FUNDAMENTALS OF DIGITAL TELEVISION TRANSMISSION

FUNDAMENTALS OF DIGITAL TELEVISION TRANSMISSION

GERALD W. COLLINS, PE
GW Collins Consulting

A Wiley-Interscience Publication
JOHN WILEY & SONS, INC.
New York • Chichester • Weinheim • Brisbane • Singapore • Toronto

This book is printed on acid-free paper.♾

Published simultaneously in Canada.

For ordering and customer service, call 1-800-CALL-WILEY.

Library of Congress Cataloging-in-Publication Data:

Collins, Gerald W.
Fundamentals of digital television transmission / Gerald W. Collins.
p. cm.
Includes index.
ISBN 0-471-39199-9 (cloth : alk. paper)
1. Digital television. I. Title.

TK6678 .C63 2000
621.388—dc21 00-035919

Printed in the United States of America.

10 9 8 7 6 5 4 3 2 1

To ***God***
who created the electromagnetic force
and
the law that governs its operation in communications systems

and

To my beautiful wife **Wilma**
who, after 39 years of marriage,
still wonders why I'm thinking about my work!

CONTENTS

PREFACE

Many engineers familiar with analog television broadcast systems are now faced with designing, operating, and maintaining digital television systems. A major reason for this introductory book is to make the transition from analog to digital television broadcasting as painless as possible for these engineers. The emphasis is on radio-frequency (RF) transmission, those elements of the system concerned with transmitting and propagating the digitally modulated signal. I begin with the digital signal as it emerges from the transport layer and end with the RF signal as it arrives at the receiver. The emphasis is on factors affecting broadcast system performance.

The scope of this book is necessarily limited; some topics, such as studio-to-transmitter links and receivers are not covered. It is intended as a self-study resource by the broadcast system engineer, as well as a reference for the design engineer, system engineer, and engineering manager. An index is included to make it a more useful resource for future reference. It may be used as a text for a formal training class.

Most people would agree that a useful engineering tool must include some mathematics. For this reason, and to make the presentation as clear as possible, concepts have been described verbally, mathematically, and in many cases, graphically. The mathematics used include algebra, trigonometry, and a small amount of calculus. For those not interested in the mathematical formulation, the charts and graphs should be sufficient to grasp the key points.

For those who wish to probe further, extensive footnotes are provided. These not only provide much more detail but are my attempt to give credit to the many workers who have brought digital television to its present state of maturity. Even with ample footnotes, I may have failed to give credit to all who deserve it. This is by no means intentional; the references included are simply those sources of which I am aware.

To the extent possible I have used the mathematical symbols most commonly used for the quantities discussed. However, the literature for the many subsystems comprising a digital television transmission system use common symbols to represent a large number of the quantities. To avoid confusion, I have added subscripts and used alternative type fonts to distinguish such quantities where necessary. When I found it necessary to use a nonstandard symbol, I attempted to make the relationship between the quantity and its symbol as intuitive as possible.

To the extent that information was available to me, I have discussed the American ATSC, the European DVB-T system, and Japan's ISDB-T system. My personal experience and library are heavily biased in the direction of the ATSC and DVB-T systems, however, a fact that will readily be apparent to the reader. The information presented should not be considered an endorsement of a specific system for any particular country or group of countries. There are many factors to be considered when selecting a transmission system, not all of which are determined by performance parameters such as transmitter peak-to-average ratio or threshold carrier-to-noise ratio. These include the type of network, program and service considerations, and the extent of the use of mobile receivers, as well as language, industrial policy, and other issues. The information presented is factual to the best of my understanding. Readers are left to draw the appropriate conclusions for their applications.

My personal design background is in antennas, analog transmitter systems, passive RF components, and propagation. When the transition to digital television began, it became necessary to educate myself with regard to digital modulation techniques, system design, and testing. This has required collaboration with many experts and the study of many reports and papers. This book is the result of that effort. If in some respect the presentation of any topic is incomplete, I take full responsibility.

The implementation of digital television is a process that will continue for many years to come. The transition periods will take up to 15 years in some countries. The process will not start in Japan until after 2003. In the United States the transition period has started and is mandated to be short. However, stations whose initial channel is outside the core spectrum will be required to move to a core channel after the transition. Those whose analog and digital channel is inside the core will be permitted to chose their permanent channel. It is hoped that this book will be helpful to those who are designing and implementing these systems, both now and in the future.

JERRY COLLINS
December 1999

ACKNOWLEDGMENTS

I most certainly do not claim originality for much of the material included in this book. In fact, the story of digital television builds on the many contributions of workers since the beginning of radio and television transmission. Rather, this book represents the result of my own attempt to understand and manage the development of digital television broadcast equipment since 1989. I am especially grateful to my former colleagues and the management of Harris Corporation Broadcast Division for their outstanding efforts. Together we participated in the process of developing digital television standards, designing equipment, and testing broadcast systems. It is to them that I owe so very much.

In naming some, I'm sure I will miss some important contributors. However, I must mention the very beginning of our work when Bob Plonka, Jim Keller, I, and others worked with Charlie Rhodes of the ATTC to develop the RF test bed by which the proponent transmission systems were tested. Bob and Jim have continued their work developing, implementing, and testing new designs and production equipment for Harris. Charlie's name is almost synonymous with DTV transmission. As soon as it was clear that the 8 VSB system would be the standard for the United States, I involved others in my R&D group in the development of the first series of 8 VSB exciters. These fine engineers included Dave Danielsons, Ed Twitchell, Paul Mizwicki, Dave Nickell, Dave Blickhan, Bruce Merideth, and Joe Seccia. The system engineering skills of Bob Davis were vital. We started the work on power amplifier development soon after the exciter. This could not have been accomplished without the able contributions of the engineers at our sister facility in Cambridge, England, under the leadership of Dave Crawford and Barry Tew. Dmitri Borodulin joined us in Quincy, Illinois for solid state PA development, along with Jim Pickard who made many contributions to the design of the IOT amplifier. I wish to emphasize the role of Harris management—especially my good friend Bob Weirather—in the development process. Without their support and encouragement we would have accomplished very little. Finally, my sincere thanks to Bob for his review of the manuscript and his constructive comments.

1

DIGITAL TELEVISION TRANSMISSION STANDARDS

A great deal of fear, uncertainty, and doubt can arise among engineers with an analog or radio-frequency (RF) background at the mere mention of digital transmission systems. Engineers sometimes fall into the trap of believing that digital systems are fundamentally different from their analog counterparts. As will be demonstrated, this is not the case. In concept, the transmission of digital television signals is no different than for analog television. The difference is in the details of implementation (hence the need for this book).

A block diagram of a typical broadcast transmission system is shown in Figure 1-1. This block diagram may, in fact, represent either an analog or a digital system. Major components include a transmitter comprising an exciter, power amplifier, and RF system components, an antenna with associated transmission line, and many receiving locations. Between the transmitter and receivers is the over-the-air broadcast transmission path. The input to the system is the baseband signal by which the RF carrier is modulated. In an analog system the baseband signal includes composite video and audio signals. In separate amplification, these modulate separate visual and aural carriers. If common amplification is used, the modulated signals are combined in the exciter and amplified together in the power amplifier. The combined signals are then transmitted together through the remainder of the link.

For a digital system, the conceptual block diagram most resembles common amplification. A single baseband signal modulates a carrier and is amplified in the transmitter, broadcast by means of the antenna, and received after propagating through the over-the-air link. The baseband signal is a composite digital data stream that may include video and audio as well as data. Since the method of modulation is also digital, the exciter used with the transmitter is also different. Beyond these details, the remainder of the system is fundamentally

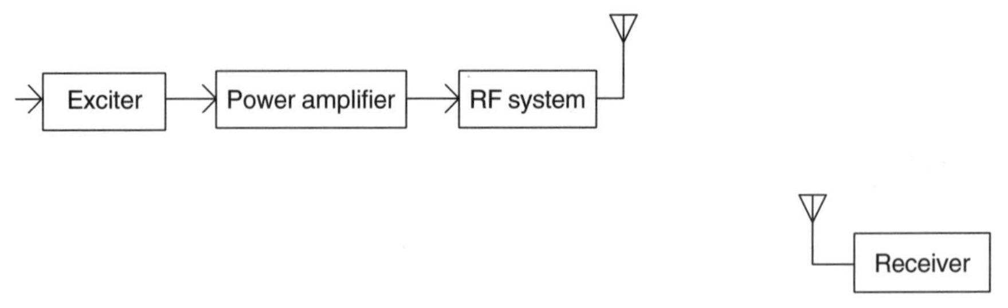

Figure 1-1. Broadcast transmission system.

the same, although there are further subtle differences in power measurement, tuning, control, and performance measurement, upconverters, power amplifiers, transmission lines, and antennas.

The similarities between digital and analog systems is also apparent when we consider the transmission channel. The ideal channel would transfer the modulated RF carrier from the modulator to the receiver with no degradation or impairment other than a reduction in the signal level and the signal-to-noise ratio. As a matter of fact, the real transmission channel is far from ideal. The signal may suffer linear and nonlinear distortions as well as other impairments in the transmitter and other parts of the channel. For analog television signals, these impairments are characterized in terms of noise, frequency response, group delay, luminance nonlinearity, differential gain, incidental carrier phase modulation (ICPM), differential phase, lower sideband reinsertion, and intermodulation distortion. For digital signals, linear distortions are also characterized in terms of frequency response and group delay. For nonlinear distortions, AM-to-AM and AM-to-PM conversion are the operative terms. In either case, the objective of good system design is to reduce these distortions to specified levels so that the channel may be as transparent as possible.

The antenna and transmission line may introduce some of the linear distortions. In most cases, these are relatively small compared to distortions introduced by the propagation path. This is especially true of matched coaxial transmission lines. Waveguides may introduce nontrivial amounts of group delay. Under some circumstances an antenna may introduce significant frequency response, nonlinear phase, and group delay distortion. Once the system design is finalized, however, there no attempt may be made to equalize distortions introduced by the transmission line or antenna.

The propagation path from the broadcast antenna to the receiver location may be the source of the most significant impairments. These impairments include noise and linear distortions resulting from reflections and other sources of multipath. Depending on specific site characteristics, the linear distortions may be severe. The impairments introduced by propagation effects vary from location to location and are also a function of time. Obviously, there is no practical means of equalizing these distortions at the transmitter. Any equalization to mitigate response and group delay introduced by the over-the-air path must be done in the receiver. The random noise introduced in the propagation path may be overcome at the transmitter only by increasing the average effective radiated power (AERP).

ATSC TERRESTRIAL TRANSMISSION STANDARD

At the time of this writing, the U.S. Federal Communications Commission (FCC), Canada, and South Korea have adopted the standard developed for digital television by the Advanced Television Systems Committee (ATSC). This standard, designated A/53, represents the results of several years of design, analysis, testing, and evaluation by many experts in industry and government. It promises to be a sound vehicle for digital television delivery for decades to come. The standard describes the system characteristics of the U.S. digital television system, referred to in this book as the ATSC or DTV system. The standard addresses a wide variety of subsystems required for originating, encoding, transporting, transmitting, and receiving of video, audio, and data by over-the-air broadcast and cable systems. The transmission system is a primary subject of this book, which is described in detail in Appendix D of the ATSC standard. The ATSC standard specifies a system designed to transmit high-quality digital video, digital audio, and data over existing 6-MHz channels. The system is designed to deliver digital information at a rate of 19.29 megabits per second (Mb/s).

The transmitter component affected most by the implementation of this standard is the exciter, although, only portions of the exciter need be affected. Figure 1-2 is a conceptual block diagram of a television exciter. As drawn, this block diagram could represent either an analog or a digital exciter. The first block, the modulator, represents composite video and audio processing and modulation in the case of analog television; for digital television, this block represents digital data processing or channel coding and modulation. (It is assumed that the reader is familiar with analog video and audio modulator functions; if not, refer to Chapter 6.2, "Television Transmitters," of the *NAB Engineering Handbook*, 9th edition.)

The second block, intermediate frequency (IF)-to-RF conversion, represents upconversion, IF precorrection and equalization, final amplification, and filtering. In principle, this block is the same for both analog and digital television signals in that the main purpose is to translate the IF to the desired RF channel. For the time being, the discussion will focus on processing the digital baseband signal prior to upconversion. To facilitate this, the nature of the input and output signals of the digital modulator block is first discussed.

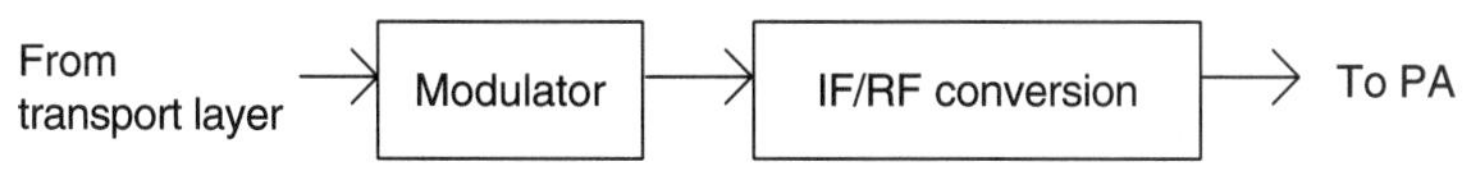

Figure 1-2. Block diagram of TV exciter.

The digital input signal to the ATSC transmission system is a synchronous serial MPEG[1]-2 transport stream at a constant data rate of 19.39... Mb/s. This serial data stream is comprised of 187-byte MPEG data packets plus a sync byte. The payload data rate is 19.2895. Mb/s. The payload may include encoded packets of digital video, digital audio, and/or data. The transport stream arrives at the exciter input on a single 75-Ω coaxial cable with a BNC input connector. The data clock is embedded with the payload data. Biphase mark coding is used. The data clock frequency error is specified to be less than ±54 Hz. The standard input level is 0.8 V ± 10% peak to peak as defined by the SMPTE Standard 310M-Synchronous Serial Interface for an MPEG-2 digital transport stream.

The output signal from the modulator block is an eight-level vestigial sideband modulated signal. Ordinarily, this is at some frequency intermediate to the baseband and RF channel frequency. The frequency, level, and other interface characteristics of the IF are generally dependent on the design choices made by the equipment manufacturer.

Figure 1-3 is a simplified block diagram of the signal processing functions required to convert the MPEG-2 transport stream to the eight-level vestigial sideband signal (8 VSB) required by the ATSC transmission system. The modulator may be viewed as performing two essential functions. The first function is channel coding. Among other things, the channel coder modifies the input data stream from the transport layer by adding information by which the receiver may detect and correct transmission errors. These are errors as a result of impairments introduced in the transmission channel. Without channel coding, the receiver would be unable to decode and display the signal properly except at receive sites with a very high signal-to-noise ratio and a minimum of multipath. The second block in Figure 1-3 is the modulator proper. It is in this block that an IF signal is modulated with the channel-coded data stream to produce the 8 VSB signal required for terrestrial over-the-air transmission.

A block diagram of the channel coder is shown in Figure 1-4. Six major functions are performed in the channel coder: data randomizing, Reed–Solomon (R/S) coding, data interleaving, trellis coding, sync insertion, and pilot signal insertion.

The incoming data from the transport stream are first randomized. This process exclusive-ORs the data bytes with a pseudorandom binary sequence locked to the data frame. The purpose of randomization is to assure that the data spectrum is uniform throughout the 6-MHz channel, even when the data are constant.

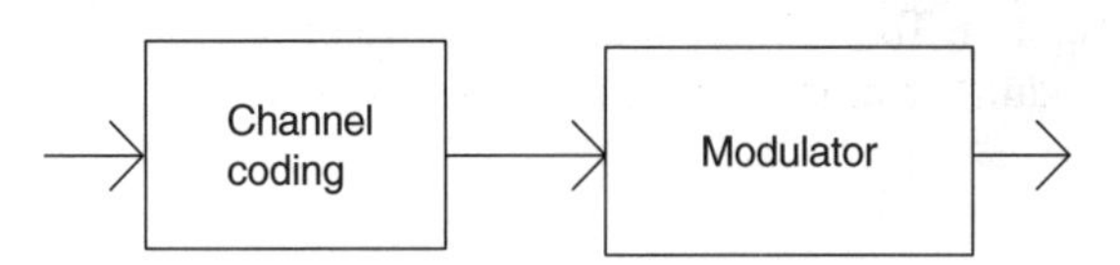

Figure 1-3. DTV modulator.

[1] Motion Pictures Expert Group.

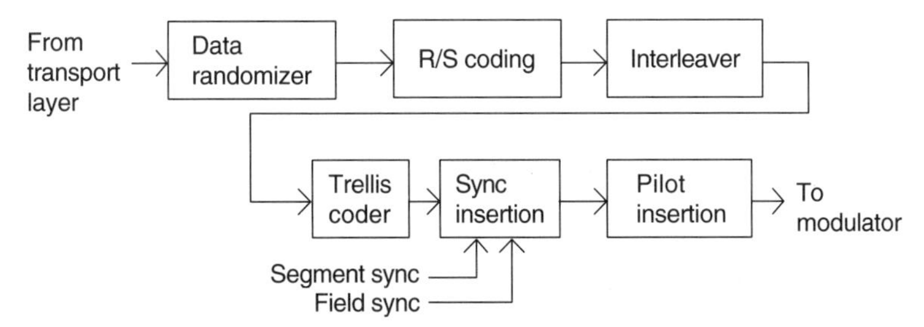

Figure 1-4. DTV channel coding. (From ATSC DTV Standard A/53, Annex D; used with permission.)

This pseudorandom sequence is generated in a 16-bit shift register with nine feedback taps. A complementary derandomizer is provided in the receiver to recover the original data sequence. Randomizing is not applied to the sync byte of the transport packet.

The next step is R/S coding. This is a forward error correction (FEC) code designed to protect against noise bursts. In this code, 20 parity bytes are added to each data block or 187-byte data packet. The R/S code selected is capable of correcting up to 10-byte errors per data block. Because of the additional bytes, the clock and data rate is necessarily increased from 19.39 Mb/s to 21.52 Mb/s. As with randomization, R/S coding is not applied to the sync bytes.

After R/S coding, the data structure is formatted into data bytes and segments, fields, and frames as defined in Figure 1-5. A data field is comprised of 312 data segments plus a sync segment, for a total of 313 segments. A data frame is comprised of two data fields, or 626 segments. The R/S coded data are interleaved to provide additional error correction. This process spreads the data bytes from several R/S packets over a much longer period of time so that a very long burst of noise is required to overrun the capability of the R/S code. A total of 87 R/S packets are processed in the interleaver.

Trellis coding, another error correction code, follows the R/S interleaver. The purpose of this code differs from the R/S code in that it has the effect of improving the signal-to-noise ratio (S/N) threshold in the presence of thermal or white noise. It is termed a $\frac{2}{3}$-rate code because every other input bit is encoded to 2 output bits; the alternate bit is not encoded. Thus the output of the trellis coder is a parallel bus of 3 bits for every 2 input bits. The trellis-coded data are interleaved with a 12-symbol code interleaver. The data rate at the output of the trellis coder is increased by a ratio of $\frac{3}{2}$, to 32.28 Mb/s. Taken together, the output bits of the trellis coder comprise the 3-bit symbols. These symbols (-7, -5, -3, -1, 1, 3, 5, 7) are the eight levels of the VSB modulator. The symbol rate is one-third that of the trellis-coded data rate, or 10.76 symbols/s.

The spectral efficiency, η_s, is the ratio of the encoded data rate to the channel bandwidth:

$$\eta_s = \frac{32.28}{6} = 5.38 \text{ bps/Hz}$$

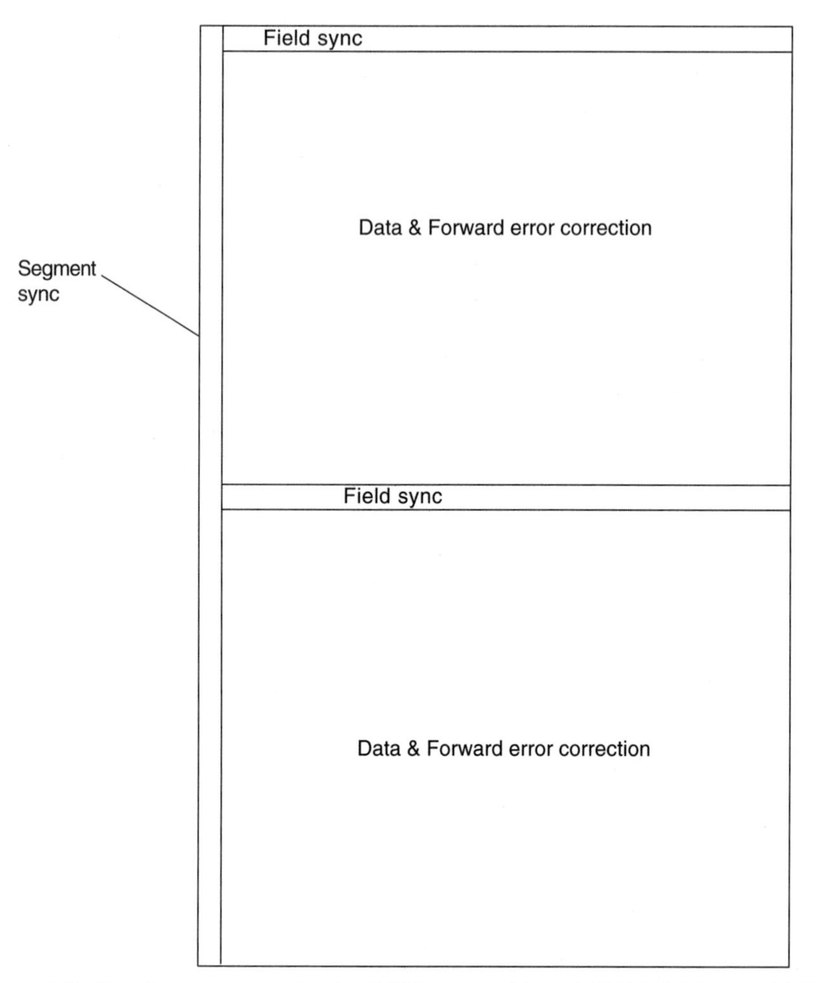

Figure 1-5. Data frame structure for the ATSC system. (From ATSC DTV Standard A/53, Annex D; used with permission.)

This is a consequence of using 3 bits per symbol to create the eight VSB levels ($M = 8$) and the excess bandwidth of the Nyquist filter ($\alpha_N = 0.1152$). Using these parameters, the spectral efficiency may be computed by

$$\eta_s = \frac{2 \log_2 M}{1 + \alpha_N} \qquad \text{bps/Hz}$$

which also results in 5.38 bps/Hz.

A data segment is comprised of the equivalent of the data from one R/S transport packet plus FEC code and data segment sync as shown in Figure 1-6. Actually, the data come from several R/S packets because of interleaving.

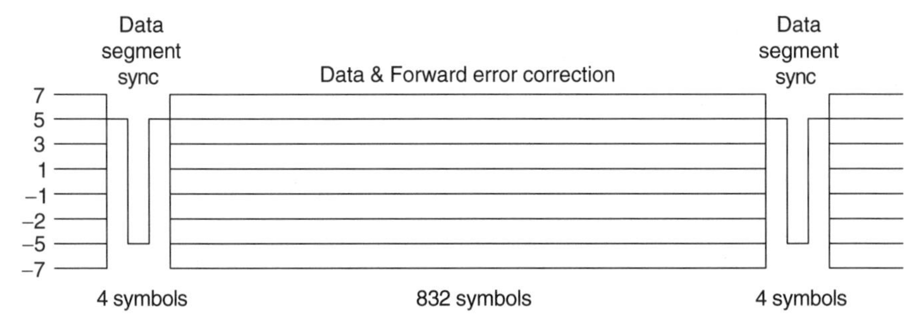

Figure 1-6. Data segment for the ATSC system. (From ATSC DTV Standard A/53, Annex D; used with permission.)

Since each R/S packet is 207 bytes in length, a data segment is 208 bytes $(207 + 1)$. At 8 bits per byte and 3 bits per symbol, the data segment is 208 bytes $\times$ 8 bits/byte $\times$ $\frac{3}{2}/3$ bits/symbol, or 832 symbols in length, 828 of which are FEC coded data; the remaining four are segment sync symbols. There are 3×832, or 2496 bits per segment, 2484 of which are data and 12 of which are segment syncs. For the data rate of 32.28 Mb/s the time per bit is 31 ns. Thus the time per segment is 2496×31, or 77.3 μs, and the segment rate, $f_{\text{seg}} = 12.94$ data segments per second. With 313 segments per field, the field time is 313×77.3 μs, or 24.2 ms, and the field rate is 41.3 kHz. The frame rate, f_{frame}, is one-half the field rate, or 20.66 kHz.

Following the trellis coding, field and segment sync symbols are inserted. The structure of the data field sync segment is defined in Figure 1-7. As with the data segments, the field sync segment is 832 symbols in length. Each symbol is binary encoded as either + or -5. Four data segment sync symbols replace the MPEG sync byte. These are followed by a series of pseudorandom number (PN) sequences of length 511, 63, 63, and 63 symbols, respectively. The PN63 sequences are identical, except that the middle sequence is of opposite sign in every other field. This inversion allows the receiver to recognize the alternate data fields comprising a frame.

The PN63 sequences are followed by a level identification sequence consisting of 24 symbols. The last 104 symbols of the field sync segment are reserved; 92 of these symbols may be a continuation of the PN63 sequence. The last 12 of these symbols are duplicates of the last 12 symbols of the preceding data segment.

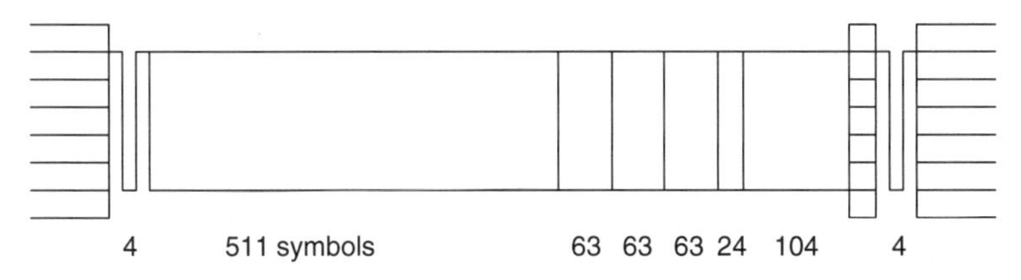

Figure 1-7. Field sync for the ATSC system. (From ATSC DTV Standard A/53, Annex D; used with permission.)

In addition to providing a means of synchronizing the receiver to the formatted data, the sync segments serve as training signals for the receiver equalizer. The equalizer improves the quality of the received signal by reducing linear distortions. This is analogous to ghost reduction due to multipath in analog systems. Since the sync sequences are known repetitive signals, the equalizer taps may be adjusted to reproduce these sequences with a minimum of distortion. The taps, thus adjusted, reduce distortion of the received data. The sync segments may also be used for diagnostic purposes.

The data field and frame structure has the familiar appearance of the field and frame structure of analog television. However, it should not be assumed that a data field corresponds to a video field. Each data field may include video, audio, or other data, so there is generally no correspondence between data fields and video fields.

VESTIGIAL SIDEBAND MODULATION

Vestigial sideband modulation may be accomplished in either the analog or the digital domain. Manufacturers have generally developed their own modulation schemes, some of which may be proprietary. Since the purpose of this book is to describe the principles of digital television transmission, a generic modulator using analog circuitry is presented.

Such a modulator is illustrated in Figure 1-8. The signal (i.e., the 3-bit multilevel symbols or pulses from the output of the trellis coder) is divided equally to form in-phase (I) and quadrature (Q) paths at the input to the modulator. The pulses are then shaped to minimize intersymbol interference. This pulse shaping is accomplished in a Nyquist filter. This is a low-pass linear-phase filter with flat amplitude response over most of its passband. At the upper and lower band edges, the filter response transitions to the stopband by means of skirts with a root-raised-cosine shape. The steepness of the skirts is determined by the shape factor, α_N. For the ATSC system, α_N is specified to be 0.1152. The Nyquist filter multiplies the shaped signals by either $\sin(\pi t/2T)$ or $\cos(\pi t/2T)$, where T is the symbol time.

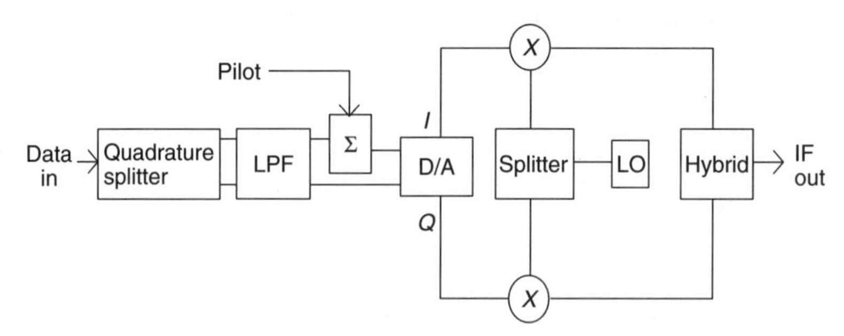

Figure 1-8. Typical digital modulator.

The shaped I and Q signals are now presented to digital-to-analog (D/A) converters in each of the I and Q channels. The I and Q signals are each multiplied by equal levels of the local oscillator (LO) signal. For the Q path, the LO signal is 90° out of phase with respect to the LO signal for the I path. These signals are then summed in a two-way power combiner to produce the IF output. The resulting spectrum contains only one of the sidebands of the modulated signals and the carrier is suppressed. Thus this modulation technique is called vestigial sideband. A pilot signal is inserted in the I path of the modulator. By adding a small direct-current (dc) offset of 1.25 V to all of the encoded symbols (including sync), a tone at the same frequency as the suppressed carrier is generated in the output of the VSB modulator. The presence of the pilot adds very little power (only 0.3 dB) to the modulated signal, but it is important in that it enables receiver tuning under conditions of severe noise and interference. It also speeds carrier recovery and, therefore, data acquisition in the receiver. It is apparent that the quality of the IF output is dependent on the stability of both the incoming data and the LO.

At this point in the system, the complete DTV signal has been generated, consisting of eight amplitude levels, four positive and four negative. The signal is often displayed in a two-dimensional I–Q or constellation diagram, as shown in Figure 1-9. This is a graphical representation of the orthogonal I and Q components of the modulated waveform, plotted in X–Y or rectangular

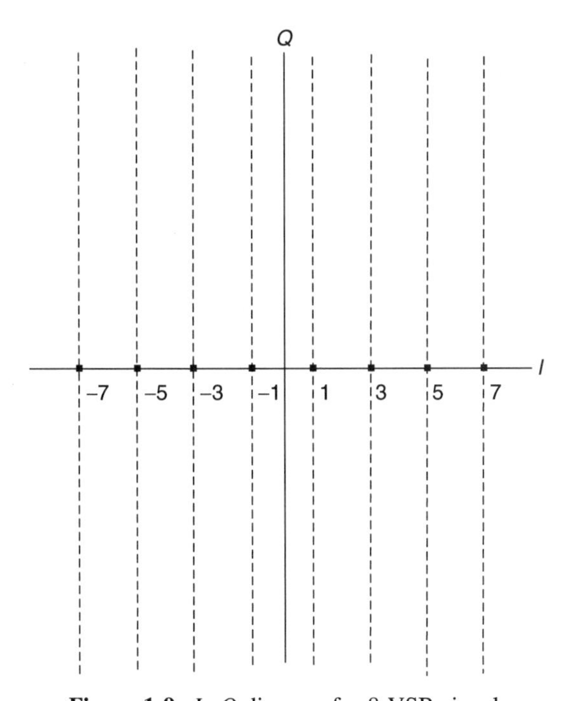

Figure 1-9. I–Q diagram for 8 VSB signal.

coordinates, where the X and Y axes are called the I and Q axes, respectively. Each point in the I–Q diagram represents a specific amplitude and phase of the RF carrier. For 8 VSB, information is carried only by the I component, for which the distinct levels of 8 VSB are plotted on the horizontal axis. Although a quadrature component is present and is displayed in the direction of the Q-axis, there are no distinct levels associated with the Q component and no information conveyed.

The modulated signal occupies 6 MHz of total bandwidth by virtue of the vestigial sideband modulation scheme. The spectrum of the modulated signal is shown in Figure 1-10. The energy is spread uniformly throughout most of the channel. At both the upper and lower band edges, the spectrum is shaped in accordance with the root-raised-cosine or Nyquist filter. A complementary root-raised-cosine filter of the same shape is included in the receiver so that the system response is a raised cosine function.

The 3-dB bandwidth of the resulting transmitted spectrum is 5.38 MHz. At the RF channel frequency, the pilot is located at the lower 3-dB point, 0.31 MHz above the lower edge of the channel. The pilot is the same frequency as the suppressed carrier. (The pilot may be at the opposite end of the spectrum at the IF.) The DTV pilot is offset from the NTSC visual carrier to minimize DTV-to-NTSC cochannel interference. The remainder of the system exists for the purposes of upconverting to the desired channel, amplifying to the required power level, and radiating the on-channel signal.

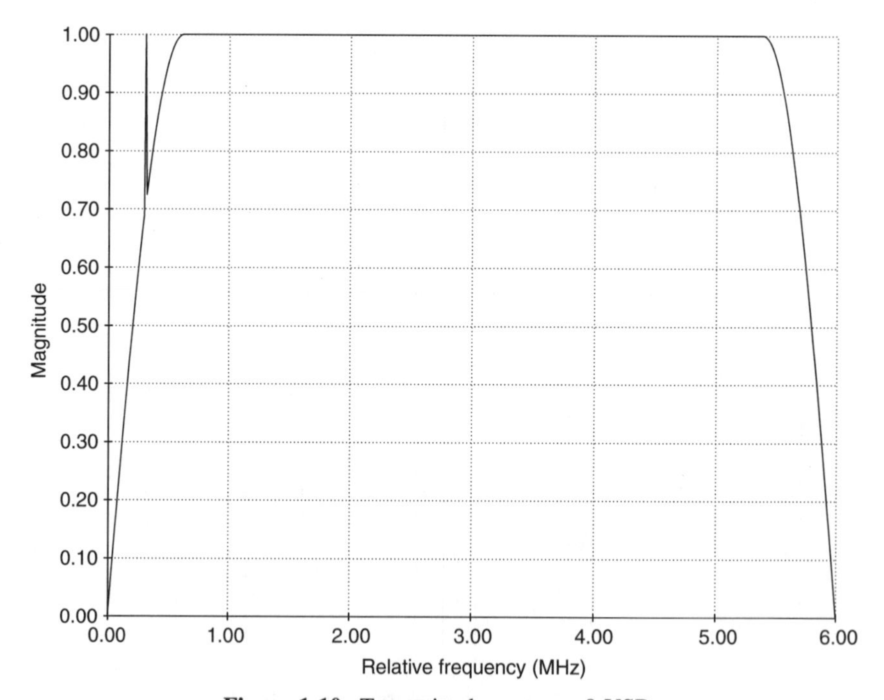

Figure 1-10. Transmitted spectrum, 8 VSB.

DVB-T TRANSMISSION STANDARD

The European Telecommunications Standards Institute has adopted a set of standards for digital broadcasting of television, sound, and data services. Standards have been adopted for satellite, cable, and terrestrial signal delivery. The standard for terrestrial transmission, ETS 300 744, is designated Digital Video Broadcast–Terrestrial (DVB-T). This standard describes a baseline transmission system for digital broadcasting of television. At the time of this writing, it has been adopted by the 15 members of the European Union, Australia, and New Zealand. It is similar in many respects to the U.S. DTV standard. However, there are also important and significant differences in both channel coding and modulation.

The DVB-T standard specifies a system designed to transmit high-quality digital video, digital audio, and data over existing 7- or 8-MHz channels. The system is designed to deliver digital information at rates from 4.98 to 31.67 Mb/s. Although there are many similarities with the ATSC standard in the transport layer and channel coding, a significant difference is in the type of modulation used. Coded orthogonal frequency-division multiplex (COFDM) has been selected for DVB-T, in part due to the unique requirements of European broadcasting stations and networks. Single-frequency networks (SFN) are used extensively in Europe to more effectively use the channels available; COFDM is seen as best suited to this requirement. In a SFN, all stations broadcasting a particular program do so on the same channel, each being synchronized to precisely the same reference signal and having common baseband timing. A receiver tuned to this channel may receive signals from one or more stations simultaneously, each with a different delay. Under multipath conditions, the signal strength from each station may vary with time. The guard intervals and equalization built into the COFDM system facilitate effective reception under these conditions. The guard interval may be selected from $\frac{1}{32}$ to $\frac{1}{4}$ the duration of the active symbol time, so that the total symbol duration is from $1\frac{1}{32}$ to $1\frac{1}{4}$ the active symbol time.

As with the ATSC standard, the transmitter assembly most affected by the transition to digital broadcast is the exciter, with the major changes required being baseband processing and modulation. Thus the focus of this discussion is on the modulator block. The nature of the input and output signals is discussed first.

In common with DTV in the United States, the digital input signal to the DVB-T transmission system is a MPEG-2 synchronous transport stream comprised of 187-byte MPEG data packets plus a sync byte. The payload may include encoded packets of digital video, digital audio, and/or data. The parallel transport stream connector at the modulator input is a DB25 female connector. The data clock line is separate from the payload data lines.

The output signal from the modulator block is a COFDM signal. Ordinarily, this is generated at some frequency intermediate to the baseband and RF channel frequency. The frequency, signal level, and other interface characteristics of the IF are generally dependent on design choices made by the equipment manufacturer.

As in the ATSC system, the modulator may be viewed as performing the functions of channel coding and modulation proper. The functions performed in the channel coder include energy dispersal or data randomization, outer or R/S coding, outer interleaving, inner or trellis coding, and interleaving. The modulator functions include mapping, frame adaptation, and pilot insertion.

The incoming data from the transport stream are first dispersed or randomized. A complementary derandomizer is provided in the receiver to recover the original data sequence. As with the DTV system, randomizing is not applied to the sync byte of the transport packet.

The next step is outer or R/S coding. The details of the code selected differ from those of the DTV standard in that it is capable of correcting up to only eight byte errors per data block. In this code, 16 parity bytes are added to each sync and data block or 188-byte packet. The R/S coded data are interleaved to provide additional error correction.

Convolutional coding and interleaving follow the R/S interleaver. The DVB-T system allows for a range of punctured convolutional codes. Selection of the code rate is based on the most appropriate level of error correction for a given service and data rate. Punctured rates of $\frac{2}{3}$, $\frac{3}{4}$, $\frac{5}{6}$, or $\frac{7}{8}$ are derived from the $\frac{1}{2}$-rate mother code. Interleaving consists of both bitwise and symbol interleaving. Bit interleaving is performed only on the useful data.

The purpose of the symbol interleaver is to map bits on to the active OFDM carriers. Detailed operation of the symbol interleaver depends on the number of carriers generated, whether 2048 (2^{11}) in the 2k mode or 8192 (2^{13}) in the 8k mode. Some of the carriers are used to transmit reference information for signaling purposes (i.e., to select the parameters related to the transmission mode). The number of carriers available for data transmission is 1705 in the 2k mode or 6817 in the 8k mode. The overall bit rate available for data transmission is not dependent on the mode but on the choice of modulation used to map data on to each carrier.

The OFDM modulator follows the inner coding and interleaving. This involves computing an inverse discrete fourier transform (IDFT) to generate multiple carriers and quadrature modulation. The transmitted signal is organized in frames, each frame having a duration of T_F and consisting 68 OFDM symbols. The symbols are numbered from 0 to 67, each containing data and reference information. In addition, an OFDM frame contains pilot cells and transmission parameter signaling (TPS) carriers. The pilot signals may be used for frame, frequency, and time synchronization, channel estimation, and transmission mode identification. TPS is used to select the parameters related to channel coding and modulation.

The many separately modulated carriers may employ any one of three square constellation patterns: quadrature-phase shift keying (QPSK, 2 bits per symbol), 16-constellation-point quadrature amplitude modulation (16 QAM, 4 bits per symbol), or 64 QAM (6 bits per symbol). By selecting different levels of QAM in conjunction with different inner code rates and guard interval ratios, bit rate may be traded for ruggedness. For example, QPSK with a code rate of $\frac{1}{2}$ and a guard interval ratio of $\frac{1}{4}$ is much more rugged than 64 QAM with a code rate of $\frac{5}{6}$ and $\frac{1}{32}$ guard interval ratio. However, the available data rate is much less. The 8k mode has the longest available guard interval, making it the best choice for single-frequency networks with widely separated transmitters.

Hierarchical transmission is also a feature of the DVB-T standard. The incoming data stream is divided into two separate streams, a low- and a high-priority stream, each of which may be transmitted with different channel coding and with different modulation on the subcarriers. This allows the broadcaster to make different trade-offs of bit rate and ruggedness for the two streams.

The power spectral density of the modulated carriers is the sum of the power spectral density of the individual carriers. The upper portion of the transmitted spectrum for an 8-MHz channel is shown in Figure 1-11, plotted relative to the channel center frequency. Overall, the energy is spread nearly uniformly throughout most of the channel. However, since the symbol time is larger than the inverse of the carrier spacing, the spectral density is not constant. The center frequency of the DVB-T channels is the same as the current European analog ultrahigh frequency (UHF) channels. The minimum carrier-to-noise (C/N) ratio is dependent, among other parameters, on modulation and inner code rate. As with the DTV system, there is no visual, chroma, or aural carrier frequencies as in analog TV.

The complete DVB-T signal has been generated at the output of the modulator. The remainder of a transmitting system exists for the purposes of upconverting

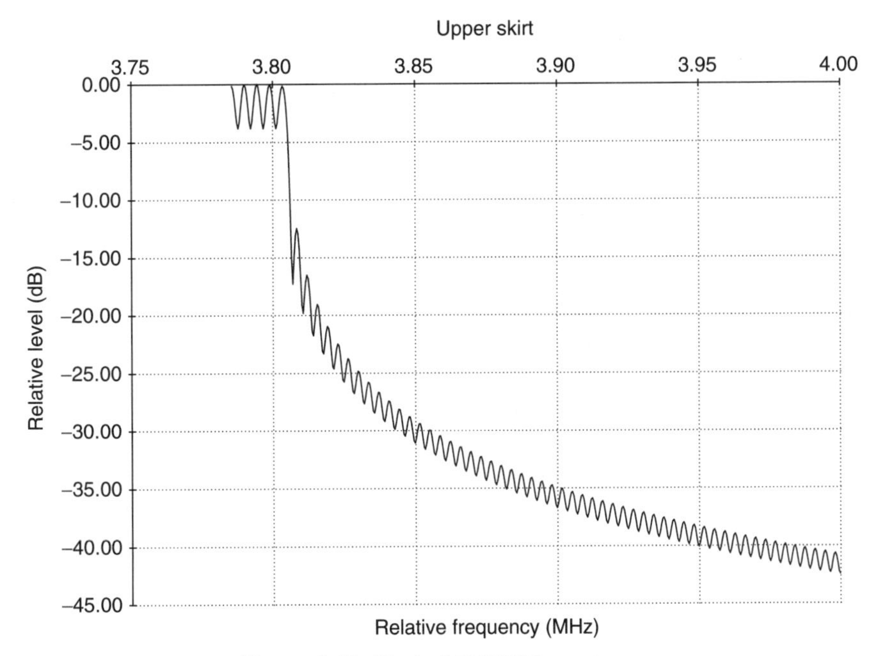

Figure 1-11. Typical COFDM spectrum.

to the desired channel, amplifying to the required power level, and radiating the on-channel signal.

ISDB-T TRANSMISSION STANDARD

Japan's Digital Broadcasting Experts Group (DiBEG) has developed a standard for digital broadcasting of television, sound, and data services, designated integrated services digital broadcasting (ISDB). Standards have been developed for delivery of satellite, cable, and terrestrial signals. These standards include a description of a baseline transmission system that provides for digital broadcasting of television, including channel coding and modulation. The transmission standard for terrestrial digital television is similar in many respects to the DVB-T standard. It is entitled Integrated Services Digital Broadcasting–Terrestrial (ISDB-T).[2] A key difference with respect to DVB-T is the use of band-segmented transmission–OFDM (BST-OFDM). This is a data segmentation approach that permits the service bandwidth to be allocated to various services, including data, radio, standard definition television (SDTV), and high-definition television (HDTV) in a flexible manner. It is planned that digital television will be launched in Japan after 2003.

The ISDB-T standard specifies a system designed to transmit over existing 6-, 7-, or 8-MHz channels. The system is designed to deliver digital information at data rates from 3.561 to 30.980 Mb/s.

In common with the other world standards, the digital input signal to the ISDB-T transmission system is a MPEG-2 synchronous transport comprised of 187-byte MPEG data packets plus a sync byte. The payload may include encoded packets of digital video, digital audio, text, graphics, and data. In addition, transmission and multiplex control (TMCC) is defined for hierarchical transmission. To make use of the band-segmenting feature, the data stream is remultiplexed and arranged into data groups, each representing all or part of a program or service. After channel coding, these data groups become OFDM segments. Each OFDM segment occupies $\frac{1}{14}$ of the channel bandwidth. This arrangement allows for both broadband and narrowband services.

For example, a single HDTV service might occupy 12 of the OFDM segments, with the thirteenth used for sound and data.[3] Alternatively, multiple SDTV programs might occupy the 12 OFDM segments. A maximum of three OFDM segment groups or hierarchical layers may be accommodated at one time. For the narrowband services, a small, less expensive narrowband receiver may be used. The OFDM segment in the center of the channel is dedicated to such narrowband or partial reception services. Obviously, a receiver decoding a single OFDM segment receives only a portion of the original transport stream.

[2] "Channel Coding, Frame Structure, and Modulation Scheme for Terrestrial Integrated Services Digital Broadcasting (ISDB-T)," *ITU Document 11A/Jxx-E*, March 30, 1999.

[3] The upper and lower channel edges occupy the bandwidth of the remaining OFDM segment.

ISDB-T has many features in common with DVB-T. Both inner and outer FEC codes are applied to the data. The resulting data stream modulates multiple orthogonal carriers. Thus both standards make use of COFDM. The guard interval may be selected from $\frac{1}{32}$ to $\frac{1}{4}$ the duration of the active symbol time. As in Europe, Japan will use this approach to increase the number of available channels by means of SFNs. The R/S code is capable of correcting up to eight-byte errors per data block. A total of 16 parity bytes are added to each sync and data block. The system also allows for a range of punctured convolutional codes. Selection of the code rate is based on the most appropriate level of error correction for a given service or data rate. Code rates of $\frac{2}{3}$, $\frac{3}{4}$, $\frac{5}{6}$, or $\frac{7}{8}$ are derived from a $\frac{1}{2}$-rate mother code.

Despite similarities, there are differences in implementation of the channel coding. As shown in Figure 1-12, the order of R/S coding and energy dispersal are interchanged with the order used in either the DTV and DVB-T systems. The R/S coding is applied to the data as they emerge from the remultiplexed transport stream, including any null packets. Energy dispersal, delay adjustment, bytewise interleaving, and trellis coding are then applied in that order to each data group separately. This permits the length of the interleaving, code rate for the inner FEC code, and signal constellation to be selected independently for each hierarchical layer. The null packets at the output of the R/S coder are removed. The delay resulting from the bytewise interleaving differs for each layer, depending on the channel coding and modulation. To compensate, a delay adjustment is inserted prior to the interleaver.

The OFDM modulator follows the inner coding and interleaving. This involves computing an IDFT to generate multiple carriers and quadrature modulation. The number of carriers available ranges from 1405 to 5617 for all channel bandwidths depending on the transmission mode. Of these, the number of carriers available for data transmission ranges from 1249 to 4993. Obviously, the carrier spacing is increased for the wider channels for a given number of carriers. The information bandwidth is approximately 5.6, 6.5, and 7.4 MHz for the 6-, 7-, and 8-MHz channels, respectively.

The transmitted signal is organized in frames. However, the frame duration is not fixed as in the DVB-T standard. Rather, the frame duration depends on the

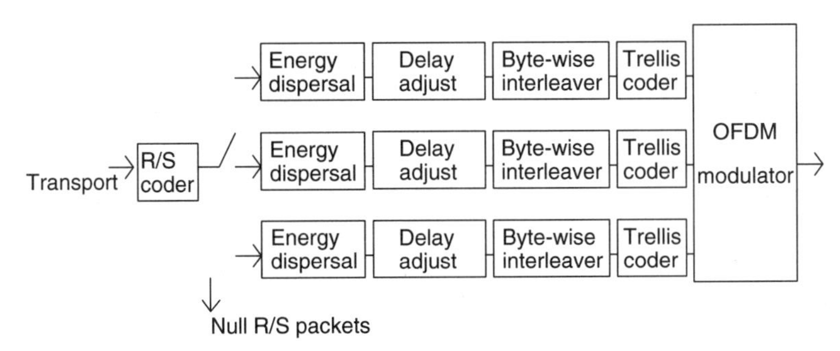

Figure 1-12. Block diagram of ISDB-T channel coding.

TABLE 1-1. Frame Duration (ms) versus Mode and Guard Interval Ratio

Guard Interval Ratio	Mode		
	1	2	3
$\frac{1}{4}$	64.26	128.52	257.04
$\frac{1}{8}$	57.83	115.67	231.34
$\frac{1}{16}$	54.62	109.24	218.46
$\frac{1}{32}$	53.01	106.03	212.06

transmission mode and the length of the guard interval. These relationships are summarized in Table 1-1. The frame duration doubles from mode 1 to mode 2 and doubles again from mode 2 to mode 3. Modes 1, 2, and 3 are defined for 108, 216, and 432 OFDM carriers per segment, respectively. The packets comprising the frames are numbered consecutively and contain the payload data as well as the information necessary for operation of the broadcast system. Scattered and continual pilot signals are available for frequency synchronization and channel estimation. The TMCC subchannels carry packets with information on the transmission parameters. Auxiliary subchannels carry ancillary information for network operation.

The many separately modulated carriers may employ the same three constellation patterns as provided in the DVB-T standard-QPSK, 16 QAM or 64 QAM. In addition, differential quadrature-phase shift keying (DQPSK) is available.

With modulation complete, the complete ISDB-T signal has been generated. The remainder of the transmitting system exists for the purposes of upconverting to the desired channel, amplifying to the required power level, and radiating the on-channel signal.

CHANNEL ALLOCATIONS

The DVB-T system operates in 7- and 8-MHz channels primarily within the existing European UHF spectrum (bands IV and V), although implementation guidelines have been published for bands I and III as well.[4] In Japan, 6-MHz channels are to be used. In general, the availability of spectrum varies from country to country; in virtually every case a scarcity exists. As in the United States, the tendency is to move terrestrial television to the UHF spectrum to free other frequencies for other uses. A common solution in Europe, Japan, and other countries is to use regional and/or national SFNs. This approach allows for the broadcast of just a few programs to a high percentage of the target area using a minimum number of channels. Generally, the digital services will coexist

[4] *Implementation Guidelines for DVB-T: Transmission Aspects*, European Telecommunications Standards Institute, April 1997.

with existing analog services for some extended period of time (say, 15 years), after which the analog service will be discontinued. Where possible, existing transmitter sites, antennas, and towers are expected to be used. To maximize the possibility of viewers using existing receiving antennas, assignment of channels near the existing analog channels with the same polarization is desirable.

In the United States, the DTV system operates in 6-MHz channels in portions of both the very high frequency (VHF) and UHF spectrum. A core spectrum is defined which includes a total bandwidth of 294 MHz, extending from channel 2 through channel 51. Because of the limited availability of spectrum and the need to minimize interference, some broadcasters are assigned spectrum outside the core at frequencies as high as channel 69 during the transition period. Use of channel 6 is minimized due to the potential for interference with the lower FM frequencies. The use of Channels 3 and 4 in the same market is minimized to facilitate use of cable terminal devices, which may operate on either of these channels. Channel 37 is reserved for radio astronomy. Where it is necessary to use adjacent channels in the same market, the NTSC and DTV stations are colocated, if possible. The licensees for the adjacent channels are required to lock the DTV and NTSC carrier frequencies to a common reference frequency to protect the NTSC from excessive interference.

The U.S. Telecommunications Act of 1996 provides that initial eligibility for an advanced television license is limited to existing broadcasters with the condition that they eventually relinquish either the current analog channel or the new digital channel at the end of the transition period. The purpose of this provision is, in part, to promote spectrum efficiency and rapid recovery of spectrum for other purposes. Consequently, DTV was introduced in the United States by assigning existing broadcasters with a temporary channel on which to operate a DTV station during the transition period, which will extend to 2006. It is planned that 78-MHz of spectrum will be recovered at the end of the transition period; it is also planned that 60 MHz in channels 60 to 69 will be recovered earlier.

If in the future channels 2 to 6 prove to be acceptable for transmission of DTV, the core spectrum may be redefined to be channels 2 to 46.[5] Those stations operating outside the core spectrum during the transition will be required to move their DTV operations to a channel inside the core when one becomes available. Broadcasters whose existing NTSC channel is in the core spectrum could move their DTV operations to this channel in the future. Broadcasters whose NTSC and DTV channels are in the core spectrum could chose which of those will be their permanent DTV channel.

It is evident that broadcast engineers in the United States will face the challenge of transmission system design and operation for the foreseeable future. Many important decisions must be made for initial systems during the transition period. Many of these systems will continue to operate without the need for major changes after the transition period ends. Others systems, however, may require major changes to accommodate the channel shifts required by the FCC.

[5] *FCC 6th Report and Order DTV Allocations*, Appendix D, April 22, 1997, p. D-11.

ANTENNA HEIGHT AND POWER

In the United States, the antenna height above average terrain (HAAT) and AERP for DTV stations operated by existing licensees is designed to provide equivalent noise-limited coverage to a distance equal to the present NTSC grade B service contour. The maximum permissible power for new DTV stations in the UHF band is 316 kW. The maximum antenna height is 2000 ft above average terrain. For HAATs below this value, higher AERP is permitted to achieve equivalent coverage. The maximum AERP is 1000 kW regardless of HAAT. The minimum AERP for UHF is 50 kW. Power allocations for VHF range from 200 W to slightly more than 20 kW.

MPEG-2

Although the source encoding and transport layer are distinct from the transmission system, they are closely associated. It is therefore important that the transmission system engineer have an understanding of MPEG-2. The following discussion is a cursory overview; for more details, the interested reader is referred to ATSC A/53 or the Implementation Guidelines for DVB-T, which point to additional documents.

In accordance with the International Telecommunications Union, Radio Sector (ITU-R) digital terrestrial broadcast model, the transport layer supplies the data stream to the RF/transmission system. This is illustrated in Figure 1-13. Since there is no error protection in the transport stream, compatible forward error correction codes are supplied in the transmission layer as already described.

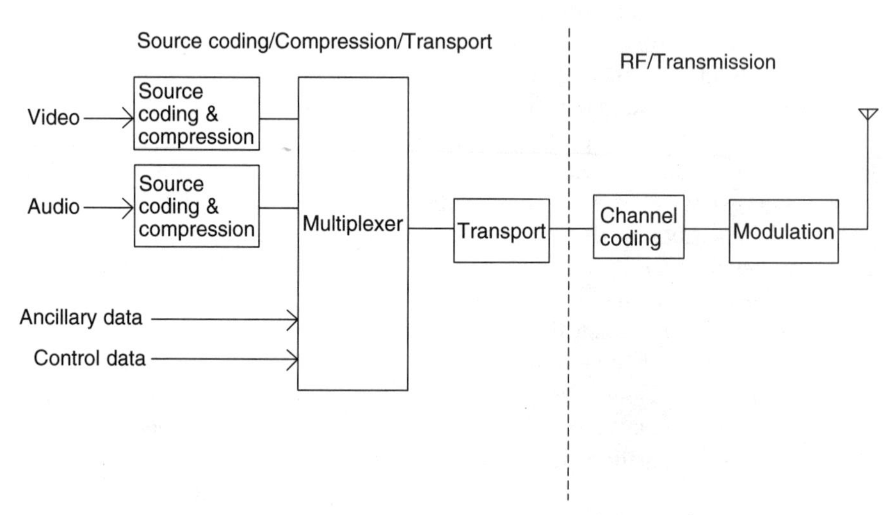

Figure 1-13. Digital television broadcast model. (From ATSC DTV Standard A/53, Annex D; used with permission.)

MPEG-2 refers to a set of four standards adopted by the International Standards Organization (ISO). Together, these standards define the syntax for the source coding of video and audio and the packetization and multiplexing of video, audio, and data signals for the DTV, DVB-T, and ISDB-T systems. MPEG-2 defines the protocols for digital compression of the video and audio data. These video coding "profiles" allow for the coding of four source formats, ranging from VCR quality to full HDTV, each profile requiring progressively higher bit rates. Several compression tools are also available, each higher level being of increased sophistication. The sophistication of each level affects the video quality and receiver complexity for a given bit rate. In general, the higher the bit rate, the higher the video and audio quality. Tests indicate that studio-quality video can be achieved with a bit rate of about 9 Mb/s. Consumer-quality video can be achieved with a bit rate ranging from 2.5 to 6 Mb/s, depending on video content.

Audio compression takes advantage of acoustic masking of low-level sounds at nearby frequencies by coding these at low data rates. Other audio components that cannot be heard are not coded. The result is audio quality approaching that of a compact disk at a relatively low data rate. The transport format and protocol are based on a fixed-length packet defined and optimized for digital television delivery. Elementary bit streams from the audio, video, and data encoders are packetized and multiplexed to form the transport bit stream. Complementary recovery of the elementary bit streams is made at the receiver.

The transport stream is designed to accommodate a single HDTV program or several standard definition programs, depending on the broadcaster's objectives. Even in the case of HDTV, multiple data sources are multiplexed, with the multiplexing taking place at two distinct levels. This is illustrated in Figure 1-14. In the first level, program bit streams are formed by multiplexing packetized elementary streams from one or more sources. These packets may be coded video, coded audio, or data. Each of these contain timing information to assure that each is decoded in proper sequence.

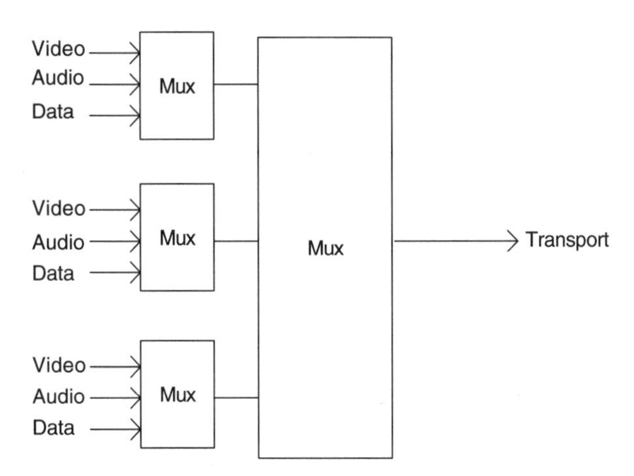

Figure 1-14. MPEG-2 multiplexing.

A typical program might include video, several audio channels, and multiple data streams. In the second level of multiplexing, many single programs are combined to form a system of programs. The content of the transport stream may be varied dynamically depending on the information content of the program sources. If the bit rate of the multiplexed packets is less than the required output bit rate, null packets are inserted so that the sum of the bit rates matches the constant bit rate output requirement. All program sources share a common clock reference. The transport stream must include information that describes the contents of the complete data stream and access control information, and may include internal communications data. Scrambling for the purpose of conditional access and teletext data may also be accommodated. An interactive program guide and certain system information may be included.

As implemented in the ATSC system, the video and audio sampling and transport encoders are frequency locked to a 27-MHz clock. The transport stream data rate and the symbol rate are related to this clock. If the studio and transmitter are colocated, the output of the transport stream may be connected directly to the transmitter. In many cases, the transport stream will be transmitted via a studio-to-transmitter link (STL) to the main transmitter site. This requires demodulation and decoding of the STL signal to recover the transport stream prior to modulation and coding in the DTV, DVB-T, or ISDB-T transmitter.

2

PERFORMANCE OBJECTIVES FOR DIGITAL TELEVISION

Characterization of the signal quality is an aspect in which digital systems differ most from their analog counterparts. With analog TV signals, engineers can readily measure the transmitted or received power at the peak of the sync pulse. The average power varies depending on picture content. Methods are available for separately measuring aural and chroma carrier power levels. Nonlinear distortions are characterized by differential gain and phase, luminance nonlinearity, and ICPM. Linear distortions are evaluated in terms of swept response and group delay.

For digital television systems, some of the familiar performance measurements are somewhat elusive. A regularly recurring sync pulse is not available for the purpose of measuring peak envelope power. The data representing video, chroma, and sound are multiplexed into a common digital stream; separate visual, chroma, and aural carriers do not exist. Because of the random nature of the baseband signal, the average power within the transmission bandwidth is constant. The quality measures of interest include average power, peak-to-average power ratio, carrier-to-noise ratio (C/N),[1] the ratio of the average energy per bit to the noise density (E_b/N_0), symbol and segment error rates (SER), bit error rate (BER), error vector magnitude (EVM), eye pattern opening, intersymbol interference (ISI), AM-to-AM conversion, AM-to-PM conversion, and spectral regrowth. Characterization of linear distortion by frequency response and group delay is common for both analog and digital systems.

[1] Reference to signal-to-noise ratio (S/N) and carrier-to-noise ratio (C/N) will be found in the literature with no distinction in meaning. In other works, C/N refers to predetection or input signal-to-noise power ratio, S/N to postdetection or output signal-to-noise power ratio. The latter convention is followed in this book.

Channel capacity is a function of carrier-to-noise ratio and channel bandwidth. Therefore, the factors affecting system noise and transmission errors at the receiver are discussed first. Following this is a discussion of factors that describe transmitter performance.

SYSTEM NOISE

Ideally, a digital television transmission system should provide an impairment-free signal to all receiving locations within the service area. Obviously, there will be some locations where this ideal cannot be achieved. In a practical system, linear distortions, nonlinear distortions, and various sources of noise and interference will impair the signal. The overall effect of these impairments is to degrade the carrier-to-noise plus interference ratio ($C/(N+I)$). In the absence of interference, this term reduces to the more familiar C/N.

Consider first the case for which there is no interference from other digital or analog signals. Knowing the received signal power and the noise power at the receiving location allows determination of the C/N and the noise-limited coverage contour in the absence of multipath and interference. Methods of determining the average power of the received signal, P_r, are discussed in Chapter 8. In the following discussion, the average carrier power, C, is considered to be equivalent to P_r after adjustment for receive antenna gain and downlead attenuation.

At distant receive locations, thermal noise should be the predominate noise source in the absence of severe multipath or interference. Thermal noise is often assumed to be additive white Gaussian noise (AWGN). The noise power spectrum of AWGN is flat over an infinite bandwidth with a power spectral density of $N_0/2$ watts per hertz.[2] The total noise power, N, in a channel of bandwidth, B, is the product of N_0 and B,

$$N = N_0 B$$

Much of the thermal noise power is due to the noise generated in input stages of the receiver. Total noise power at the receiver input may be expressed as

$$N = kT_s B \qquad \text{watts}$$

where k is Boltzmann's constant (1.38×10^{-23} Joules/Kelvin) and T_s is the receive system noise temperature in Kelvins. This formula may be written in terms of decibels above a milliwatt (dBm)

$$N(\text{dBm}) = -198.6 + 10 \log B + 10 \log T_s$$

[2] The assumption of white noise is not strictly true for all sources of noise. For example, noise from galactic sources decreases with increasing frequency. However, for all practical purposes over the bandwidth of one channel, the noise spectrum may be considered to be flat.

For DTV transmission in the United States, the channel bandwidth is 6 MHz, so that the thermal noise limit for a perfect receiver at room temperature, N_t, is

$$N_t = 1.38 \times 10^{-23} \times 290 \times 6 \times 10^6 = 24.01 \times 10^{-15} \text{ W}$$

Converting to dBm, the thermal noise limit is -106.2 dBm. For the 7- and 8-MHz channels provided for in the DVB-T and ISDB-T standards, the thermal noise limit is -105.7 and -105.2 dBm,[3] respectively.

To determine the threshold receiver power, P_{mr}, required at the receiver, the threshold carrier-to-noise ratio and receiver noise figure, NF, must be added to the thermal noise limit. That is,

$$P_{\text{mr}} = N_t + C/N + \text{NF}$$

To determine the threshold power at the antenna, the line loss ahead of the receiver must be added and the receive antenna gain subtracted from the threshold receiver power:

$$P_{\text{ma}} = P_{\text{mr}} - G_r + L$$

For planning purposes in the United States, the FCC Advisory Committee on Advanced Television Service has recommended standard values for receiver noise figure, the loss of the receiving antenna transmission line, and antenna gain at the geometric mean frequency of each of the RF bands.[4] These planning factors are shown in Table 2-1. The resulting threshold received power at the antenna and receiver terminals is also shown in the last two lines of this table. Satisfactory reception is defined in terms of the threshold of visibility (TOV). For the U.S. DTV system this is set at a threshold C/N value of 15.2 dB.

A similar table for the DVB-T system using 8-MHz channels is constructed in Table 2-2. For this system, the theoretical threshold C/N for nonhierarchical transmission in a Gaussian channel ranges from 3.1 to 29.6 dB.[5] For Table 2-2,

TABLE 2-1. FCC Planning Factors and Threshold Power

Component	VHF		UHF
	Low	High	
Receiver antenna gain, G_r (dB)	4	6	10
Line loss, L (dB)	1	2	4
Noise figure, NF (dB)	10	10	7
Threshold C/N (dB)	15.2	15.2	15.2
Threshold power at antenna, P_{ma} (dBm)	−84.0	−85.0	−90.0
Threshold power at receiver, P_{mr} (dBm)	−81.0	−81.0	−84.0

[3] The equivalent noise bandwidth for an 8-MHz channel is actually 7.6 Mhz.

[4] *FCC Sixth Report and Order*, April 3, 1997, p. A-1.

[5] *ETS 300 744*, March 1996, pp. 38–41.

TABLE 2-2. DVB-T Minimum Receiver Signal Input Levels for 8-MHz Channels

	Band			
Component	I	III	IV	V
Receiver antenna gain, G_r (dB)	3	7	10	12
Line loss, L (dB)	1	2	3	5
Noise figure, NF (dB)	5	5	5	5
Threshold C/N (dB)	13.9	13.9	13.9	13.9
Threshold power at antenna, P_{ma} (dBm)	−88.3	−91.3	−93.3	−93.3
Threshold power at receiver, P_{mr} (dBm)	−86.3	−86.3	−86.3	−86.3

a $\frac{7}{8}$ inner code rate, Δ/T_u of $\frac{1}{8}$ and 16 QAM are assumed, yielding a threshold C/N of 13.9 dB needed to achieve a BER of 2×10^{-4} before R/S decoding. The corresponding payload data rate is 19.35 Mb/s. Since this is just one of many possible scenarios, the entries in this table should not be construed as planning factors. A significant difference between this table and Table 2-1 is the much lower receiver noise figure. In addition, different values of antenna gain and line loss are assumed for the upper and lower portions of the UHF band.

For the ISDB-T system, the theoretical minimum C/N required to achieve a BER of 2×10^{-4} is 16.2 dB, using the same channel bandwidth, modulation, inner code rate, and guard interval ratio[6] as assumed previously for DVB-T. The corresponding payload data rate is 18.93 megabytes per second (MB/s). Thus, in this example the DVB-T system is capable of better performance than the ISDB-T system by about 2.3 dB while achieving a somewhat higher data rate. In fact, the performance difference ranges from 1.4 to 2.7 dB for all possible inner code rates and modulation types. In the hierarchical mode, the DVB-T system requires higher C/N thresholds and achieves lower data rates.

At the time of this writing, an implementation loss of up to 1 dB has been measured on ISDB-T; for DVB-T the measured implementation loss is currently 2.7 dB.[7] As hardware and software developments proceed, performance improvements should be expected. At present, actual performance of both systems is about equal, but the greater potential for improvement is in favor of DVB-T.

EXTERNAL NOISE SOURCES

Although it is standard practice to make calculations as presented in Tables 2-1 and 2-2, this may not tell the complete story. These results represent the minimum power required in an environment limited to random noise, due to the receiver. To obtain the total system noise, the effect of antenna noise temperature, T_a,

[6] "Transmission Performance of ISDB-T," *ITU-R Document 11A/Jyy-E*, May 14, 1999.

[7] Yiyan Wu, "Performance Comparison of ATSC 8-VSB and DVB-T COFDM Transmission Systems for Digital Television Terrestrial Broadcasting," *IEEE Trans. Consumer Electron.*, August 1999.

and the noise contribution of the antenna-to-receiver transmission line must be included. The result is a fictitious temperature that accounts for the total noise at the input to the receiver. When the effects of antenna and line on total are included, the total noise power available at the receiver is

$$N = \frac{kT_aB}{\alpha_r} + (\alpha_r - 1)kT_0B + kT_rB$$

where α_r is the line attenuation factor, T_0 is the ambient temperature, and T_r is the receiver noise temperature. The antenna noise power is attenuated by the transmission line; the noise contribution of the line is added directly to the receiver noise. The receiver noise temperature is related to the noise factor, F, by

$$F = 1 + \frac{T_r}{T_0}$$

Receiver noise factor is related to noise figure by

$$NF = 10 \log F$$

Transmission line loss, L, is related to the attenuation factor by

$$L = 10 \log \alpha_r$$

With the inclusion of these factors, system noise temperature, referenced to the receiver input, is given by

$$T_s = \frac{N}{kB}$$

To illustrate the impact of the external noise sources, the equivalent noise temperature and noise power contributions for each of these components are listed in Table 2-3 for an assumed ambient temperature of 290 K. The receiver noise temperatures are computed from the noise figures given in Table 2-1 for the U.S. DTV system. The sum of all contributions is shown as the receive system noise floor. Two cases are shown. The first is a good approximation for rural areas, based on the curve labeled "rural" in Figure 2-1. The second is based on the curve labeled "suburban." These curves show the increasing effect of impulse noise at the lower frequencies. The antenna noise temperature is assumed to be equal to the values on these curves. The threshold signal required at the input to the receiver under the assumed conditions is also shown in Table 2-3. Since the total system noise already includes the receiver contribution, the threshold receiver signal is determined simply by adding the threshold C/N to the total noise floor.

The results shown for threshold signal level in Table 2-3 are higher than those in Table 2-1 and those normally published in DTV receiver noise budgets. This is because estimates of system noise are often published considering only the receiver noise figure and neglecting the contributions of the external sources through the receive antenna and transmission line-to-system noise.

TABLE 2-3. Antenna, Line, and Receiver Contributions to Noise in U.S. DTV Systems

Component	VHF Low	VHF High	UHF
Case 1: Rural			
Receiver temperature (K)	2610	2610	1450
Line temperature (K)	75	170	440
Antenna temperature (K)	3000	250	24
System noise temperature (K)	5070	2940	1900
System noise floor (dBm)	−93.8	−96.1	−98.0
Minimum receiver power (dBm)	−78.6	−80.9	−82.8
Case 2: Suburban			
Receiver temperature (K)	2610	2610	1450
Line temperature (K)	75	170	440
Antenna temperature (K)	189,000	15,700	1500
System noise temperature (K)	153,000	13,000	2490
System noise floor (dBm)	−79.0	−89.8	−96.9
Minimum receiver power (dBm)	−63.8	−74.6	−81.7

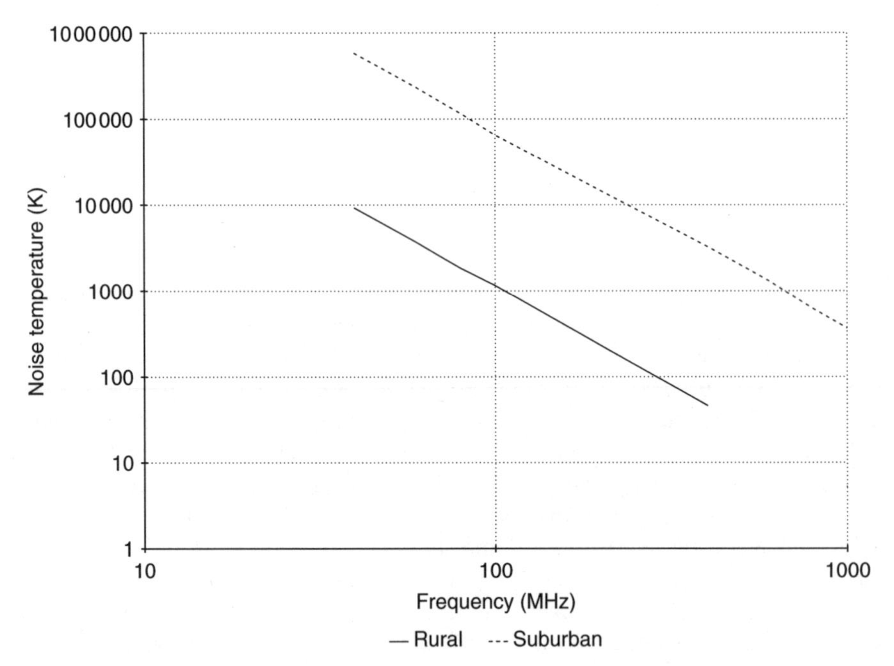

Figure 2-1. External noise temperature. (From *Reference Data for Radio Engineers*, 6th ed., Howard W. Sams, Indianapolis, Ind., 1977, p. 29-2; used with permission.)

Figure 2-1 and the calculations in Tables 2-1 and 2-3 show that the contribution of natural and man-made noise to the antenna and system noise temperature is highly dependent on location, whether in an urban, suburban, or rural environment. In suburban areas the system noise floor may be degraded by external sources by more than 2 dB at UHF; at low-band VHF, the degradation may be over 20 dB. Noise in urban areas may be 16 dB higher than in suburban locations. Rural areas may be quieter than suburban areas by 18 dB or more. Since urban and suburban receivers are more likely to be in areas of high signal strength, there is some justification for using the lowest values for antenna noise temperature to estimate the limits of coverage in many cases. UHF stations may expect to enjoy a 3- to 20-dB noise advantage over low-band VHF stations and a 3- to 6-dB advantage over high-band stations. The advantage due to lower noise level tends to compensate for the higher propagation losses experienced at the higher frequencies.

In practice, the line loss varies with receiver installation as well as frequency. The receiver noise figure varies depending on manufacturer, production tolerances, and frequency. In the tables it is assumed that outside antennas will be used. In those locations where an inside antenna is used, the minimum receive power is increased by the difference in antenna gain. This, too, varies from site to site. The antenna gain varies with manufacturer, production tolerances, and frequency. Thus the threshold receiver power must be understood for what it is — an estimate whose actual value in any given location depends on many site-specific variables.

The higher system noise level due to external sources is qualitatively consistent with field measurement in the United States. In the Charlotte, North Carolina, DTV field tests[8] there were six sites for which no cochannel interference was noted on Channel 6. The average noise floor recorded at these sites was −67.9 dBm; the minimum was −73 dBm and the maximum was −64 dBm. Adjusting these values for the VHF system gain of 25.5 dB results in an average noise floor of −93.4 dBm, a minimum of −98.5 dBm, and a maximum of −89.5 dBm. The equivalent receiver input noise power for the receiver used (NF = 6 dB) was −100.2 dBm, 1.7 dB below the minimum measured value (after adjustment for system gain). The minimum value was evidently measured at a rural location some 21 miles northeast of the transmitter site. Most (but not all) of the locations at which higher noise floors were observed appear to be at more urban or suburban sites. The location at which maximum noise was measured was a part of the Charlotte grid.

For UHF, the average noise floor recorded at the Charlotte field test sites was −71.0 dBm; the minimum was −71.9 dBm and the maximum was −68.2 dBm. Adjusting these values for the UHF system gain of 29.4 dB results in an average noise floor of −100.4 dBm, a minimum of −101.3 dBm, and a maximum of −97.6 dBm. The equivalent receiver input noise power for the receiver used

[8] *Field Test Results of the Grand Alliance HDTV Transmission System*, Association of Maximum Service Television, Inc., September 16, 1994.

(NF = 7 dB) was −99.2 dBm, 2.1 dB above the minimum measured value, 2.4 dB below the maximum measured value, and 1.2 dB above the average measured value (all after adjustment for system gain). From these data it may be concluded that use of only receiver input noise power is a much better predictor of noise floor at UHF. Variation in noise power from location to location is much less at UHF.

The impact of man-made noise at VHF is recognized in the Implementation Guidelines for DVB-T. Noise power is assumed to increase by 6 dB in band I and 1 dB in band III. No allowance is made for man-made noise in bands IV and V.

TRANSMISSION ERRORS

At least three different methods may be used to count transmission errors: segment error rate, bit error rate, and symbol error rate. Symbol error rate is defined as the probability of a symbol error before forward error correction coding. This quantity is often plotted as a function of C/N or the related quantity, E_b/N_0. The relationship between E_b/N_0 and C/N may be derived as follows.

The average carrier power may be written as[9]

$$C = \frac{E_s}{T}$$

where E_s is defined as the energy per symbol and T is the symbol time. The average energy per bit is therefore

$$E_b = \frac{C}{R_b} = \frac{E_s}{TR_b} = \frac{E_s}{\log_2 M}$$

where R_b is the transmission rate in bits per second and M is the number of levels. For example, for 8 VSB, $M = 8$, so that $E_b = E_s/3$. Dividing both sides by N_0, we see that E_b/N_0 is related to C/N by

$$\frac{E_b}{N_0} = \frac{C}{RN_0} = \frac{C}{N}\frac{B}{R_b}$$

Both quantities are usually expressed in decibels, so the latter expression is often written

$$\frac{E_b}{N_0}(\text{dB}) = \frac{C}{N}(\text{dB}) - 10\log\left(\frac{B}{R_b}\right)$$

[9] David R. Smith, *Digital Transmission Systems*, Van Nostrand Reinhold, New York, 1985, pp. 240–241.

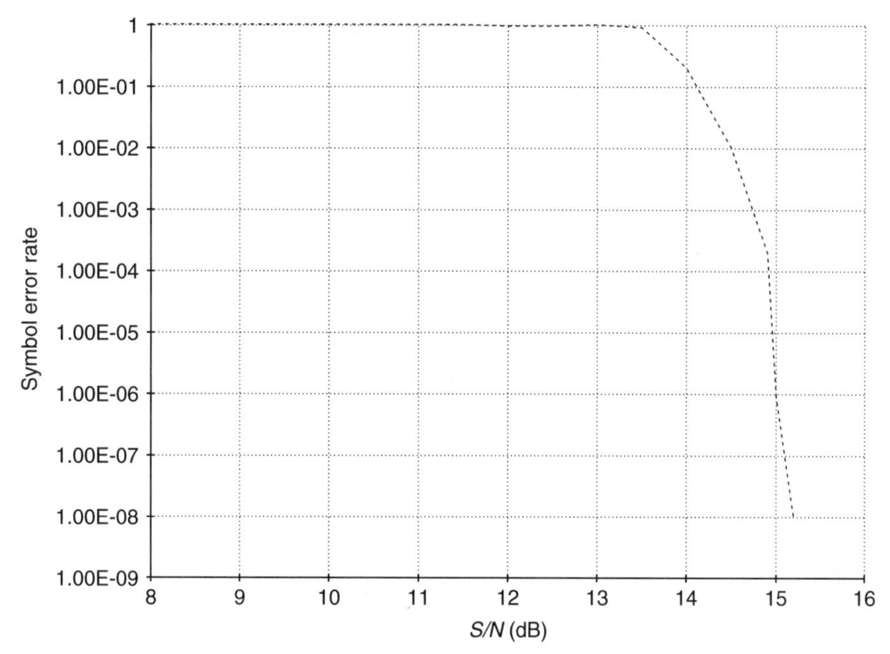

Figure 2-2. Symbol error rate versus S/N. (From Advanced Television Systems Committee, "Guide to the Use of the ATSC Digital Television Standard," *Document A/54*, ATSC, Washington, D.C., Oct. 4, 1995; used with permission.)

The receiver noise bandwidth is assumed equal to the channel bandwidth. This expression allows fair comparison of the relative performance of different systems with differing C/N thresholds and data rates on the basis of E_b/N_0, provided that the error rates are equivalent. If error rates are not equivalent, further adjustment is required. Bit error rate, the probability of a bit error before FEC, is also usually plotted as a function of E_b/N_0.

Segment error rate refers to the probability of an error in a data segment, after forward error correction. Measurements of the segment error rate versus S/N for the 8 VSB terrestrial broadcast mode is shown in Figure 2-2. It is apparent that the system is quite robust until the threshold level is approached. The TOV has been determined to occur for a segment error rate of 1.93×10^{-4}. Recalling that the segment length is 832 symbols and the symbol rate is 10.76 Msymbols/s, it is evident that the TOV corresponds to 2.5 segment errors per second. The equivalent BER is 3×10^{-6} after R/S decoding.

It is instructive to compare the DVB-T and ATSC systems using E_b/N_0. Such a comparison has been made by Wu,[10] who concludes that the ATSC system holds a theoretical advantage over DVB-T of about 1.3 dB. This advantage can be accounted for entirely by the more powerful R/S and convolutional codes

[10] Wu, op. cit., p. 3.

used for the ATSC system. As presently implemented, the advantage is 3.6 dB in the AWGN channel. Measurements on the ATSC system resulted in only a 0.4-dB implementation loss. With improvements, the implementation loss of both systems will be reduced, the DVB-T system having the greater potential improvement.

ERROR VECTOR MAGNITUDE

The quality of the in-band signal may also be expressed in terms of error vector magnitude. This a useful quantitative measure defined as the root-mean-square (RMS) value of the vector magnitude difference between the ideal constellation points, D_i, and the actual constellation points, D_a, of the $I-Q$ diagram, expressed in percent. An error signal vector e_i, may be computed at each symbol time:

$$e_i = D_i - Y_i$$

EVM is usually computed as an average over a large number, N_s, of samples, so that

$$\text{EVM} = \left(\frac{1}{N} \sum_{n=1}^{N_s} |e_i|^2 \right)^{1/2} \times 100\%$$

A perfect digital transmission system would exhibit an EVM of 0%.

The inverse relationship between EVM and C/N may be seen by considering the error signal to be noise. The C/N is simply the ratio of the RMS value of the desired constellation points to the RMS value of the noise:

$$\frac{C}{N} = 10 \log \frac{\sum_{n=1}^{N_s} D_i^2}{\sum_{n=1}^{N_s} |e_i|^2}$$

The relationship between C/N and EVM is illustrated in Figure 2-3, which is a plot of measured in-band performance at the output of an 8 VSB DTV transmitter. Over most of the range of measurements, EVM is inversely proportional to C/N. Only at low values of EVM is a more complex relationship evident.

Overall, EVM may be considered the best overall measure of in-band DTV performance. It takes into account all impairments that contribute to intersymbol interference (ISI), the underlying cause of symbol and bit errors. ISI is caused by any energy within one symbol time that would interfere with reception in another symbol time. In addition to noise, this energy may be due to dispersion within the channel due to linear distortion or timing errors caused by bandlimiting in the system. The channel response smears and delays the transmitted signal at the receiver. When ISI becomes sufficiently severe, the receiver mistakes the value of the transmitted symbols.

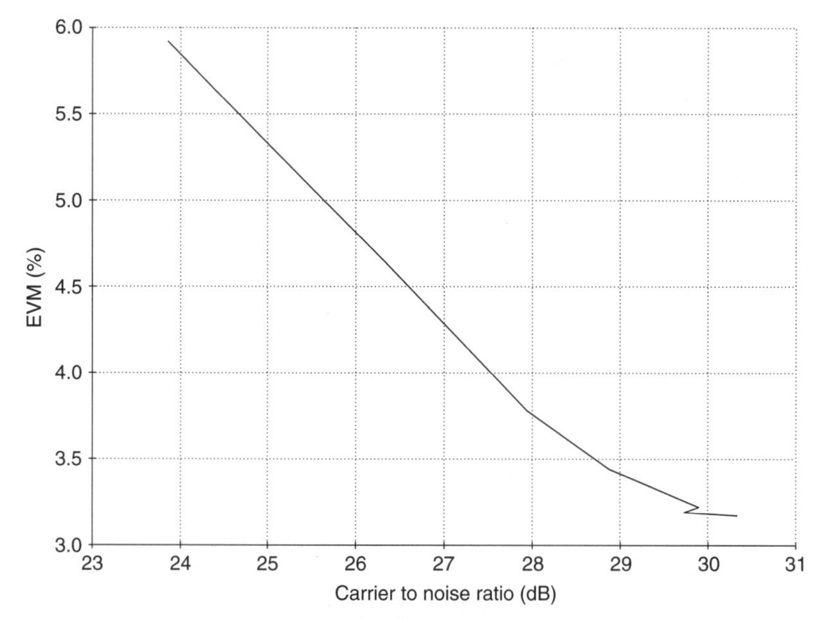

Figure 2-3. EVM versus C/N.

To minimize the effect of dispersion and maximize noise immunity and the resultant ISI in the ATSC system, the square pulses at the input to the 8 VSB modulator are shaped by means of a Nyquist filter. This low-pass linear-phase filter has a flat amplitude response over most of its passband, approximating an ideal low-pass filter. In practice, the ideal low-pass filter with infinitely steep skirts is not physically realizable. Therefore, the response of the Nyquist filter is actually made somewhat more gradual.

The pulse shape at the output of the Nyquist filter is very nearly described by the familiar sinc function,

$$\frac{\sin \pi t/T}{\pi t/T}$$

This function has the property that it is equal to zero at $t = \pm T$, $t = \pm 2T$, $t = \pm 3T$, and so on, but is equal to unity for $t = 0$. Thus pulses occurring at symbol times other than $t = 0$ do not contribute to received symbol power and there is no ISI. The sinc function may be multiplied by any other function without changing the timing of the zeros, thus preserving the property of no ISI. The usual choice is to multiply by a function having a root-raised-cosine response characteristic. The resulting pulse shape is a modified sinc function:[11]

$$\frac{\sin \pi t/T}{\pi t/T} \frac{\cos \alpha_N \pi t/T}{1 - (4\alpha_N^2 t^2/T^2)}$$

[11] Smith, op. cit., p. 210.

This pulse shape also has the property that it is equal to zero at $t = \pm T$, $t = \pm 2T$, $t = \pm 3T$, and so on, but is equal to unity for $t = 0$. As with the sinc function, pulses occurring at symbol times other than $t = 0$ do not contribute to received symbol power and there is no ISI, provided that the half-power bandwidth is $1/2T$. The tails of this pulse decrease at a faster rate than the ideal low-pass filter, so that timing precision is not as critical.

The half-power bandwidth of this filter is often referred to as the Nyquist bandwidth, which for the U.S. DTV system is 5.38 MHz. The full filter bandwidth is greater by a factor of $1 + \alpha_N$, so that the channel bandwidth is 1.1152×5.38 or 6 MHz. Outside this frequency band, the response is zero.

EYE PATTERN

The eye pattern is a convenient method for visually and qualitatively assessing the ISI and C/N performance of a digital transmission system. The signal is displayed on an oscilloscope set to trigger at the symbol time. The persistence of the scope creates a composite of all possible waveforms. At each level of the signal, the overlapping waveforms produce a pattern that resembles the human eye. The degree to which the eyes are open is a measure of the ISI and hence signal quality. The eye openings should be greatest at the sampling time. Eyes open 100% correspond to an EVM of 0%.

Approximate quantitative measure of C/N may be made from visual assessment of the eye pattern.[12] The signal amplitude is represented by the center-to-center distance between symbol levels, V. Noise is represented by the accumulated thickness of the intersecting lines at each symbol level, ΔV. The log ratio of these distances, $20 \log(V/\Delta V)$, is an estimate of C/N ratio. For example, if $\Delta V = 0.1$ V, the C/N value is 20 dB. Similar closing of the eyes in the horizontal dimension may be an indication of timing jitter.

INTERFERENCE

Although the noise floor is a useful concept for estimating the maximum extent of coverage, in the real world interference is often present, tending to place further limits on coverage. Interfering signals may originate with cochannel and adjacent channel stations. Signals further removed in frequency may be either harmonics or intermodulation products (IPs). In any case, the level of these signals at their source is usually outside the control of the stations with which they interfere. During the transition period, the interference may come from both analog and digital TV signals. Only the effects of interference on the digital television signal will be discussed.

[12] Luobin and K. Cassidy, "Analyze QAM Signals with Oscilloscope Eye Diagram," *Microwaves and RF*, January 1998, p. 115.

The geographical locations, channels, and effective radiated power (ERP) of existing analog stations are presently fixed. Digital stations will generally be located at or near analog sites with sufficient power to replicate analog service. These factors and the channel assignments of the digital services are the starting point for interference analysis. Propagation of interfering signals is dependent on the same factors that affect the desired signal. Given the AERP, tower-siting parameters, and operating channel, the time-varying signal level at a specific location depends on distance, topographical factors, atmospheric conditions, and sources of multipath. The methods described in Chapter 8 may be used to estimate these levels using the parameters of the interfering station(s).

COCHANNEL INTERFERENCE

Cochannel signals are the desired signals for stations in other markets. The digital TV receiver detects a cochannel digital signal as just another source of noise. When the C/N value is less than threshold due to the combination of thermal sources and interference, reception will fail. The thermal noise and interfering signal powers are additive, so that the $C/(N+I)$ threshold is increased in inverse proportion to the cochannel interference. This is illustrated in Figure 2-4, which shows the C/N threshold as a function of carrier-to-interference ratio, C/I, for a pair of 8 VSB DTV stations operating on the same channel, assuming an

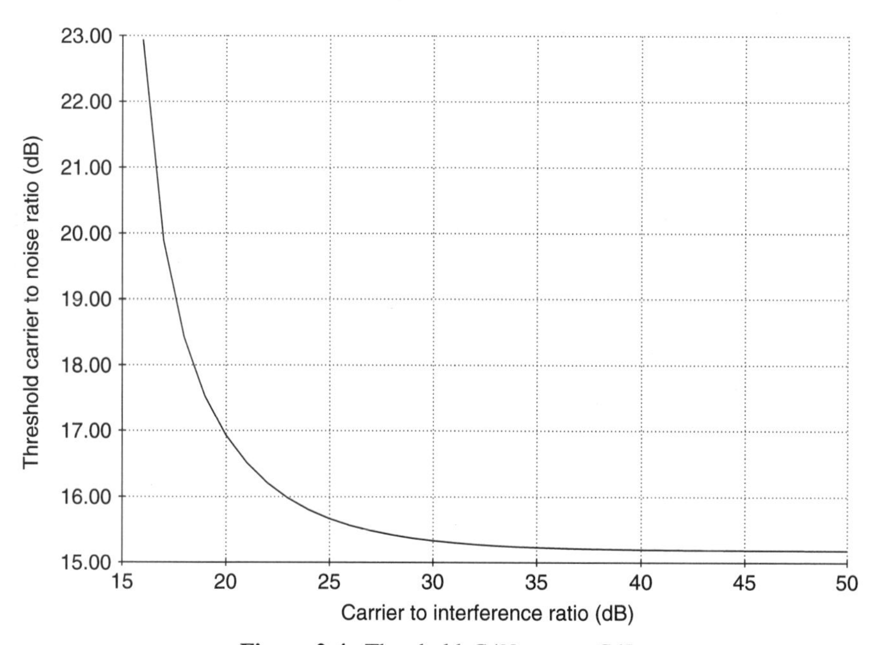

Figure 2-4. Threshold C/N versus C/I.

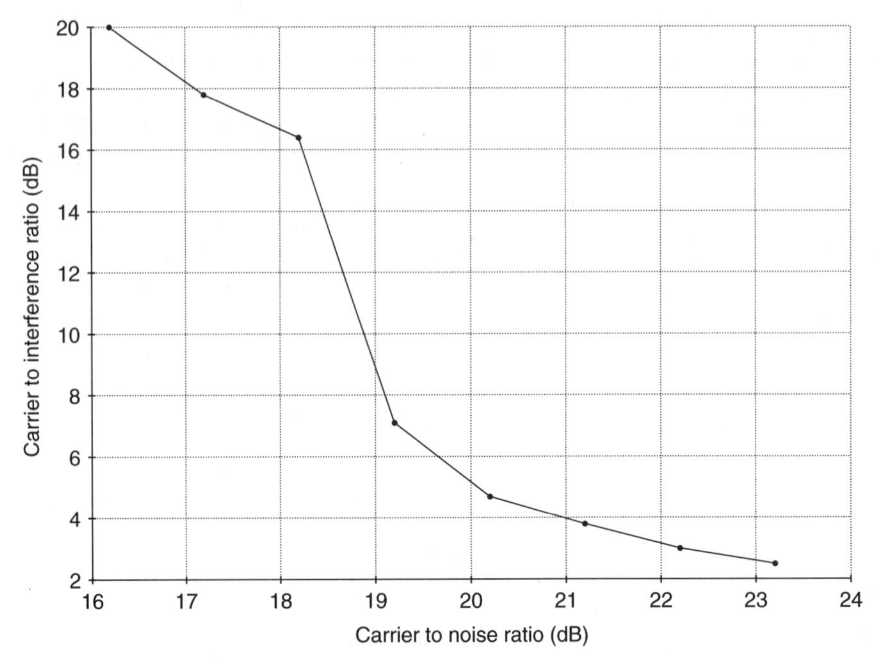

Figure 2-5. C/I versus C/N. (From *DTV Express Training Manual*; used with permission.)

omnidirectional receiving antenna. When the C/I value is high, say greater than 35 dB, the threshold C/N approaches 15.2 dB, the interference-free value. As the interference increases, the threshold C/N increases correspondingly. It is apparent that near the edge of noise-limited coverage, the minimum C/I is much higher than in areas where C/N is high. At the edge of noise-limited coverage, no interference can be tolerated.

During the transition period, cochannel analog stations may also affect digital TV service. Results of tests at the Advanced Television Test Center (ATTC)[13] on the 8 VSB system using a noncommercial laboratory DTV demodulator are shown in Figure 2-5. From these tests it is apparent that for low DTV carrier-to-noise ratios ($C/N < 19.2$ dB) the critical desired-to-undesired ratio, D/U, is 7 dB. That is, the desired signal must exceed the undesired signal by 7 dB for acceptable DTV reception. At the critical D/U value the DTV picture was impaired with black squares frozen in time and the audio was muted. At lower D/U, the picture, sound, and data failed completely. The laboratory DTV demodulator was equipped with a NTSC cochannel rejection filter, which automatically switched in when the C/I value fell below 16.5 dB. At high carrier-to-noise ratios ($C/N > 23$ dB), the critical D/U is 2 dB. At the time of this writing, no data are available with consumer-grade receivers. A

[13] Charles Rhodes, *DTV Express Training Manual*, Harris Corporation, Melbourne, FL, p. 2–9.

NTSC cochannel rejection filter may not necessarily be present in all commercial receivers.

The referenced test data were the basis of channel allocations in the United States, where a cochannel analog-to-digital protection ratio of 1.8 dB is used. For DVB-T, the corresponding protection ratio is 4 dB. In general, no improvement should be expected from the use of precise carrier offset by analog transmitters since interfering signals may come from any one of many stations. Some analog stations may offset the visual carrier by 10 kHz, with a tolerance of up to ± 1 kHz.

In the absence of a NTSC offset, offsetting the DTV pilot frequency by 28.615 kHz above its normal frequency has the effect of placing the NTSC visual, chroma, and aural signals near the nulls of the NTSC reject filter in DTV receivers[14] equipped with these filters.

ADJACENT CHANNEL INTERFERENCE

An adjacent channel signal may be the desired signal for another station or the result of third- or higher-order intermodulation products generated in the power amplifier of other transmitters. Whatever the source, these signals appear as spurious sidebands in the adjacent channel just outside the desired channel. When generated by digital TV transmitters, these components appear as noise. The FCC and DVB-T masks define strict limits on this interference at the output of the transmitter system. The noise remaining after application of the mask adds to the noise from other sources.

There is very little difference between the effect of interference from upper or lower digital TV sidebands. The effect is much the same as that of digital-to-digital cochannel interference, except that the levels are offset by 45 dB. This is approximately equivalent to the total noise power relative to the average in-band power resulting from use of an emissions mask.

ANALOG TO DIGITAL TV

For analog transmitters on adjacent channels, the major concern is for the visual, color, and aural carriers interfering with the digital TV signal via the adjacent channel reject bands of the receiver. There are no artifacts produced until a critical D/U value is reached. The critical D/U is -48 dB for a lower adjacent NTSC and -49 dB for an upper adjacent NTSC signal. Thus, the undesired signal may be as much as 49 dB greater than the desired signal. At a D/U above this level, reception fails abruptly — there is no picture, sound, or data.

The sidebands of an adjacent analog station may also interfere with the in-band digital TV signal. For example, the specification for lower sideband reinsertion

[14] C. Eilers and G. Sgrignoli, "Digital Television Transmission Parameters — Analysis and Discussion," *IEEE Trans. Broadcasting*, Vol. 45, No. 4, December 1999, pp. 365–385.

for NTSC transmitters is only 20 dB. To the writer's knowledge, no tests have been conducted to measure the magnitude of this effect.

TRANSMITTER REQUIREMENTS

To provide the desired C/N at the receiver, it is necessary that the transmitter system produce a signal free of noise and of both linear and nonlinear distortions at sufficiently high power to cover the service area. The average power is the parameter to which the performance of a digital TV transmitter system is referenced. A typical time-domain signal envelope for an 8-VSB modulated signal is shown in Figure 2-6. Note the great variability in the envelope peaks. Obviously, there is no regularly recurring peak corresponding to the familiar sync pulse of analog television. Neither is there a predictable peak envelope signal level. However, the average power level is constant. A similar envelope could be plotted for OFDM signals. Therefore, a digital TV station's power is normally stated in terms of the average transmitter output power (TPO) or average effective radiated power.

Aside from being a constant for any combination of video, audio, and data, there are many advantages to using average power to characterize a digital television transmission system. Techniques for measurement of average power are well developed. For high-power systems, use of a calorimeter provides a useful means of calibrating average system output power. Other average reading instruments using methods that depend on development of a dc level due to heating of the sensing element may also be used. Because the average power level is constant in time and the spectrum is constant over most of the channel bandwidth, relative power levels may readily be determined by integration of the radiated spectrum.

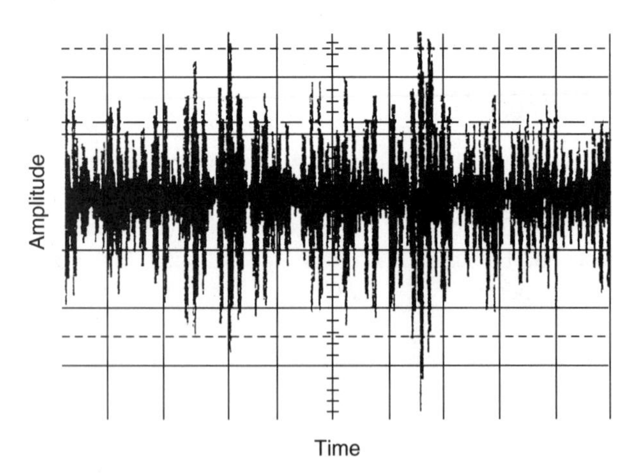

Figure 2-6. Time-domain envelope of RF DTV signal. (Photographed by Bob Plonka, Harris Corporation; used with permission.)

Although average power is used to establish transmitter and station ratings, knowledge of the peak power is not unimportant. Sufficient headroom must be allowed in the intermediate and final amplifiers of the transmitter as well as any other nonlinear components in the transmission chain to avoid excessive compression — thus the requirement for very linear power amplification. Moreover, the peak power rating of transmission lines, filters, RF systems, and antennas must be specified properly to avoid voltage breakdown.

The model numbers of most transmitters are usually assignal in terms of a peak rating. This seems to be due more to the tradition of rating analog transmitters in terms of their peak sync rating than of any definite measurement. Nevertheless, measurement of peak power is much more difficult and in many cases not necessary. It is common practice to estimate the peak-to-average ratio (PAR) based on reasonable assumptions, and to multiply this factor by the average power to obtain an estimate of peak power. PAR may be estimated on the basis of the cumulative distribution function (CDF). This is illustrated using data from 8 VSB tests in Figure 2-7. Note that PAR does not exceed 7.5 dB about 99.99% of the time. For the COFDM signal, the corresponding PAR is about 10 dB. These figures are considered to be adequate estimates for specifying the allowable compression in a power amplifier. Peak-to-average ratios in 8 VSB transmitters of up to 11 dB have been reported. The higher values might be used to select a transmission line conservatively based on the breakdown voltage or

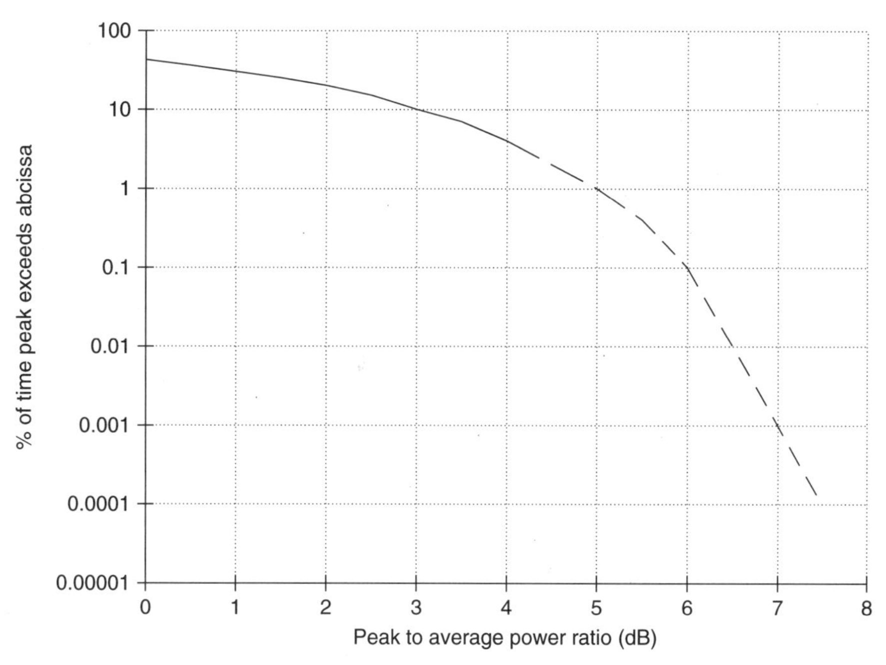

Figure 2-7. Typical CDF of 8 VSB signal. (From *DTV Express Training Manual*; used with permission.)

to specify the headroom in components that must be extremely linear, such as the exciter.

The large signal peaks relative to the average signal level result in compression in the output stages of practical power amplifiers. The resulting nonlinearity produces higher-order intermodulation products, which are observed as spectral spread. Figure 2-8 shows the spectrum of an 8 VSB signal measured at the output of a typical power amplifier. The signal within the channel bandwidth is essentially constant except for a spike in the spectrum on the left side. This spike is due to the presence of the pilot signal. Outside the channel bandwidth there is evidence of spectral regrowth. In an ideal linear amplifier, the energy in this region would be limited to the levels generated in the exciter. The measured level in a practical amplifier depends on the extent to which the amplifier is driven into compression during the signal peaks as well as the specific nonlinear characteristic of the amplifier.

Spectral spread is an extremely important parameter, in that it determines the level of interference to digital and analog stations allocated to adjacent channels. From the viewpoint of the adjacent channel, the out-of-band energy simply becomes another source of noise. Even if there is no adjacent channel allocation, the FCC and DVB-T specifications require stringent limits to out-of-band signals. For example, the mandated FCC radiation mask is shown in Figure 2-9. The 0-dB reference is set at the average in-band power level. In practical terms the mask requires that all signals outside the 6-MHz channel allocation be 36.7 dB below the average in-channel level, decreasing in linear fashion to 99.7 dB below the in-channel level at frequencies 6 MHz above and below

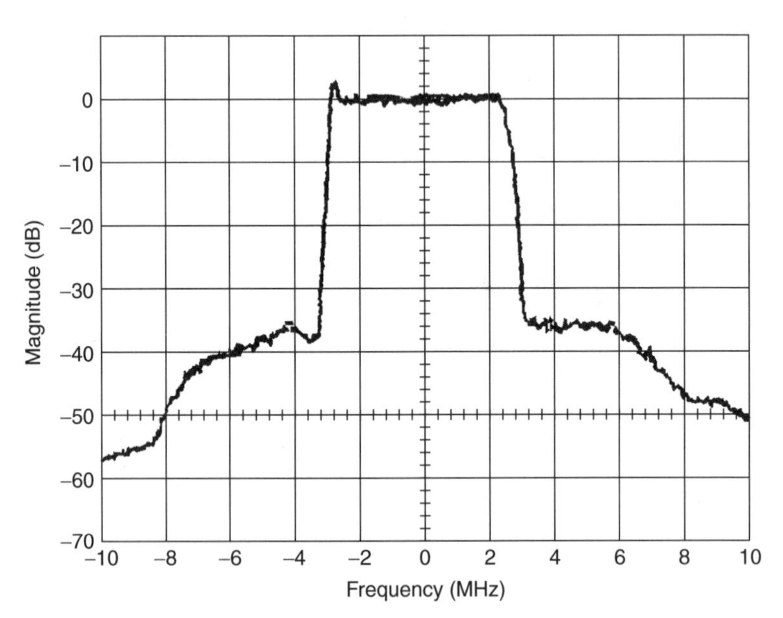

Figure 2-8. Typical transmitter output spectrum.

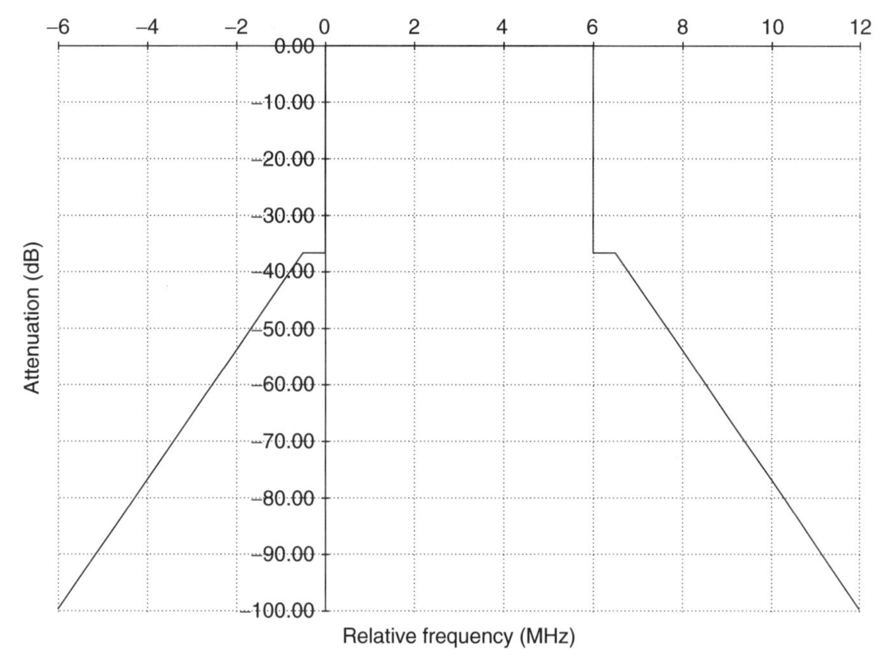

Figure 2-9. FCC emissions mask.

the channel edge. As the mask is plotted here, it is assumed that the resolution bandwidth of the spectrum analyzer is fixed to a sufficiently small value (say, 30 kHz) to properly present the in-band and out-of-band measurements. Similar masks for the DVB-T system are shown in Figures 3-10 and 3-11. Needless to say, to achieve the levels required by the applicable masks requires a power amplifier with adequate headroom, precise power control, stable and precisely controlled precorrection circuits, and a well-designed output bandpass filter.

AM-to-AM conversion is the primary mechanism by which spectral regrowth occurs. This term is used to describe the degree to which the transmitter output voltage is directly proportional to the input. Output phase may also be a function of input level. The deviation from ideal linear phase is described as AM-to-PM conversion. A perfectly linear transmitter would produce no AM/AM or AM/PM.

Consider a practical amplifier (or any other quasilinear component). In the time domain, the output, S_o, of a third-order nonlinearity may be described as a function of the input, S_i, as follows:

$$S_o = gS_i + g_3S_i^3$$

where g is the gain of the amplifier in the linear region of the transfer function and g_3 represents the degree of third-order nonlinearity. Thus, if g_3 is zero, the amplifier is ideal, having no AM/AM or AM/PM conversion. If g_3 is nonzero, the amplifier is nonlinear. The resulting AM/AM and AM/PM will give rise to third-order intermodulation products and spectral regrowth.

The nonlinear transfer function is illustrated in Figure 2-10 for the case of $g = 10$ (20 dB) and $g_3 = -1$. Over most of the input range, the output increases in direct proportion to the input. Note that when the input signal is about unity (0 dB), the compression is approximately 1 dB. As the input signal increases to a value of 1.8 (5 dB), no further increase in output occurs. The compression at this point is about 3.4 dB. Beyond this point, increases in the input produce less output signal. Some practical amplifiers, such as klystrons, actually exhibit this type of nonlinear characteristic, in which the output is reduced when the input signal increases beyond a limiting value.

AM-to-PM conversion may be illustrated by considering the third-order term in the transfer function to be complex. In this case we simply write

$$g_3 = g_{3I} + jg_{3Q}$$

where g_{3I} is the in-phase component of the nonlinear term and g_{3Q} is the quadrature component. Nonlinear phase and amplitude are illustrated in Figure 2-11 for the case of $g_{3I} = g_{3Q} = -1$. As the input signal increases, the output phase lags. At the 1-dB compression point, the incidental phase shift is about $-7°$, typical of many practical amplifiers.

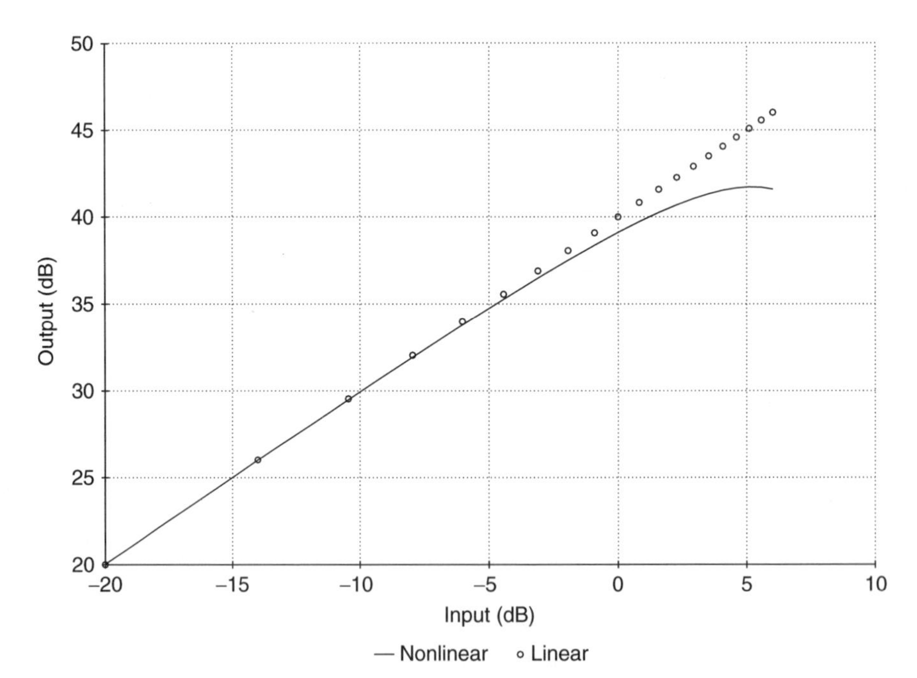

Figure 2-10. Nonlinear amplification.

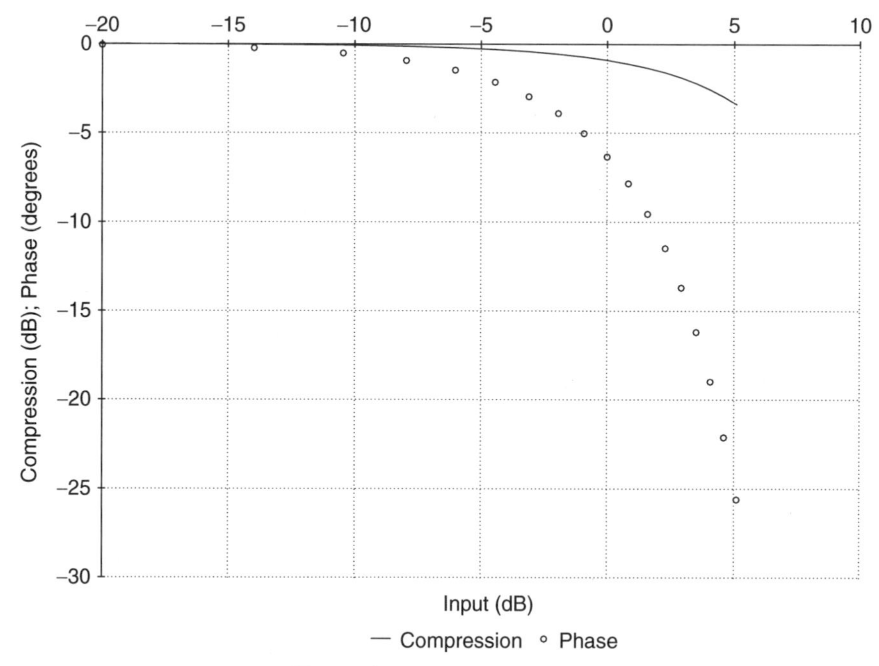

Figure 2-11. Nonlinear phase.

The relationship between AM/AM and spectral regrowth may be illustrated by considering the input signal, S_i, to be composed of two tones, S_1 and S_2, at angular frequencies ω_1 and ω_2, respectively; that is,

$$S_i = S_1 + S_2$$

When these signals are inserted into the expression for S_o, the result is

$$S_o = g(S_1 + S_2) + g_3(S_1 + S_2)^3$$

Expanding the cubic term and after some algebra, we obtain

$$(S_1 + S_2)^3 = S_1^3 + S_2^3 + 3S_1^2 S_2 + 3S_1 S_2^2$$

It is apparent that the fundamental signals have been preserved and amplified. However, additional signals, S_1^3, S_2^3, $3S_1^2 S_2$, and $3S_1 S_2^2$, have been generated. It is well known that these new signals are, among others, at frequencies $3\omega_1$, $3\omega_2$, $2\omega_1 - \omega_2$, and $2\omega_2 - \omega_1$. The third harmonic signals at $3\omega_1$ and $3\omega_2$ are well outside the channel and may be filtered. However, the difference signals at $2\omega_1 - \omega_2$ and $2\omega_2 - \omega_1$ are in and near the channel and represent intermodulation products (IPs) or spectral regrowth. Recognizing that the digital signal may be described as a continuous spectrum, it is apparent that the continuous spectral regrowth due to nonlinearity shown in the measured data is to be expected. In fact, the amplitude of the intermodulation products could be computed using this type of model.

In the absence of other noise sources, spectral regrowth places a minimum value on the transmitted in-band noise floor. Consider the case when ω_1 and ω_2 are closely spaced and near the center of the channel, say at 2.5 and 3.5 MHz above the channel edge. In this case the IPs will be 1 MHz above and below the fundamental tones, well within the channel bandwidth. These intermodulation products represent noise with respect to the desired signal. Again considering the digital signal as a continuous spectrum, it is evident that transmitter system nonlinearity places an upper limit on the carrier-to-noise ratio.

There are noise sources other than thermal and third-order products due to nonlinear distortions within a transmitter system. For example, quantization noise in digital-to-analog converters and oscillator phase noise contribute to total noise power. In a well-designed and well-maintained transmitter system, these sources should be small.

3

CHANNEL CODING AND MODULATION FOR DIGITAL TELEVISION

As discussed in Chapter 1, the purpose of the exciter is to convert the digital input signal from the transport layer to an on-channel RF signal for use as a drive signal for the power amplifier. Two major exciter functions are required to accomplish this task: (1) channel coding and modulation, and (2) upconversion and linear amplification. Further details of the these functions are now discussed. These functions are further subdivided into data synchronization, channel encoding, sync and pilot insertion, Nyquist filtering or spectral shaping, and quadrature modulation.

For the ATSC system, the on-channel RF signal is a single-carrier 8 VSB signal; for the DVB-T and ISDB-T systems, the modulated RF signal is a coded orthogonal frequency-division multiplex signal. There are many features common to these systems, all of which are designed to provide robust delivery of the digital signal in the presence of noise, multipath, and interference. The similarities are seen most clearly in the channel coding, which includes data synchronization, randomization, forward error correction, and interleaving. Even the modulation processes are similar, in that multiple bits are arranged in multivalued symbols to modulate the RF carrier(s). However, the modulation methods also present fundamental differences. The 8 VSB modulation is a single-carrier approach in which the spectrum of the serial data symbols fill the full channel bandwidth. COFDM is a multicarrier technique in which the data symbols are transmitted in parallel over many carriers. The symbol rate for each modulated carrier is much less than the total symbol rate, so that the spectrum of the parallel data symbols fill only a small part of the channel.

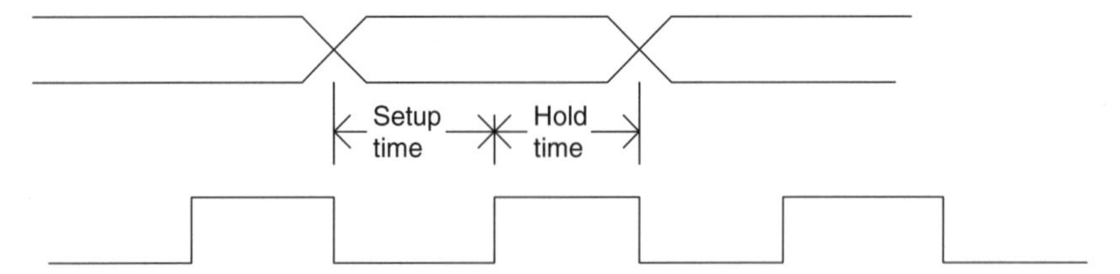

Figure 3-1. Input signals from transport data stream. (From Harris CD-1 Exciter Technical Manual, 7/7/97; used with permission.)

DATA SYNCHRONIZATION

At the exciter input, the DTV, DVB-T, or ISDB-T transmitter functions as a baseband receiver for the transport data stream. The exciter input interface must perform the functions of clock and data recovery, synchronization, and stabilization.

For the ATSC system, the DTV transport signal arrives at the input to the exciter with a bit rate clocked at a frequency derived from the studio clock, which is nominally at 27.000 MHz. This data clock is required to be accurate to within ±54 Hz of its nominal value of 19.392659 MHz. It is essential that any variation in the data clock rate be removed by the exciter clock and data recovery circuits.

Use of an embedded clock as required by the SMPTE 310M standard is a simple form of clock and data distribution. The timing signal is distributed as a part of the data stream. Timing is recovered from the incoming data using the bit transitions.

The timing relationship of the SMPTE 310M clock and data signals is shown in Figure 3-1. The period of one clock cycle is 51.6 ns. After recovery of the clock signal, the relative timing of the clock and data signal must be held as shown. The setup time should be greater than 20 ns; the hold time should be greater than 10 ns. Data acquisition time should be less than 1 ms.

To accommodate the data-rate expansion due to adding R/S parity bytes and convolutional coding, phase-locked-loop (PLL) frequency synthesizers must be used to convert the 19.39-MHz clock to 10.76 MHz and multiples thereof. The required clocks are distributed to the R/S and trellis coders.

RANDOMIZATION/SCRAMBLING

Periodic bit patterns can occur in the transport stream which, if not corrected by scrambling, would create discrete spectral lines in the modulated RF signal. Scrambling of the data is thus used to minimize the length of strings of 0's or 1's, suppressing the discrete spectral components. Scramblers thus "whiten" the data by producing data streams that contain bits in sequences in which the state of each bit is independent of adjacent bits, all pseudorandom sequences being equiprobable.

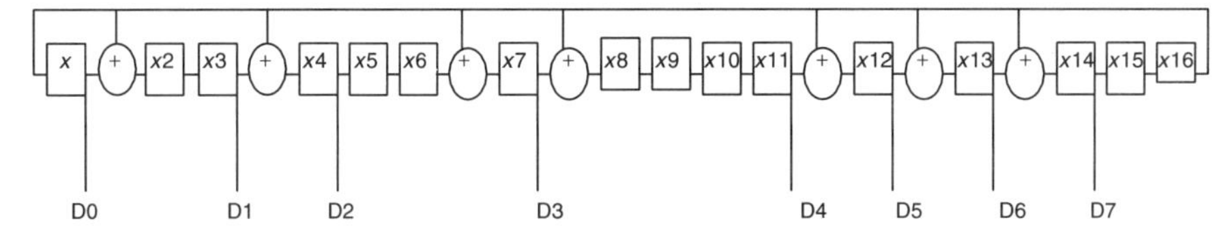

Figure 3-2. ATSC data randomizer.

Operation of the randomizer for the ATSC system is illustrated in Figure 3-2. The input data are logically summed, modulo 2, with a pseudorandom sequence produced in accordance with a generator polynomial. The pseudorandom sequence is generated by using a shift register with specified feedback connections to the adders. (A shift register is a cascade of several flip-flops.) When a shift register receives a clock pulse, the binary state of each flip-flop is transferred to the next flip-flop in the cascade. The feedback connections are located at eight positions along the cascade. The tapped signal is added at each connection and fed to the next flip-flop. The sequence can be predicted from knowledge of the shift register length and tap locations. Thus the resulting sequence is known as pseudorandom. The length of the pseudorandom sequence is determined by the length of the shift register, the position of the feedback taps, and the initial states of the flip-flops, which are preloaded to 1 during the data segment sync interval. The randomizer generator polynomial is

$$G_{(16)} = x^{16} + x^{13} + x^{12} + x^{11} + x^7 + x^6 + x^3 + x + 1$$

The bit rate at the output of the randomizer is the same as the input data rate. The shift register is clocked at the frequency recovered from the transport data stream.

The pseudorandom bit stream at the output of the randomizer may be represented in the frequency domain by its Fourier transform. For the non-return-to-zero (NRZ) transport bit stream, the Fourier transform is the familiar $\sin x/x$ or sinc function. This function is maximum at the bit sample time, with amplitude zeros occurring at multiples of the bit time. The bandwidth of the sinc function is infinite. Further processing is therefore required to limit the bandwidth to that of the channel.

FORWARD ERROR CORRECTION

Forward error correction is used to detect and correct errors in the transmitted data at the receiver. Extra bits are added to the scrambled baseband data for this purpose. The encoder accepts data from the randomizer at the payload data rate, f_p, and introduces redundancy into the bit stream so that the receiver can use it to detect and correct errors in the coded data. The output of the coder is at some higher data rate.

FEC enables the achievement of a desired BER with a significantly lower E_b/N_0, allowing a low-level high-bit-rate signal to be received in a higher noise environment than would otherwise be possible. The improvement in the value of the threshold E_b/N_0 is referred to as coding gain. For example, if the threshold value of E_b/N_0 is 15 dB without FEC and 12 dB with FEC, the coding gain is 3 dB. By implementing FEC, the effect of transmitting a much higher power level is achieved but at greatly reduced expense for equipment and prime power. Since the transmitted bit rate is higher with FEC, coding gain is achieved at the expense of increased channel bandwidth or the need to transmit a more complex symbol constellation. The digital channel bandwidth is fixed by regulatory agencies; thus the choice must be made for the more complex constellations.

There are two types of FEC used in the DTV, DVB-T, and ISDB-T systems. Although differing in implementation details, each system uses a combination of block codes and convolutional codes. Since these codes are linked in a cascade or series configuration, they are said to be concatenated. The block code is the outer code and is encoded first; the outer code is followed by the inner code. By using complementary inner and outer codes, very large coding gains may be achieved. This is especially important for systems such as space communications and digital television, in which the data are compressed; such systems are especially susceptible to transmission errors and thus require low SER or BER at low C/N. This is illustrated in Figure 3-3, which shows the typical performance of a space communications channel for three cases: with concatenated codes, trellis

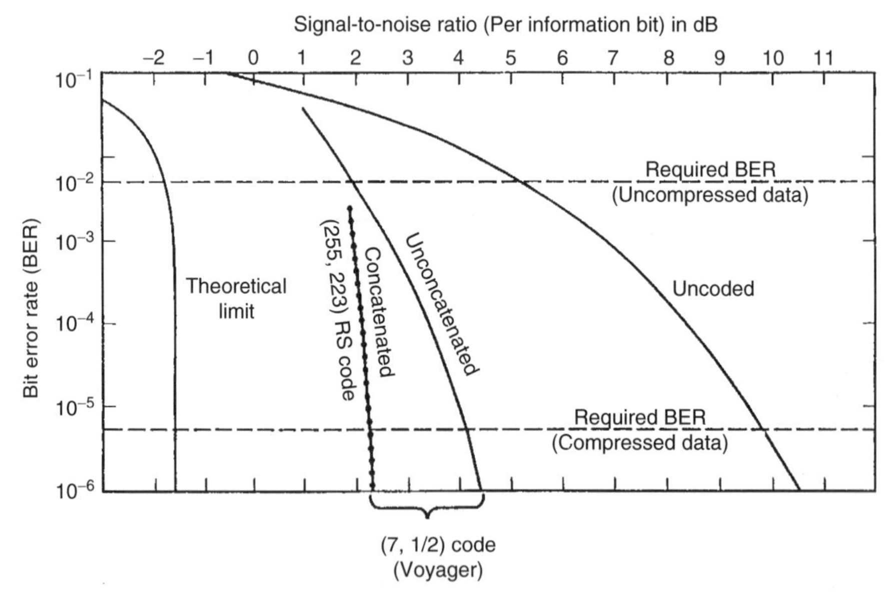

Figure 3-3. Typical performance curves for concatenated and unconcatenated coding systems for the space channel. (From Ref. 1 © 1994 IEEE; used with permission.)

code only (unconcatenated), and no FEC.[1] Use of the trellis code alone results in coding gains of 3 to 6 dB relative to the uncoded curve. (On this graph, coding gain is the difference in C/N between coded and uncoded curves at a specific BER.) Concatenating a block code with a trellis code results in over 2 dB improvement in coding gain for the compressed data but almost no improvement for the uncompressed data. The required BER for the compressed data is more than three orders of magnitude better than the uncompressed.

In the receiver, the order of the block and trellis codes is reversed. The block decoder is used to correct errors due to impulse noise and analog cochannel interference as well as short burst errors generated in or otherwise remaining after the convolutional decoder. As the name implies, block codes divide the data sequence into blocks, processing these blocks independently by adding the redundancy dictated by the desired code. The block codes used in the DTV, DVB-T, and ISDB-T systems are known as Reed–Solomon codes, named for their discoverers, Irving Reed and Gustave Solomon.[2] R/S codes are linear codes based on the mathematics of fields that can be described completely by their size. These finite fields are often called Galois fields after the French mathematician who discovered them. Finite fields are sets of numbers over which all calculations are performed. The input to the calculations and the their results must be numbers contained within the field.[3]

In R/S encoding, the randomized input data are divided into blocks, each block having a dimension of k_b bytes. A code word of n_b bytes in length is constructed by adding $n_b - k_b$ redundancy or error correction bytes to each block. The R/S code notation is therefore (n_b, k_b). To implement a R/S code, the clock rate must be increased by the ratio of the coded word length to the payload block dimension. When all bytes are encoded, the block code data rate, f_b, is

$$f_b = \frac{n_b f_p}{k_b}$$

For the ATSC system, n_b/k_b is 207/187 and the input data rate is 19.39 Mb/s, so that the output data rate (less syncs) is

$$f_b = (1.106952)(19.392659) = 21.47108 \text{ Mb/s}$$

A R/S code of length n_b and dimension k_b is capable of correcting up to t_b byte errors,[4] where

$$t_b = \frac{k_b - n_b}{2}$$

[1] S. B. Whicker and V. K. Bhargava, *Reed–Solomon Codes and Their Applications*, IEEE Press, New York, 1994, p. 27.

[2] Ibid., p. 18.

[3] A. D. Houghton, *The Engineer's Error Correcting Handbook*, Chapman & Hall, London, 1997, p. 14.

[4] Whicker and Bhargava, op. cit., pp. 4–7, 61.

Thus the R/S code used in the ATSC system is capable of correcting up to 10 byte errors per block. For the DVB-T and ISDB-T systems, n_b/k_b is 203/187, or 1.085562. This R/S code is capable of correcting up to eight byte errors per block. This accounts, in part, for the higher-threshold C/N or E_b/N_0 required by the latter systems.

Although R/S coding increases the bit rate by some 10%, this increase is not sufficient to increase the complexity of the transmitted constellation. For example, in the ATSC system, the bit rate at the output of the R/S coder could be transmitted at 2 bits/symbol within the Nyquist bandwidth of 5.38 MHz. The unencoded data rate of 19.3 Mb/s would also require 2 bits/symbol within this bandwidth. Thus no appreciable penalty is paid to obtain the coding gain of the R/S code.

INTERLEAVING

Interleaving and complementary deinterleaving in the receiver is a process for decorrelating burst errors, extending the power of block encoding to correct a larger number of errors. By interleaving a code of a given length, the code can correct a quantity of errors that would require a much longer code without interleaving. The error-correcting power of a longer code is obtained without the potential spectral efficiency penalty of a higher code rate.

There are many ways to interleave the encoded data. In general, the data are read into a memory in the order in which they are output from the FEC encoder and read out in a different order. For example, the blocks of data may be written into a memory as rows of a matrix and read out as columns, thus reordering the data. As a result, consecutive data bytes are spread out over a longer period of time. Should the data be corrupted in transmission, burst errors will be reordered when deinterleaved in the receiver and thus distributed over a similar long period. The block interleaver in the ATSC system is a diagonal byte interleaver that operates conceptually as described. A key difference is that the data are read into the channel as ordered by the matrix diagonals rather than columns.

INNER CODE

Trellis codes are most effective for coping with random errors such as those due to white noise. They are not very effective in coping with large consecutive losses of data, such as might occur with analog television cochannel interference or impulse noise. In fact, when the trellis code capacity is exceeded, a burst error is generated at the output. For these reasons, the trellis code is concatenated with the R/S block code to obtain coding gain for both types of a data loss and to obtain the synergy resulting from both codes and associated interleaving.

Unlike block codes, trellis codes operate on the data sequence without dividing it into large, independent blocks. Instead, the data are processed continuously. The

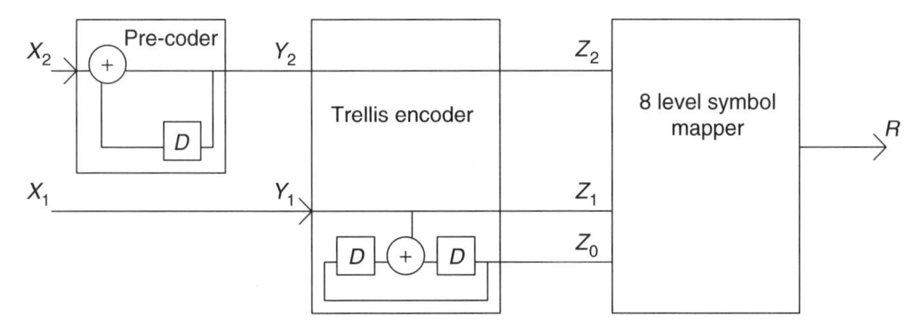

Figure 3-4. Block diagram of ATSC precoder, trellis encoder, and mapper. (From ATSC DTV Standard A/53, Annex D; used with permission.)

encoder divides the data into short blocks and outputs a new sequence of greater length. For these linear codes the coder output is a modulo-2 sum of present and previous inputs. The name is derived from the graphical representation of the encoder states as a function of symbol time which resembles a trellis.[5] Trellis codes are also called convolutional codes. The process resembles the mathematical process called convolution—hence the name.

The encoding process is illustrated by reference to the ATSC trellis encoder shown in Figure 3-4. This trellis coder is a $\frac{2}{3}$-rate device in which the two input bits are encoded to three output bits. The serial data stream from the R/S interleaver is divided into 2-bit blocks. One redundant bit is added for each pair of R/S-coded data bits. At the input to the encoder, the two input bits, Y_1 and Y_2, are encoded to three parallel output bits, Z_0, Z_1, and Z_2. This is accomplished by encoding Y_1 into a pair of output bits, Z_0 and Z_1. Output bit Z_1 is equal to Y_1, but Z_0 is the output of a $\frac{1}{2}$-rate convolutional coder, a shift register operating on Y_1. Output bit Z_2 is equal to Y_2.

In the ATSC implementation, Y_2 is actually precoded for the receiver cochannel interference filter. This is accomplished by modulo 2 adding the input bit X_2 with Y_2 delayed by 12 symbol clock cycles. Since the precoder encodes the input bit to only one output bit, the overall trellis code rate remains at $\frac{2}{3}$. The unencoded bit is $X_1 = Y_1 = Z_1$.

The 12-symbol delay, D, in the precoder and trellis encoder accounts for the intrasegment interleaver employed, shown schematically in Figure 3-5. Every twelfth symbol is processed as a group in trellis encoder and precoder 0; every next twelfth symbol is processed in coder 1, and so on, until 12 groups have been processed. The outputs of the trellis encoders and precoders are then multiplexed to produce the completed sequence for input to the modulator.

After trellis encoding and interleaving of each data segment, the state of the output multiplexer is advanced by four symbol times without advancing the state of the trellis encoders. This allows time for insertion of the data segment

[5] Wesley W. Peterson and E. J. Weldon, Jr., *Error-Correcting Codes*, MIT Press, Cambridge, Mass., 1972, pp. 413–421.

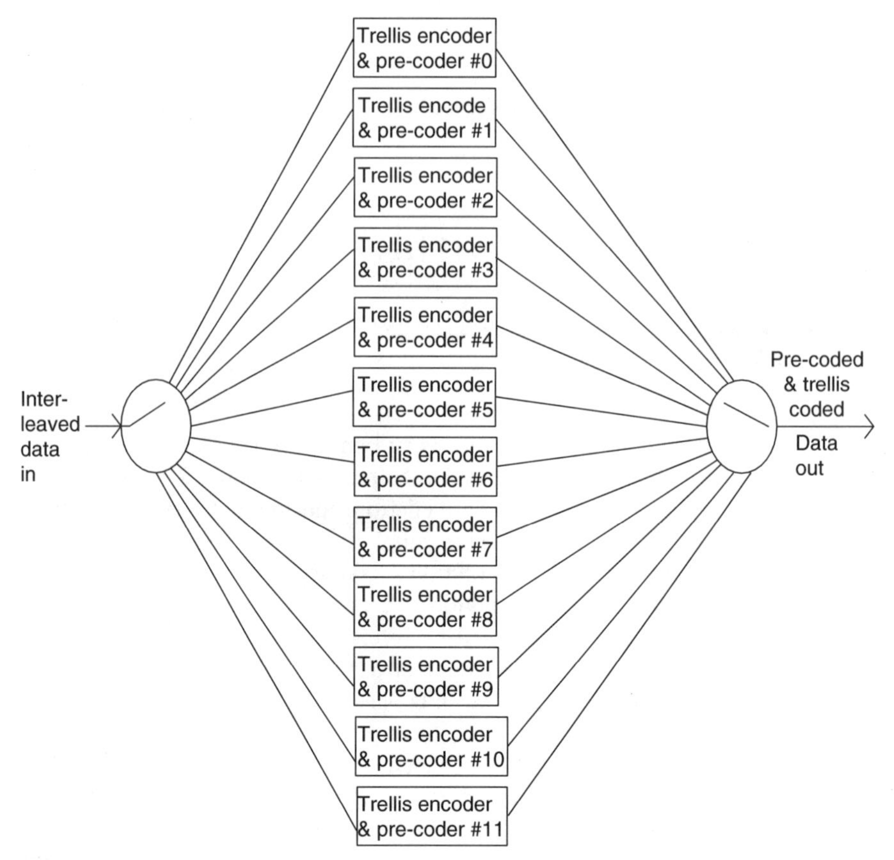

Figure 3-5. Trellis code interleaver. (From ATSC DTV Standard A/53, Annex D; used with permission.)

sync, a four-symbol sequence. Thus the next segment is processed with encoders 3 through 11 followed by encoders 0 through 3. The result is illustrated in Table 3-1 for the first three segments of a frame. In segment 0, blocks 0 through 68 contain 12 data bytes each for a total segment length of 828 bytes. The remaining segments comprising the frame follow. Given their location in the data processing chain, it is apparent that the data segment sync bytes are not subject to either R/S or trellis coding.

For the trellis coder, the encoded data rate is

$$f_t = \frac{n_t f_b}{k_t} \qquad \text{where} \quad n_t = k_t + 1$$

For the ATSC trellis coder, $k_t = 2$, so that the transmission rate is now $(\frac{3}{2})$ (21.47) or 32.20 Mb/s. The trellis coder outputs are then mapped into 2^{n_c} constellation points in signal space. For the ATSC system, $2^{n_c} = 8$, the eight levels required

TABLE 3-1. Interleaving Sequence

	Block			
Segment	0	1	···	68
0	D0 D1 ··· D11	D0 D1 ··· D11		D0 D1 ··· D11
1	D4 D5 ··· D3	D4 D5 ··· D3		D4 D5 ··· D3
2	D8 D9 ··· D7	D8 D9 ··· D7		D8 D9 ··· D7

TABLE 3-2. Map of 8 VSB Constellation Points

Z_2	Z_1	Z_0	R
0	0	0	−7
0	0	1	−5
0	1	0	−3
0	1	1	−1
1	0	0	+1
1	0	1	+3
1	0	1	+5
1	1	1	+7

for VSB modulation. These 3-bit symbols are clocked at a symbol rate $\frac{1}{3}$ of the trellis-coded data rate. The mapping of the constellation points is shown in Table 3-2, labeled with their binary and decimal equivalents. This is the unfiltered baseband 8 VSB signal.

The trellis coding and mapping process has the effect of expanding the constellation from 2 bits per symbol or four levels, to 3 bits per symbol or eight levels. Doubling the number of constellation points increases the power required at the threshold of detection, assuming no change in the separation between points and fixed noise and interference power. Fortunately, this effect is more than offset by the increase in the minimum distance, d_m, between sequences of the encoded signal. This is a measure of the difference between sequences or the number of bits that must be changed to construct one sequence from the other. The overall C/N gain due to coding and modulation[6] is given by

$$\text{gain (dB)} = 10 \log \left(\frac{d_m^2}{4 - \Delta P} \right)$$

Bingham shows that for a four-state trellis code as used in the ATSC system, $d_m = 6$; ΔP will be shown later to be 6.2 dB. Thus the overall gain in C/N is 3.3 dB.

[6] John A. C. Bingham, *The Theory and Practice of Modem Design*, Wiley, New York, 1988, pp. 341–345.

FRAME SYNC INSERTION

In the ATSC system, the trellis-coded and interleaved data are next multiplexed with the frame sync signals, a full data segment inserted at the start of each field. A fixed pseudorandom data sequence is transmitted in the first 511 symbols after the segment sync.

QUADRATURE MODULATION

The processes considered to this point convert the serial transport data stream to a pseudorandom sequence and add the parity bits needed for forward error correction. The output of these processes is parallel, multilevel symbols at a rate consistent with the expanded data rate. This signal must now modulate an RF carrier for transmission on one of the many channels allocated for digital television.

Just as for analog signals, there are three fundamental methods of digital modulation: amplitude, frequency, and phase. If the symbols are applied to the modulator as square pulses, these modulation methods are known as amplitude-shift keying, frequency-shift keying, and phase-shift keying, indicating that the value of the appropriate parameter is shifted instantaneously as a function of the value of the symbol. For the COFDM system, pulse shaping is not used; thus keying is the more appropriate descriptor, even though this term is often used interchangeably with modulation. In the single carrier 8 VSB system square pulses are not used. The pulses representing the symbols are shaped to limit the bandwidth. Therefore, it is appropriate to describe the process as modulation rather than keying.

Various combinations of amplitude and phase modulation or keying are used for each of the digital transmission systems. The ATSC system may be considered as digital amplitude modulation since the data are conveyed by discrete levels of the RF waveform. The DVB-T and ISDT-T systems convey the data by discrete values of both amplitude and phase and thus produce constellations with both in-phase and quadrature components. The instantaneous amplitude of the waveform in the time domain is determined by both the value of the symbol and, if pulse shaping is applied, by the transition path from symbol to symbol.

8 VSB

For the 8 VSB system, the output of the processes converting the serial transport data stream to a pseudorandom sequence and adding forward error correction consists of parallel multilevel symbols at a rate of 32.28 Mb/s, including sync symbols and a dc offset. The symbol rate is one-third of the encoded data rate, or 10.76 MHz. The symbols are assigned numeric values at each of eight equally probable, equally spaced levels: ± 1, ± 3, ± 5, ± 7. This is a one-dimensional constellation providing maximum immunity to noise. Symbols occur at regularly spaced intervals — the symbol time.

8 VSB is single carrier modulation format, one of a broad class of M-ary modulation schemes. There are $m = 3$ bits transmitted for every symbol giving rise to the $M = 8$ levels in accordance with

$$m = \log_2 M$$

Aspects of the waveform shape, modulator block diagram, probability of error, and bandwidth are now discussed.

At the input to the modulator, the average power, P_a, is the mean of the sum of the squares of the symbol values multiplied by the symbol rate. That is,

$$P_a = \frac{2f_s(1^2 + 3^2 + 5^2 + 7^2)}{8}$$

This is identical to the result obtained from the general equation for signal power in a single-dimensional M-ary system[7]

$$P_a = \frac{f_s(M^2 - 1)}{3}$$

Dividing this expression by the symbol rate, the energy in a single pulse, E_s, is

$$E_s = \frac{M^2 - 1}{3}$$

Ignoring the transition paths between constellation points as a result of pulse shaping, the peak power is

$$P_p = f_s(M - 1)^2 = 49f_s$$

so that the minimum peak-to-average power ratio is

$$\begin{aligned} \frac{P_p}{P_a} &= \frac{3(M-1)^2}{M^2 - 1} \\ &= \frac{3(M-1)}{M+1} = \frac{3(7)}{9} \end{aligned}$$

or 3.7 dB. Due to the Nyquist filter, the transitions cannot be ignored and the peak-to-average ratio after modulation is in excess of 6 dB.

In double-sideband amplitude modulation, the carrier and both sidebands are transmitted. Since there is no information in the carrier, and both sidebands contain identical information, the carrier and one of the sidebands may be suppressed with no loss of data. The result is a significant improvement in both power and bandwidth efficiency. In practice, complete removal of one sideband is not feasible, and vestigial sideband modulation is used. In VSB, a portion of

[7] Ibid., p. 85.

the unwanted sideband is transmitted along with the complete desired sideband. To facilitate recovery and regeneration of the carrier in the receiver, a low-level pilot at the carrier frequency is retained. Thus most, but not all of the advantages of single-sideband suppressed carrier (SSB-SC) modulation are enjoyed.

Vestigial sideband modulation can be generated by filtering a double-sideband signal or by processing the baseband signal. The latter method is preferred. The baseband signal, $x(t)$, is the sequence of the eight level symbols at the output of the trellis coder. This may be written as

$$x(t) = \sum_i d_i \delta(t - iT)$$

where d_i is the series of pulses representing the symbols and δ is the Dirac delta or impulse function, which is nonzero only when $t = iT$. This signal is applied to the Nyquist or baseband shaping filter, which has an impulse response of $h_0(t)$ and frequency response of $H_0(\omega)$, centered on zero frequency. The baseband filter impulse and frequency responses are related by the Fourier transform and its inverse. To preserve one sideband while suppressing the other, the Nyquist filter response is offset from zero frequency by one-fourth of the symbol rate, or 2.69 MHz. This is accomplished by splitting the baseband signal into two signals that are equal in magnitude but with a 90° phase relationship. This is equivalent to multiplying the impulse response of the shaped symbol pulse by $e^{j\pi t/2T}$. It is appropriate to describe the Nyquist filter with its offset response as low pass since its passband extends from 0 to 5.38 MHz. For the ATSC system, the lower sideband is discarded, so that only the upper sideband is retained.

The splitting and phase-shifting operations implement a complex mathematical operation called a Hilbert transform. Ideally, a Hilbert transform preserves the amplitude spectrum but shifts the phase of one component relative to the other by 90° at all baseband frequencies. Although the ideal Hilbert transform is physically unrealizable, it can be approximated by a signal splitter and a pair of all-pass networks that produce the 90° phase difference.

As a result of Nyquist filtering and application of the Hilbert transform, the unmodulated signal may be represented in the time domain as the convolution of the baseband impulse response with the offset Nyquist filter impulse response. Thus, the in-phase signal, $x_i(t)$, plus the quadrature signal, $x_q(t)$, may be written

$$x_i(t) + x_q(t) = \sum_i d_i \delta(t - iT) \otimes [h_0(t) e^{j\pi t/2T}]$$

where $\otimes$ represents convolution. Applying Euler's formula, the in-phase and quadrature components of the shaped baseband signal may be written

$$x_i(t) = b + \sum_i d_i \delta(t - iT) \otimes \left[h_0(t) \cos \frac{\pi t}{2T} \right]$$

and

$$x_q(t) = \sum_i d_i \delta(t - iT) \otimes \left[h_0(t) \sin \frac{\pi t}{2T} \right]$$

A DC offset, b, has been added to the in-phase component to generate the pilot. The in-phase and quadrature signals may be applied separately to the baseband inputs of a quadrature modulator. This is indicated by the signals labeled I and Q in Figure 1-8. A local oscillator operating at the intermediate carrier frequency is split into equal quadrature components with the outputs applied to the inputs of the modulator. The resulting IF signals are combined in the hybrid to form the desired VSB signal. This signal may be represented mathematically as

$$S_v(t) = \left\{ b + \left[\sum_i d_i \delta(t - iT) \right] \otimes \left[h_0(t) e^{j\pi t/2T} \right] \right\} e^{j\omega_c t}$$

The in-band spectrum is the Fourier transform of $S_v(t)$ and is identical to that of the shaped baseband signal except that it is translated upward in frequency. The spectrum is thus dependent only on the shape of the baseband signal in the frequency domain and the frequency response of the modulator, which ideally would be constant. Except at the pilot frequency, the spectrum is smooth, since the randomizer is used to assure random data at the input to the Reed–Solomon coder and hence to the modulator. From the expression for S_v, it is readily seen that the amplitude of the pilot is constant and the pilot frequency is the same as the carrier frequency. The spectrum is centered at a frequency one-fourth of the symbol rate, or 2.69 MHz above the pilot.

As noted earlier, the pilot amplitude is determined by the dc component added to the baseband signal. For the ATSC system, b is specified to be 1.25. As a result, the pilot level is 11.3 dB below the average in-band power level, which was shown earlier to be 21. That is,

$$\text{pilot amplitude} = \frac{10\ \log(1.25)^2}{21} = -11.3\ \text{dB}$$

BANDWIDTH

There are two common definitions of the modulated signal bandwidth. In both, the bandwidth is defined in terms of power spectral density. The half-power or 3-dB bandwidth is defined as the difference between frequencies at which the power spectral density is half the peak value. For the ATSC system, this is required to be 5.38 MHz.

Bandwidth may also be defined in terms of the spectral mask. In this case the power must be attenuated to the levels specified by the FCC as shown in Figure 2-9. Under this definition, the signal bandwidth is also dependent on the nonlinear characteristics of the transmitter. This is discussed in detail in Chapters 2 and 4. The channel bandwidth is inherent in this definition. For the ATSC system, the power spectral density at the upper and lower edges of the 6-MHz channel is required to be 36.7 dB below the average in-band value.

ERROR RATE

The symbol error rate for multilevel signals such as 8 VSB is determined primarily by the number of and minimum distance between the constellation points. In the presence of noise, the most likely error is that a symbol will be mistaken for the closest adjacent symbol. For convenience of calculation, the error rate is evaluated in the presence of an additive white Gaussian noise distribution even though there are other important sources of noise, such as impulse noise and interference from other signals. Under the conditions of an AWGN distribution, the symbol error rate is proportional to the error rate for binary signals.[8] The probability of error for binary signals is the area under the normal distribution integrated from the carrier noise to infinity. This function is plotted versus C/N in Figure 3-6. The error rate for multilevel signals may be obtained by scaling this curve to the right by an amount equal to the change in C/N. The scale factor, $\Delta C/N$, is

$$\frac{\Delta C}{N} = 10 \log \frac{M^2 - 1}{3}$$

For 8 VSB, $M = 8$, so that $\Delta C/N = 13.2$ dB. The usual practice is to plot error rate versus E_b/N_0. Recall from Chapter 2 that

$$\frac{E_b}{N_0} = \frac{C}{N}\frac{B}{R_b}$$

where B/R_b is the inverse of the bandwidth efficiency. This has the effect of moving the curve back to the left by $10 \log(5.38)$, or 7.3 dB. The result is a plot of symbol error rate versus E_b/N_0, as shown in Figure 3-7.

Without the data-rate expansion resulting from trellis coding, a 4 VSB system ($M = 4$) could have been implemented. In this case, $\Delta C/N = 7.0$ dB. Thus, in the absence of the error-correcting capability of the trellis code, the additional required received power, ΔP, for 8 VSB would be 6.2 dB for the equivalent symbol error rate.

COFDM

Coded orthogonal frequency-division multiplex is a multicarrier modulation method in which many closely spaced carriers occupy the channel. The channel is thus divided into many overlapping subbands. Frequency-division multiplex (FDM) refers to the set of evenly spaced carriers. Coded refers to the use of channel coding to combat frequency-dependent fading and degradation of the symbol or bit error rate, similar to that used for 8 VSB. Orthogonal refers to the relationship between the multiple carriers. The integral of the product of the

[8] Ibid.

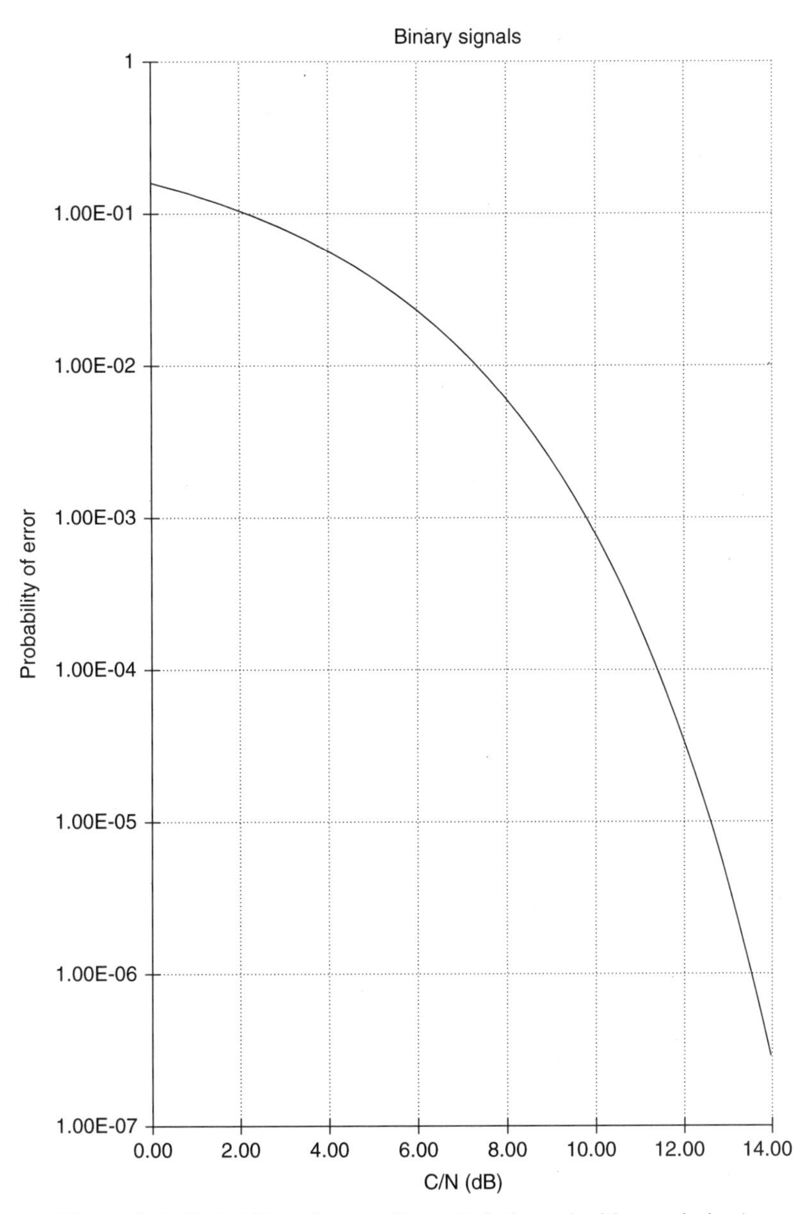

Figure 3-6. Probability of error. (From Ref. 6; used with permission.)

time-domain signal of any carrier and the time-domain signal of any other carrier over the active symbol time is ideally zero. This arrangement assures that the sidebands overlap in such a way that they can be received without significant intercarrier interference. The carriers are spaced in frequency by an amount equal to the inverse of the active symbol interval.

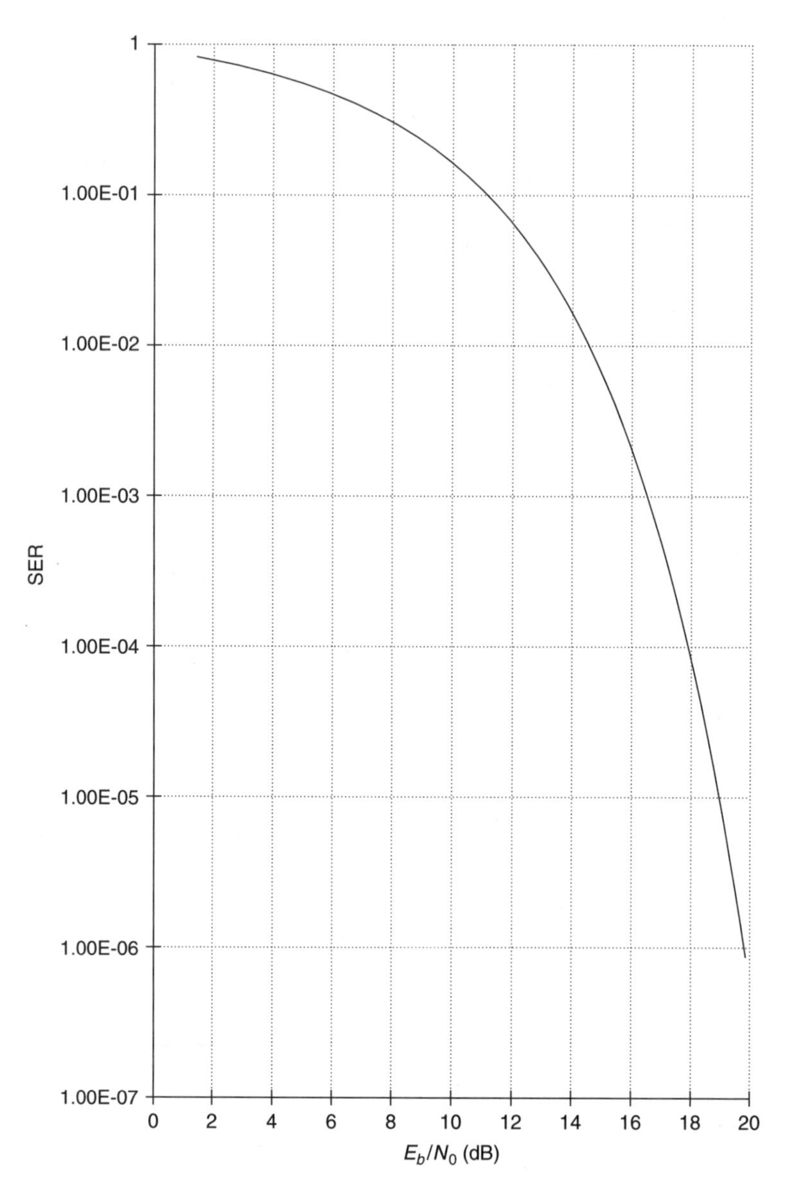

Figure 3-7. 8 VSB SER versus E_b/N_0.

COFDM is very robust in the presence of interference and linear distortions due to multipath. For similar reasons, COFDM is useful for single-frequency networks. Depending on the specific modulation format employed on each carrier, the DVB-T and ISDB-T systems support a wide range of payload bit rates. The average transmitter power is much less than that required for analog TV transmission.

Unlike 8 VSB, which is a single-carrier system, COFDM is a parallel system in which the high-speed serial data stream is transmitted as a multiplexed set of lower-speed data streams. The spectrum of any one modulated subband occupies only a small part of the available channel bandwidth, unlike 8 VSB, in which the spectrum of each data symbol occupies the entire available bandwidth. As a result, relatively few of the COFDM carriers and associated data symbols are affected by frequency-dependent fades. Since the variation of attenuation and delay across each subband is greatly reduced, burst errors due to fading or interference may distort some but not all of the data transmitted. The complete transmitted data stream may be reconstructed from the symbols received on the less-affected carriers. Since the distortion within each subband is small, equalization of the subbands is relatively simple.

The set of closely spaced carriers is generated by an Inverse Fast Fourier Transform (IFFT). The carrier phases and data timing of the separate subbands are arranged to maintain a relatively flat power spectrum for the composite signal and to permit separation of the overlapping subbands without significant intersubband interference. Nearly ideal performance can be achieved if the number of carriers is large enough.

The symbol period is divided into an active interval, T_u, and a guard interval, Δ. The total symbol interval is the sum of T_u and Δ. Data are transmitted only during the active interval. The purpose of the guard interval is to overcome the effects of multipath signals that are delayed less than Δ. All received signals with a delay less than Δ add constructively with the direct signal. Cochannel signals in single-frequency networks combine in a similar manner.

Since the data symbols are not subject to Nyquist shaping, the power spectral density of each carrier, $P_{kc}(f)$, is the familiar sinc function

$$P_{kc}(f) = \left[\frac{\sin \pi(f - f_{kc})T}{\pi(f - f_{kc})T}\right]^2$$

where $f_{kc} = f_c + k'/T_u$ and $k' = k_c - (K_{\min} + K_{\max})/2$; $K_{\min} < k_c < K_{\max}$. The channel center frequency is f_c and k_c is the carrier number, an integer.[9] The power spectral density for a few subbands with zero guard interval is illustrated in Figure 3-8. Obviously, the individual carrier spectra are not band limited. However, the sum of these spectra, shown in Figure 3-9, is bandlimited.

A plot of the highest five carriers of an 8-MHz channel in the DVB-T 2k mode (1705 carriers) using a guard interval ratio Δ/T_u of $\frac{1}{4}$ is shown in Figure 1-11. Since the total symbol time is greater than the inverse of the carrier spacing, the main lobe of the power spectral density of each carrier is slightly narrower than twice the carrier spacing, and the transmitted subbands are not strictly

[9] "Framing Structure, Channel Coding and Modulation for Digital Terrestrial Television (DVB-T)," *ETS 300 744*, p. 35.

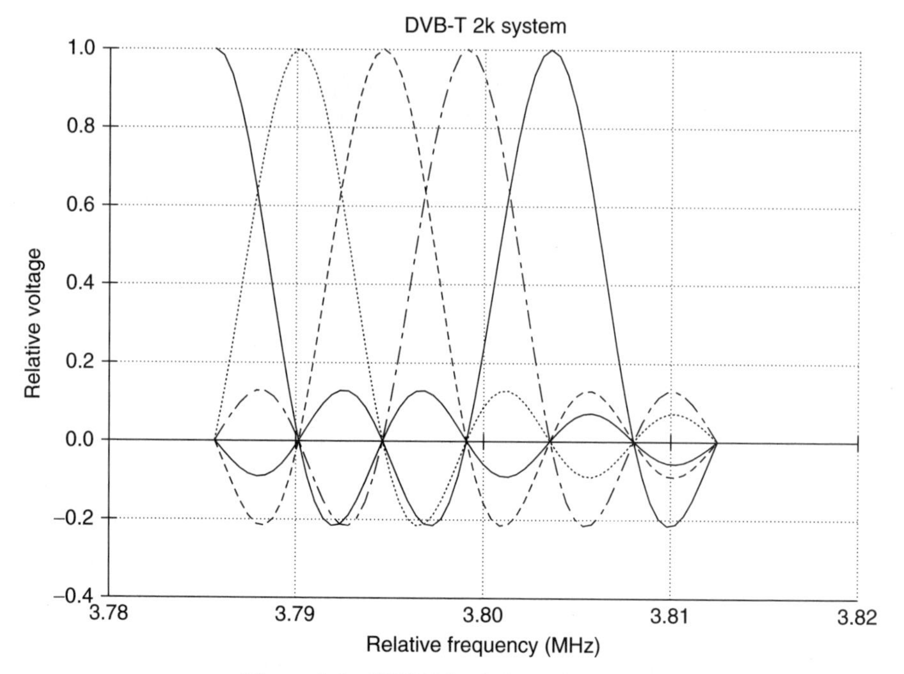

Figure 3-8. COFDM subchannel spectra.

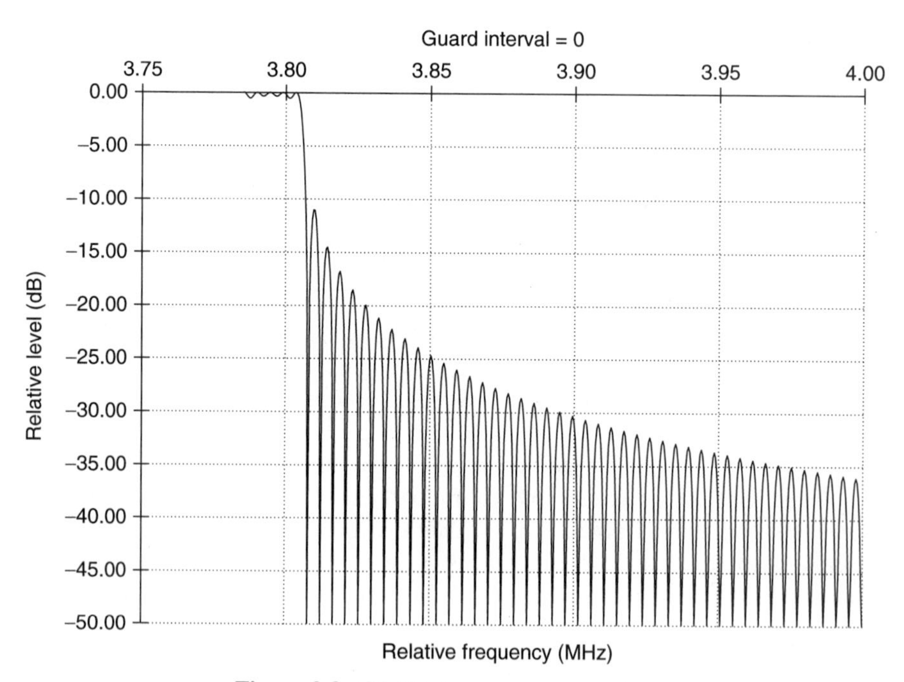

Figure 3-9. COFDM spectrum, upper skirt.

orthogonal.[10] Therefore, the power spectral density is not constant within the channel bandwidth. Above the highest-frequency carrier, the sidelobes of the sinc functions add to produce a power spectral density whose peaks decrease steadily. The highest carrier frequency is sufficiently below the channel edge so that the ideal spectrum is more than 35 dB below the in-band spectrum. A similar relationship occurs at the lower band edge.

FLEXIBILITY

There are many degrees of freedom in the COFDM signal, permitting a large amount of flexibility in the structure of the signal. For example, it is not necessary that all carriers be transmitted at the same amplitude. Under some circumstances, such as the occurrence of high levels of interference or severe multipath, unequal carrier amplitudes may provide the potential for higher overall data rate.

The optimum power allocation to each subband may be calculated by the water filling theorem[11] of communications theory. This theorem states that the power spectral density of the transmitted signal, $S_x(f)$, should be chosen so that

$$S_x(f) = P_t - S_n(f)$$

where $S_n(f)$ is the power spectral density of the noise or interference normalized to the power transfer function of the channel and P_t is the transmitted power. When the power spectral density is selected in accordance with this expression, the channel capacity is given by

$$C_c = \frac{1}{2} \int \log_2 \left[1 + \frac{S_x(f)}{S_n(f)} \right] df \qquad \text{bits/second}$$

The outcome of this kind of calculation is determined by the nature of the dominating impairment (i.e., whether the impairment is due to white noise, impulse noise, interference from analog TV stations, or multipath). For example, if only white noise is considered, the power distribution across the band should be constant. In the presence of cochannel interference from an analog TV signal, the system may be made more robust by not transmitting in the COFDM subbands around the vision carrier, color subcarrier, and sound carrier frequencies, where there are peaks in the analog signal. All other subbands would use the same number of bits per symbol in the same constellation and at the same power.[12] An analogous unequal allocation could be made on the basis of multipath or impulse noise. In practice, cochannel interference, impulse noise, and multipath are dependent on location and time, so that use of a nonuniform distribution is not usually feasible

[10] Orthogonality is restored in the receiver by integrating the demodulated signal over the useful symbol interval. For echoes of duration less than Δ, the receiver can find an interval of length T_u in which there are no symbol transitions.

[11] T. M. Cover and J. A. Thomas, *Elements of Information Theory*, J Wiley, Inc, New York, 1991.

[12] Anders Vahlin, and Nils Holte, "OFDM for Broadcasting in Presence of Analogue Co-channel Interference," *IEEE Trans. Broadcast.*, Vol. 41, No. 3, September 1995.

in broadcast applications. The constant power distribution associated with the assumption of white noise is used in the DVB-T and ISDB-T systems.

BANDWIDTH

As with 8 VSB, there are multiple definitions of the modulated signal bandwidth of the COFDM signal. The bandwidth may be defined as the difference between the highest and lowest carrier frequencies. The spacing between these carriers is 6.66 and 7.61 MHz for 7- and 8-MHz DVB-T channels, respectively. For ISDB-T the signal bandwidth is slightly less than that of DVB-T; for a 6-MHz channel it is 5.57 MHz. The channel bandwidth is inherent in this definition.

Bandwidth may also be defined in terms of the spectral mask. In this case, the power spectral density must be attenuated to the levels specified in the standard. Several spectral masks are defined in the DVB-T, most depending on whether or not the digital station is cosited with and operating on a channel adjacent to an analog TV transmitter with specified characteristics. These masks assume that the analog and digital services are copolarized and that the peak sync transmitter power output (TPO) of the analog service is equal to the total power of the digital service. If the power output of the transmitters are not equal, proportional correction may be applied. Figure 3-10 shows the spectral mask

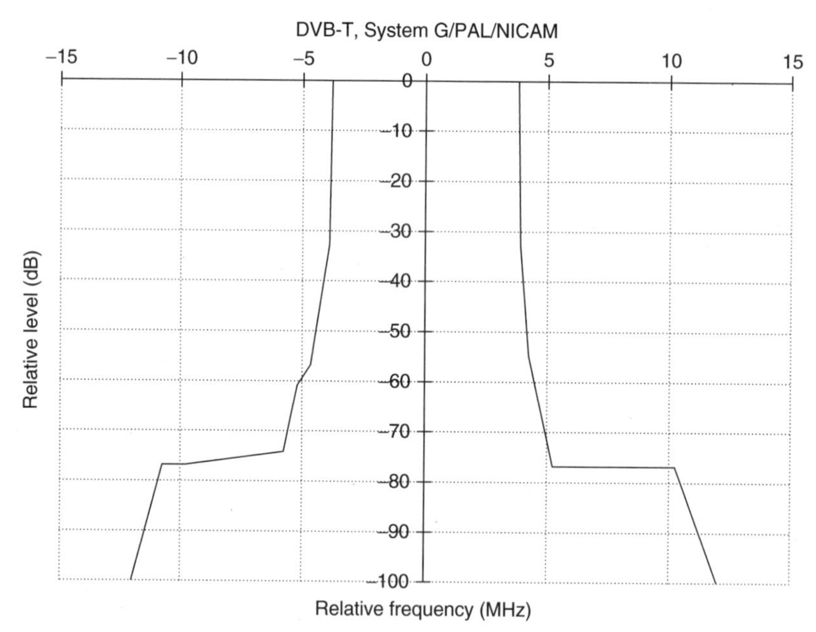

Figure 3-10. Minimum protection mask. (© ETSI 1997, © EBU 1997, ETS 300 744 is the property of ETSI and EBU. Further use, modification, redistribution is strictly prohibited and must be the subject of another copyright authorization. The above mentioned standard may be obtained from ETSI Publication Office, publications@etsi.fr, Tel: +33(0)4 92 94 42 41; used with permission.)

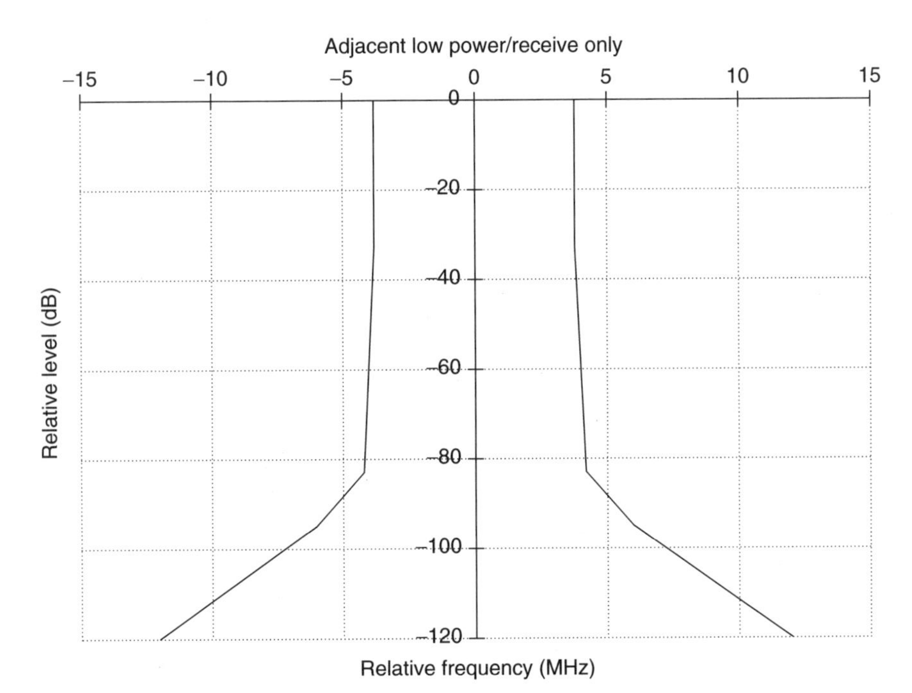

Figure 3-11. DVB-T critical mask. (© ETSI, 1997, © EBU 1997, ETS 300 744 is the property of ETSI and EBU. Further use, modification, redistribution is strictly prohibited and must be the subject of another copyright authorization. The above mentioned standard may be obtained from ETSI Publication Office, publication@etsi.fr, Tel: +33(0)4 92 94 42 41; used with permission.)

for the case where the DVB-T transmitter is cosited with and operating on a channel adjacent to a System G/PAL/NICAM analog transmitter. The resolution bandwidth is specified as 4 kHz. The 0-dB reference level corresponds to the total output power. An even more severe mask, specified to provide protection to other critical services, such as a low-power transmitter or receive-only sites, is plotted in Figure 3-11. This mask actually reduces the in-band signal bandwidth by 200 kHz.

MODULATION

The serial data are converted to parallel data and mapped into 2-, 4-, or 6-bit groups to form complex numbers representing data symbols. Choice of the number of bits per symbol is determined by the desired constellation, whether DQPSK (ISDB-T), QPSK, 16QAM, or 64 QAM (DVB-T and ISDB-T). This process is illustrated in Figure 3-12 for the QPSK constellation in the DVB-T nonhierarchical transmission mode. The data input to the demultiplexer, $x(t)$, is mapped into a pair of substreams at the outputs, b0 and b1. Each substream

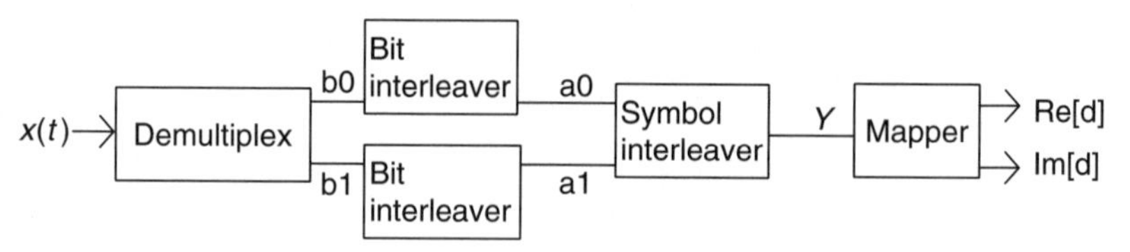

Figure 3-12. Block diagram of inner interleaver and mapper for DVB-T QPSK. (© ETSI 1997, © EBU 1997, ETS 300 744 is the property of ETSI and EBU. Further use, modification, redistribution is strictly prohibited and must be the subject of another copyright authorization. The above mentioned standard may be obtained from ETSI Publication Office, publications@etsi.fr, Tel: +33(0)4 92 94 42 41; used with permission.)

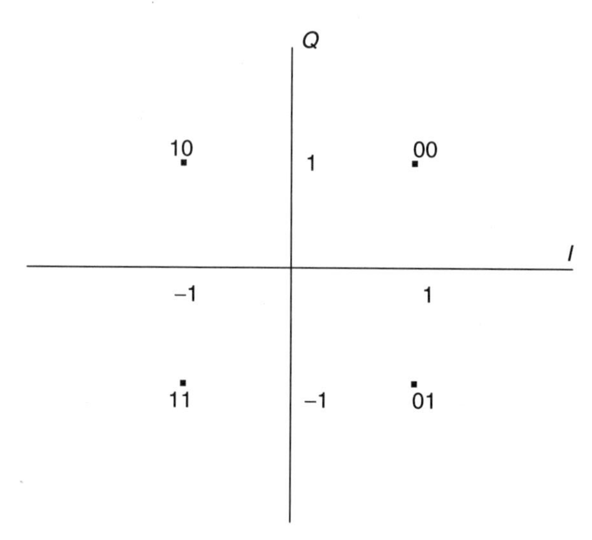

Figure 3-13. QPSK mapping and bit pattern.

is processed by a separate bit interleaver to produce the output vectors a0 and a1. Only the useful data are interleaved. The block size is 126 bits for each interleaver, but the interleaving sequence differs. The output of the interleavers is grouped to form the data symbols, each consisting of one bit from each interleaver. The symbol interleaver maps these symbols to the output vector, Y, which in turn is mapped into the signal constellation, d. This mapping is in accordance with a Gray code, as shown in Figure 3-13; that is, the value of only one bit is changed for adjacent symbols. Organization into the COFDM frame structure follows the mapper (see the discussion of the DVB-T standard in Chapter 1).

The complex data symbols, d_i, at the output of the mapper are applied to an IFFT and filtered to form the frequency-division multiplex time-domain signal approximated by[13]

$$S_f(t) = \sum_{k_c=K_{\min}}^{K_{\max}} (\mathrm{Re}[d_i]\cos\ 2\pi f_{kc}t + \mathrm{Im}[d_i]\sin\ 2\pi f_{kc}t)$$

where $f_{kc} = k_c/(K_{\max}T)$, $\mathrm{Re}[d_i]$ and $\mathrm{Im}[d_i]$ are the real (I) and imaginary (Q) parts of the data sequence. Obviously, the signal is very complex in the time domain, being the sum of multiple carriers whose amplitudes and phases depend on random complex coefficients.

From the viewpoint of the transmission system, the peak and average signal powers and their ratio are of interest. Ideally, the peak-to-average ratio of a multicarrier system is 10 log(N/4).[14] Thus for the DVB-T 2k mode, the peak-to-average ratio would be 10 log(1705), or 26.3 dB; for the 8k mode it would be 32.3 dB. However, since the data are scrambled, these theoretical values rarely occur in practice, especially for the larger constellations. The probability of exceeding 12 dB is approximately 0.01%. Even for a peak-to-average ratio of 12 dB, the signal will be susceptible to nonlinear distortion in high-power amplifiers operating near saturation. Some clipping can be tolerated without significant loss of data since the data are subject to quantization, rounding, and truncation limitations during FFT computation. Tests have shown that clipping 0.1% of the time degrades C/N by less than 0.2 dB. Nevertheless, it is important to optimize the transmitter operating point and exciter precorrection to prevent excessive levels of third-order IMD. Transmitter backoff in the range of 10 dB is required for satisfactory out-of-band performance.

The acceptable bit error rate for the DVB-T and ISDB-T systems has been set at 2×10^{-4} at the output of the convolutional decoder. This is the minimum error rate for proper operation of the R/S decoder. The C/N value required to achieve this error rate is dependent on the parameters selected for transmission, such as the constellation and convolutional code rate. Performance has been evaluated by simulation for several channel models, including the AWGN or Gaussian model. Other models evaluated are Ricean channels for fixed receivers, with multipath and Raleigh channels for portable receivers. To illustrate the performance for the AWGN distribution, the minimum C/N for DVB-T nonhierarchical transmission as a function of convolutional code rate ($\frac{1}{2}$, $\frac{2}{3}$, $\frac{3}{4}$, $\frac{5}{6}$, and $\frac{7}{8}$) for each constellation is plotted in Figure 3-14. The average C/N penalty for 16-QAM and 64-QAM relative to QPSK is 6.3 and 12 dB, respectively. The payload data rate expands as the constellation complexity increases, as shown in Figure 3-15. This increase is in proportion to the additional bits transmitted per symbol. The maximum spectral efficiency is 3.96 bits/Hz.

[13] William Y. Zou and Yiyan Wu, "COFDM: An Overview," *IEEE Trans. Broadcast.*, Vol. 41, No. 1, March 1995, pp. 1–8.
[14] Ibid., p. 6.

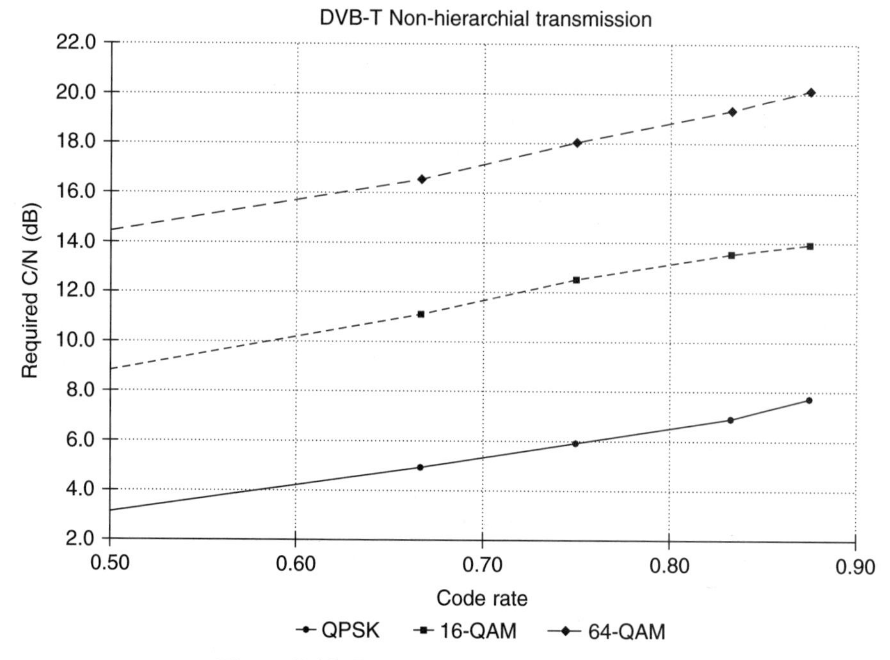

Figure 3-14. Required C/N versus code rate.

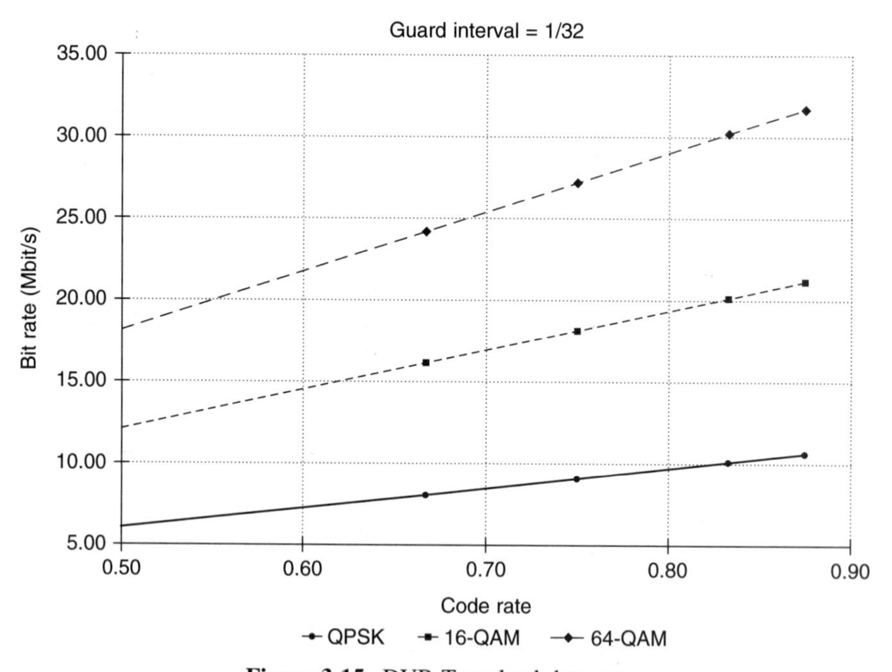

Figure 3-15. DVB-T payload data rate.

4

TRANSMITTERS FOR DIGITAL TELEVISION[1]

The introduction of digital TV has continued to drive important advances in TV transmitter technology. This continues the stream of new technology and ideas that have provided high-quality analog TV transmission while improving reliability, reducing maintenance, and lowering overall cost of ownership. These new technologies include solid-state high-power amplifiers, improvements in UHF tube transmitters, and use of digital signal processing and computerized control. An environment of technical deregulation allows continued flexibility in transmitter design and system operation. In this chapter we discuss the relevant technology and provide information needed for selection, installation, operation, and maintenance of digital TV transmitters.

A digital television transmitter may be considered to comprise two essential components, the exciter and the RF power amplifier (PA). The exciter signal processing functions required to convert the baseband digital signal into a modulated IF signal were discussed in Chapter 3. The exciter also performs other essential functions subsequent to the modulator. These may include precorrection and equalization, upconversion, bandlimiting, and amplification to a relatively low power RF signal.[2] Traditionally, these have been implemented in analog TV transmitters as analog functions. However, with the availability of a digital input signal and the recent advances in digital signal processing (DSP)

[1] Portions of this chapter were previously published in the following and reprinted with permission: "TV Transmitters," Chap. 6.2 in *NAB Engineering Handbook*, 9th ed., National Association of Broadcasters, Melbourne, Fla, 1999; "Transmitters for Analog Television," in *Encyclopedia of Electrical and Electronics Engineering*, Wiley, New York, 1999; and "Transmitter Considerations for Digital Television," in *DTV Handbook*, DTV Express, Harris Communications, Melbourne, Fla, 1998.

[2] Because the output of the exciter is a bandlimited modulated RF signal, most commercially available units may be considered low-power transmitters.

technology, several of these functions may be implemented in digital circuitry. The possibility of performing the modulation and bandlimiting functions in DSP was noted in Chapter 3. It is also now feasible to perform equalization and precorrection using DSP. By sampling the transmitter output at appropriate points, these functions may be made adaptive. Depending on processing speed and designer choices, it may be possible to produce the bandlimited, modulated carrier on channel without the need for upconversion. Thus the time is rapidly approaching when it may be possible to perform all functions except high-power amplification using DSP.

In the following paragraphs, precorrection and equalization are first discussed as if they occurred subsequent to modulation and without regard to the technology in which they are implemented. This approach is followed for several reasons. By confining the digital processes to baseband functions, the description of channel coding is not cluttered with the details of adaptive precorrection and equalization or techniques for using DSP for modulation and bandlimiting, much of which is proprietary. Many engineers are already familiar with analog methods of performing these functions; thus the focus may be on application of system-level concepts to digital television. Specific implementations depend on design choices made by the various manufacturers. These choices will no doubt tend to favor more functions being in DSP as technology advances. However, the objective of this discussion is to understand the system-level issues rather than any specific implementation.

PRECORRECTION AND EQUALIZATION

For this discussion, assume that the output of the modulator is at the IF and that precorrection of nonlinear distortions and equalization of linear distortions is done using analog circuits operating in the IF frequency band. Although these are separate functions that are adjusted independently, they must be considered together for proper system operation. The precorrection and equalization concepts described herein are termed feedforward,[3] meaning that the correctors and equalizers are upstream in the RF line for the purpose of correcting distortions farther downstream. Traditionally, adjustment of these circuits for optimum equalization has been manual; an operator observes the system response on appropriate test equipment and makes the necessary adjustments. When implemented as an automatic adaptive process, the operator is replaced with the digital processing circuitry and software.

EQUALIZATION

Equalization is introduced in the system to compensate for linear distortions produced in various bandlimiting downstream components. The objective is to

[3] This should not be confused with a feedforward loop, which is described later as a technique for linearizing the intermediate-power amplifier.

provide a complementary complex transfer function that when multiplied by the complex frequency response of the bandlimiting components will reduce linear distortions to an acceptable level. The ideal system frequency response is constant, independent of frequency. The most likely sources of linear distortions are PA input-matching networks and cavities, output networks and cavities, and RF filters.

In the absence of nonlinear distortions, the system block diagram may be represented by the two blocks shown in Figure 4-1. The first block is the equalizer with complex frequency response, $H_{\text{eq}}(\omega)$; the second block is the PA and filters with complex frequency response, $H_0(\omega)$. Assuming that the networks are independent, the transfer function for the combination of these blocks is

$$H_s(\omega) = H_0(\omega)H_{\text{eq}}(\omega)$$

When the system is equalized, this product is equal to unity, so that

$$H_{\text{eq}}(\omega) = [H_0(\omega)]^{-1}$$

The frequency-response functions of the PA and the equalizer are complex, meaning that both amplitude and phase must be equalized. If the response of the blocks is separated into amplitude and phase components, the transfer function of the second block may be written

$$H_0(\omega) = A(\omega)e^{-j\phi(\omega)}$$

so that

$$H_{\text{eq}}(\omega) = \frac{1}{A(\omega)}e^{j\phi(\omega)}$$

where $\phi(\omega)$ is a nonlinear phase function. The product of these functions is the desired

$$H_s(\omega) = 1$$

It is not unusual for the unequalized response of a PA and filters to exhibit tilt, quadratic, cubic, and other higher-order distortions. In general, the response may be represented as a constant plus a polynomial, $f(\omega)$, or

$$H_0(\omega) = 1 + f(\omega)$$

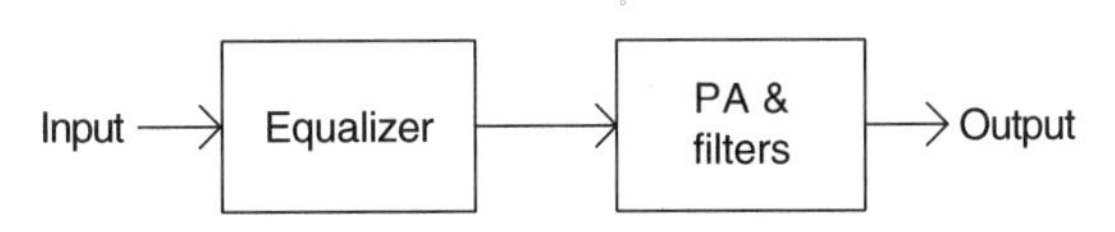

Figure 4-1. System block diagram of linear distortion and equalizer.

Consequently, the equalizer must be adjusted so that its frequency response is the inverse of the distortion, or

$$H_{\mathrm{eq}}(\omega) = \frac{1}{1 + f(\omega)}$$

For proper operation of precorrection circuits, this relationship must hold over a frequency band that includes the RF channel and the upper and lower adjacent channels.

If there is no amplitude distortion,

$$H_{\mathrm{eq}}(\omega) = Ke^{j\phi(\omega)}$$

where K is the constant amplitude or gain. This equation describes the response of an all-pass filter network.[4] The phase response for a second-order all-pass filter is given by

$$\phi(\omega) = -2\tan^{-1}\frac{a\omega_0}{b\omega_0^2 - \omega^2}$$

where a and b are constants equal to the transfer function coefficients at resonance ($\omega = \omega_0$). Traditional analog techniques used to accomplish group delay equalization at IF have included both active and passive networks of this type. Adjustment of the resonant frequency and quality factor, $\boldsymbol{Q}$, of multiple all-pass networks in cascade is used to equalize a wide variety of linear distortions.

The linear distortions due to phase versus frequency are often characterized in terms of group delay, which is the nonuniform delay of different frequencies over the signal bandwidth. Mathematically, group delay is the negative first derivative of phase with respect to angular frequency, or

$$\text{group delay} = \frac{-d\phi}{d\omega}$$

In general, a rapid change in amplitude as a function of frequency in tuned circuits implies large amounts of group delay distortion.

When implemented in an adaptive system using DSP, the equalizer transfer functions must be realized as a digital filter. The process is shown conceptually in Figure 4-2. The output of the transmitter is sampled at a point downstream of all components contributing to linear distortions. After demodulation, the sample, I_s and Q_s, is compared to the ideal signal, I_i and Q_i to produce an digital error signal. The error signal, e_i and e_q, is then used to adjust the tap coefficients on a digital filter. This process is iterated until the error function is driven to a sufficiently small value. As operating conditions and the amount of linear distortions change,

[4] *Reference Data for Radio Engineers*, 6th ed., Howard W. Sams, Indianapolis, Ind., 1977, pp. 10-19 to 10-21.

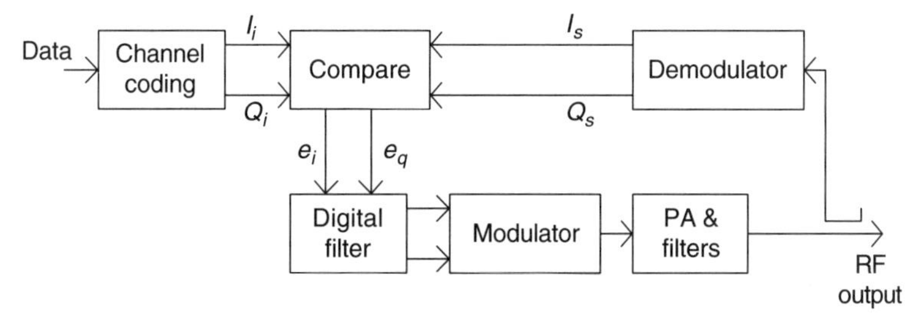

Figure 4-2. Conceptual block diagram of adaptive equalization.

the adaptive filter coefficients are constantly adjusted to maintain desired system performance.

PRECORRECTION

Most nonlinear distortions are generated in the high-power RF amplifier by means of the mechanism described in Chapter 2. The out-of-band intermodulation products sufficiently removed from the channel of operation are attenuated by the high power-output filter. However, in-band IPs and those close to the channel can be removed only by making the transmitter sufficiently linear to reduce these signals to specified levels. Given sufficient bandwidth in the stages succeeding the modulator, the IF corrector predistorts the modulated signal, generating spectral energy to cancel the unwanted intermodulation products.

In the absence of linear distortions, the system block diagram may be represented by two blocks, as shown in Figure 4-3. The second block is the PA, for which the complex output voltage, v_o, may be written in terms of its input as

$$v_o = g[v_c + f_o(v_c)]$$

where g is the amplifier gain and $f_o(v_c)$ is a polynomial representing the nonlinearities. The first block is the precorrector for which the output voltage, v_c, is a sum of linear and nonlinear functions,

$$v_c = g_c[v_i + f_i(v_i)]$$

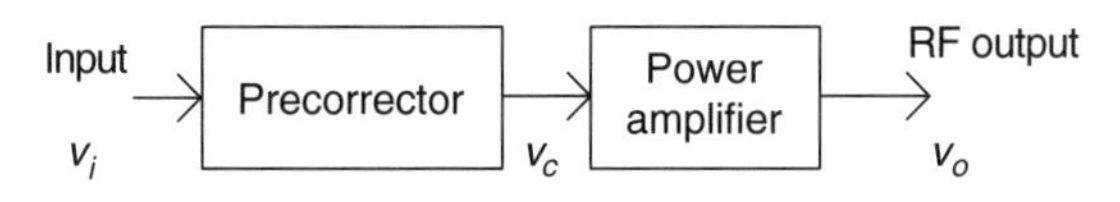

Figure 4-3. Precorrection of amplifier nonlinearity.

In terms of the input to the precorrector, v_i, the system output voltage is

$$v_o = gg_c[v_i + f_i(v_i)] + gf_o(v_c)$$

Since the desired output voltage is the first term, $v_o = gg_c v_i$, it follows that the precorrection is properly adjusted when

$$gg_c f_i(v_i) + gf_o(v_c) = 0$$

or

$$g_c f_i(v_i) = -f_o(v_c)$$

If both sides of this equation are polynomials of order N, the precorrection is properly set up when the coefficient of each term of order n on the left-hand side is equal in magnitude but opposite in sign to the corresponding term on the right-hand side. For example, if the amplifier output voltage as a function of input is

$$v_o = gv_c + g_3 v_c^3$$

it follows that

$$gf_o(v_c) = g_3 v_c^3$$

and

$$g_c f_i(v_i) = \frac{-g_3}{g} v_c^3$$

Thus a properly scaled cubic function is required in the precorrector to correct a third-order amplifier nonlinearity. Since there is no change in average power as the picture changes, the required transfer curve remains constant.

Traditionally, simple analog IF linearity correctors have been used to approximate the desired polynomial by means of a piecewise linear curve. In these circuits, diodes are set to conduct at a specific level of the IF modulated signal. When the diodes conduct, the gain or attenuation is reduced as needed to create a discontinuity at the proper level in the transfer function. A nonlinear transfer function can be approximated to the desired precision by cascading several of these circuits, thereby producing a complementary match to the nonlinear PA transfer function.

Phase distortions in high-power amplifiers produce spectral components in quadrature with the modulation signal, resulting in unbalance between the upper and lower sidebands. Traditionally, these distortions have also been corrected in analog transmitters by means of direct AM/PM correctors at IF. The signal is split into two paths, which are in phase quadrature across the entire band of interest. The quadrature signal is modified with level-dependent diode expansion or compression circuits of the same type used for linearity correction.

The vector diagram of Figure 4-4 illustrates nonlinearities introduced by the PA and operation of the AM/AM and AM/PM correctors. The undistorted input

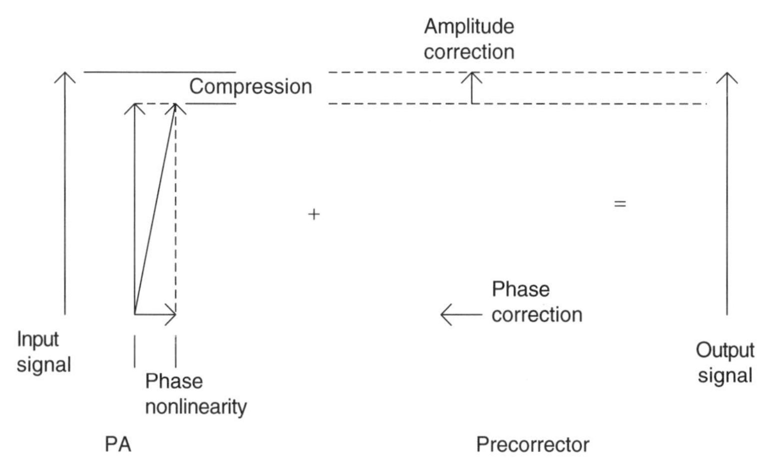

Figure 4-4. Vector diagram illustrating AM/AM and AM/PM correction.

signal is represented by the vector on the left. The transmitter introduces phase shift and compression. To compensate, the signal in the exciter is expanded in amplitude and a quadrature correction is added. When the resultant signal is amplified, the output signal is the desired linear digital television signal.

In the typical transmitter, both linear and nonlinear distortions are present simultaneously and thus must be precorrected and equalized together. For example, a tube amplifier comprises (1) an input cavity which, due to bandwidth limitations, introduces linear distortions; (2) the amplifier tube, which introduces nonlinear distortions; and (3) output cavities for extracting the output power that introduce additional linear distortions. For satisfactory performance, each of these distortions must be compensated. Solid-state PAs may introduce similar combinations of distortions. For wideband solid-state amplifiers, linear distortions may not be as severe as with tubes. However, some linear distortion will be present, due to input and output matching networks. In addition, channel filters at the output of the PA may introduce a significant amount of linear distortion.

The combination of linear and nonlinear distortion in a typical PA may be represented as shown in Figure 4-5. The nonlinear distortions due to the tube or transistor are sandwiched between the linear distortions due to the input/output cavities and/or matching networks and output filter. For satisfactory operation, the combined effect of these distortions must be compensated. The first step is to equalize the linear distortions of the first block over a bandwidth of at least three times the channel bandwidth. Once this is accomplished and assuming that the linearizer has equivalent bandwidth, there should be sufficient bandwidth to precorrect the nonlinear distortions, thereby reducing both in-band and out-of-band IPs. Finally, the linear distortions at the output may be equalized over the channel bandwidth. In general, the distortions in the system must be compensated in the reverse order in which they occur, as illustrated in Figure 4-6. This order must be followed whether the compensation is done at IF or baseband.

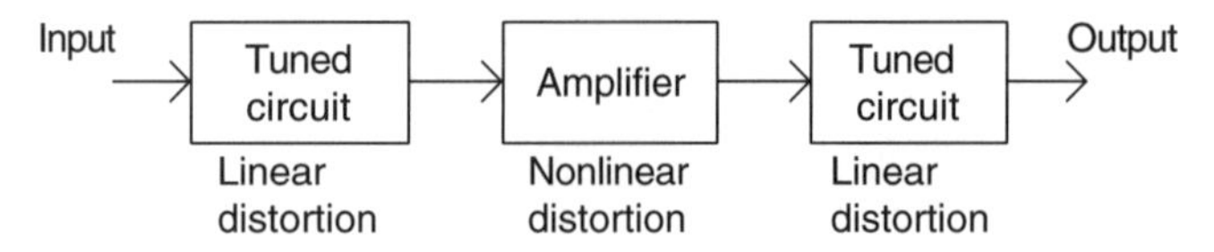

Figure 4-5. Block diagram of PA linear and nonlinear distortions.

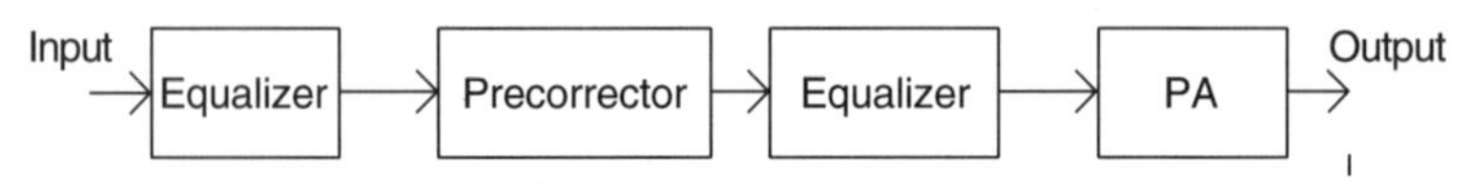

Figure 4-6. Precorrection and equalization of the PA.

UP CONVERSION

The exciter frequency control circuits generate the required local oscillator signals to translate the modulated IF to the final RF carrier frequencies. These frequencies are generated from a stable reference oscillator in low-noise phase-locked-loop (PLL) frequency synthesizers so that a single standard frequency may be used for all TV channels. A typical exciter may use an IF frequency in the neighborhood of 45 MHz. The important performance characteristics of the reference oscillator and synthesizer(s) are low phase noise, frequency stability over time and temperature extremes, and low levels of microphonics. Synthesizer performance should be properly maintained to prevent degradation of phase noise and spurious frequency levels.

PRECISE FREQUENCY CONTROL

For DTV transmitters operating in the United States on channels immediately adjacent to and within 88 km of a transmitter on a NTSC channel, the pilot frequency must be maintained at 5.082138 MHz above the NTSC visual carrier frequency with a tolerance of ± 3 Hz. To determine whether or not a particular station is required to operate with precise frequency control (PFC), the FCC Rules Part 73, Radio Broadcast Services, should be consulted.

To maintain the relative precision offset within 3 Hz requires maintenance of each carrier frequency within 1.5 Hz. Maintenance of the carrier frequency to such precise tolerance for long periods of time is beyond the capability of quartz crystal technology. If the analog and digital stations are co-sited, the NTSC visual carrier and DTV pilot frequency may be locked to a common, highly stable source such as an atomic frequency standard. If they are not co-sited, different reference frequency sources can be employed provided that they are capable of the required precision and stability.

Loran-C and GPS receivers are commercially available at modest cost to provide the 10.0-MHz reference signal.[5] The frequency accuracy and stability claimed for these sources is 1 part in 10^{12}, much better than FCC requirements. The frequency is maintained by a cesium beam or hydrogen maser and is monitored by an agency of the U.S. government. Any out-of-tolerance equipment is replaced promptly. These signals may be received continuously throughout the United States at sufficient signal-to-noise ratio for use as reference sources. The precise reference is simply connected to the external reference signal input of the exciter. The synthesizer reference oscillator is compared to the precise frequency standard by means of a phase-locked loop. The resultant error voltage is used to adjust the synthesizer oscillator. By phase-locking the carrier to the stable reference, the master oscillator acquires the stability of the reference. The PLL should be equipped with digital memory to hold the last estimate of the frequency correction. Should the reference signal be lost, the memory will continue to supply the error voltage.

RF AMPLIFIERS

The last active stage in the exciter is the RF amplifier. For digital signals it is important that this amplifier have a linear amplitude and phase transfer characteristic, flat, symmetric frequency response, and minimum group delay variation across the modulation passband. For the ATSC and ISDB-T systems, at least 6-MHz bandwidth is required. For the DVB-T system, 7- or 8-MHz bandwidths are required. Typical average output power is 250 mW.

POWER AMPLIFIERS

The PA provides the "muscle" to amplify the modulated RF signal to the desired level for transmission. The power amplifier technology to perform this function is key and has far-reaching consequences on the operation, maintenance, and cost of the digital transmission system. Both solid-state and tube devices are available in commercial digital transmitters. For VHF channels, solid-state devices predominate. For UHF, both solid-state and tube devices are used. The choice between these technologies hinges on several factors in addition to output power. These factors include system performance parameters related to the quality of the transmitted signal, system redundancy, floor loading, amount of floor space required, system efficiency and power consumption, cooling considerations, and cost. Key functions in the PA common to all amplifier technologies include ac distribution, ac-to-dc power conversion, cooling, and control.

TV transmitters are unique in that no other application requires such high levels of linear RF power generation, especially at UHF, while operating virtually uninterrupted. This has led to the development of specialized techniques to assure

[5] Charles W. Rhodes, *DTV Express Training Manual*, Harris Corporation, Melbourne, Fla., 1998, pp. 4-3, 7-1 to 7-4.

highly efficient and reliable operation. The need for high efficiency has led to the nearly universal use of partially saturated class AB final power amplifiers with the precorrection techniques required to compensate for nonlinearities inherent in this class of operation. To achieve the levels of reliability required, redundant system architectures that minimize single-point failures are used.

Since ERP is assigned or computed on the basis of average power, transmitter output is also defined in terms of average power. There is no regularly recurring sync pulse by which to determine peak power as with analog television. Nevertheless, the peak power rating of the transmitter is important. Sufficient headroom must be available to amplify the random peaks of the digital signal. Even though ERP and TPO are defined on the basis of average power, the tradition of transmitter power rating according to approximate peak power capability has been carried forward by transmitter and tube manufacturers. However, average power rating is the correct quantity to be used for purposes of discussing TPO. For the ATSC system, 6 to 7 dB of headroom in the final stage of the PA is required. With this amount of headroom, the nonlinear distortions of the RF chain can be properly precorrected. Typically, the PA will be driven 0.5 dB into compression by the signal peaks. The peak-to-average power ratio versus the sideband level at the spectral breakpoints for a solid-state transmitter is plotted in Figure 4-7. A peak-to-average ratio of about 6.6 dB yields a sideband level at

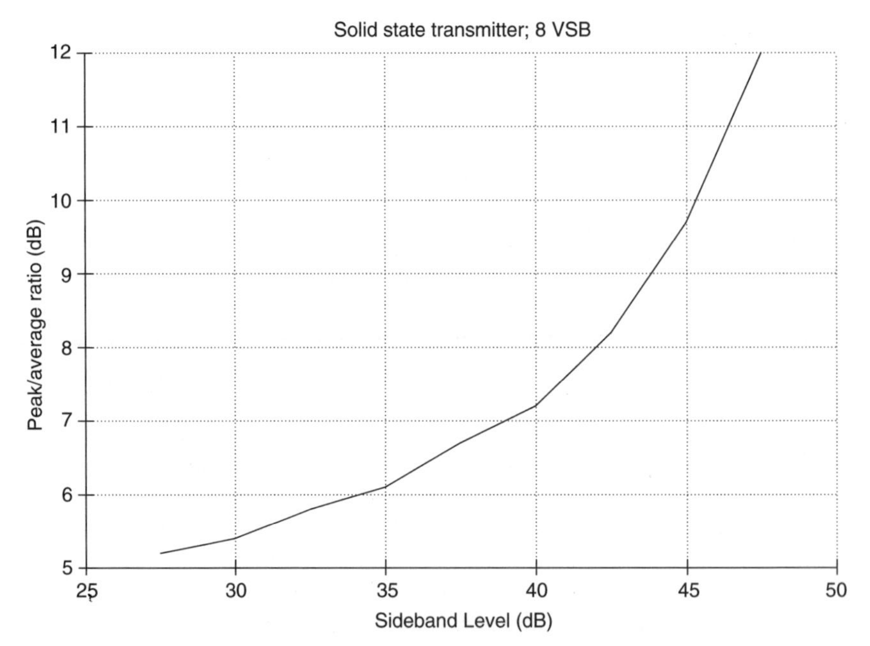

Figure 4-7. Peak-to-average ratio versus sideband level. (From M.J. Horspool, "A New Solid State UHF Television Transmitter Design for Digital Terrestrial Broadcasting," *Conference Proceedings of the International Conference and Exhibition on Terrestrial and Satellite Broadcasting*, New Delhi, India, January 9–21, 1998.)

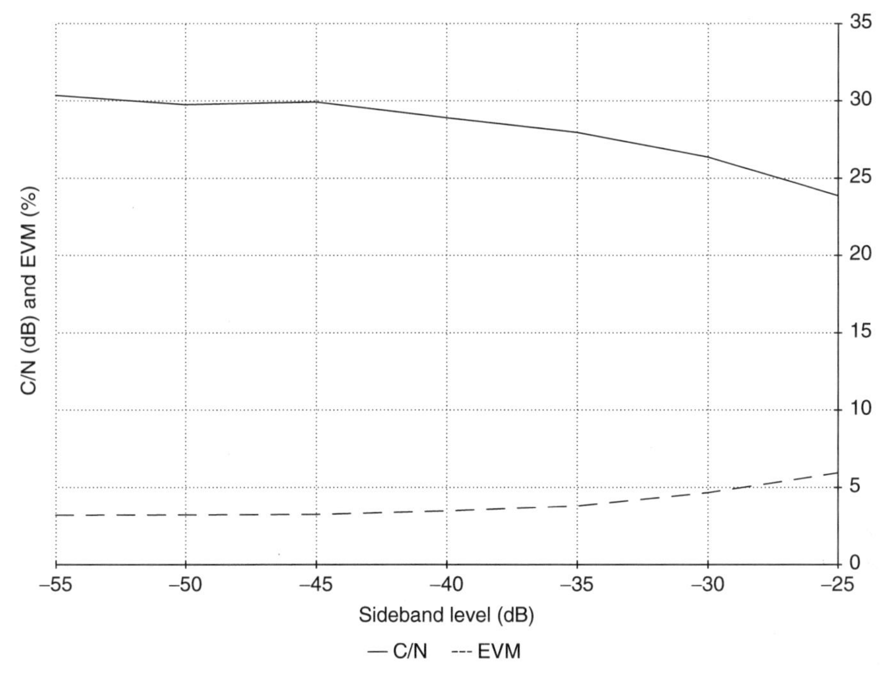

Figure 4-8. C/N and EVM versus sideband level.

the breakpoints of about 37 dB, which is consistent with FCC requirements. The combination of good precorrection and the stopband characteristics of the output filter make it possible to meet the total requirements of the FCC emissions mask. For systems using COFDM, about 2.5 dB of additional headroom is required.[6]

In-band performance is also affected by PA headroom. Output carrier-to-noise ratio and error vector magnitude for the 8 VSB signal are plotted as a function of sideband level at the spectral breakpoints in Figure 4-8. The IMD sideband at the breakpoints extends across the digital channel. The C/N and EVM improve as sideband level is reduced. The inverse relationship between C/N and EVM is shown in Figure 2-3.

SOLID-STATE TRANSMITTERS

Technological advances in field-effect transistors (FETs) have made the development of solid-state high-power linear amplifiers for digital applications both practical and cost-effective. By combining RF modules, it is practical to manufacture digital transmitters for any power range up to 25 kW for both UHF and

[6] Yiyan Wu, "Performance Comparison of ATSC 8-VSB and DVB-T COFDM Transmission Systems for Digital Television Terrestrial Broadcasting," *IEEE Trans. Consumer Electron.*, August 1999.

VHF. The final stage of a solid-state power amplifier is operated in class AB for the best trade-off of efficiency, linearity, reliability, and cost.

There are several advantages to the use of high-power solid-state technology. Due to the absence of tuning controls and degradation of filament emissions, performance is maintained over extended periods of time. No warm-up time is required; full-rated power is available within seconds of activation. Air cooling has the advantage of eliminating any chance of coolant spills, corrosion, or concern for cooling system freezing in cold climates. Safety is also enhanced. Dc operating voltages are 65 V or lower compared to tens of kilovolts for tube amplifiers. There is no need for crowbars to protect solid-state devices in the event of an arc or a short circuit. Maintenance cost is also reduced. This is due in part to the absence of high-voltage components, cooling liquids, and costly replacement tubes. Modular soft-fail architectures are employed by virtue of the large numbers of RF and power supply modules operated in parallel. Failure of any one of these units has only a minor effect on TPO. Thus immediate corrective action is not as critical as in tube transmitters, in which there may be but a single output device. Simple diagnostics displays make the identification of a failed unit easy. Hot pluggable designs and an inventory of spare modules make it possible to remove and replace the failed unit while the transmitter remains on-air. Repair of the failed unit may be done off-line.

Vertical MOSFETs have been the devices of choice for VHF transmitters. Recently, lateral diffused MOSFETs (LDMOS) have proven capable of developing cost-effective linear power for UHF. Although both bipolar transistors and FETs have merit, FETs have some advantages over bipolar devices. FETs have a higher amplification factor than bipolar transistors, are not subject to thermal runaway, are more rugged, are more linear, and have a less abrupt saturation characteristic. Higher gain means a reduction in the number of required driver stages. The fewer the drive stages, the lower the manufacturing cost and the better the linearity since there are fewer parts contributing to cost and nonlinearity. Simple bias circuitry for FETs minimizes parts count and amplifier production cost.

New developments in high-power solid-state devices are on the horizon. Silicon carbide (SiC) materials may make it possible to produce a variety of devices based on this technology. Linear high-power UHF transistors are under continuing development.[7] A 10-W MESFET device with a gain of 12 dB has been announced.[8] These transistors promise to operate at higher voltages (48 V) and temperatures (maximum junction temperature of 250°C) than silicon UHF devices, thereby increasing available power output, system efficiency, and cooling effectiveness. As this technology matures, higher-power devices will probably be available. Depending on device costs, these developments could have the effect of lowering transmitter system cost.

[7] Carlton Davis, Jack Hawkins, and Charles Einoff, Jr., "Solid State Transmitters," *IEEE Trans. Broadcast.*, September 1997, pp. 261–267.

[8] "First Silicon Carbide Microwave Power Products Are Announced," *Appl. Microwaves Wireless*, August 1999, p. 104.

As with analog transmitters, standard power ratings for solid-state transmitters are dependent on the choices made by the manufacturers in response to perceived customer needs and available technology. Typical standard maximum powers offered for UHF and VHF are listed in Table 4-1. In every case, these transmitters may be operated at lower than their maximum rating. The choice of power level must be made on the basis of desired AERP and headroom. However, it must be recognized that operating at the maximum rating results in maximum efficiency. Experience with analog solid-state transmitters has demonstrated that operation at rated power is not detrimental to product life.

Other technical factors related to UHF solid-state transmitter power levels are listed in Table 4-2. These figures may be used to estimate floor space

TABLE 4-1. Standard Digital Transmitter Power Ratings (kW)

UHF		
Tube	Solid State	VHF
100		
75		
50		
35		
25	25	
17.5	20	
15.0	15	
12.5	10	
	7.5	7.5
	5.0	5.0
	3.75	
	2.5	2.5
		1.25

TABLE 4-2. Typical Solid-State UHF Transmitter Characteristics

Average Power Rating (kW)	Size (in.)	Weight (lb)	Power Consumption (kW)
1.25	57.5 W × 61 D		8.2
1.88	57.5 W × 61 D		11
2.50	57.5 W × 61 D		14
3.75	57.5 W × 61 D		20
5.00	57.5 W × 61 D	2,500	26
7.50	91.5 W × 61 D		39
10.0	91.5 W × 61 D	4,500	50
15.0	126 W × 61 D	6,500	75
20.0	160 W × 61 D	8,500	100
25.0	194 W × 61 D	10,500	125

required, floor loading, cooling load, and operating power costs. Power supplies, ac line control components, and cooling equipment are built in. These transmitters require from one to five PA cabinets plus an exciter/control cabinet.

RF AMPLIFIER MODULES

Combining several RF power modules to achieve the desired transmitter output power increases the parallel redundancy and on-air availability. Average output power of 250 to 500 W per module has been adopted by nearly all manufacturers based on cost, weight, and size limitations.

Self-protection of each PA module against various fault conditions is common practice. By using self-protecting modules, the cabinet control logic and overall transmitter control logic is simplified, thus improving overall reliability. Diagnostic indications as a part of the module minimize time to repair. Protection from overvoltage, overdrive, VSWR, overtemperature, and assurance of equal load sharing among devices is essential for maximum amplifier life. The failure of one subassembly should not cause another subassembly to fail.

Modular amplifiers that can be removed while the transmitter is operating improve on-air availability. If an amplifier module fails, the transmitter continues to function without disrupting on-air operation. A spare PA amplifier can be used while the failed unit is repaired.

POWER SUPPLIES

Power supply design is critical to the reliability of a solid-state transmitter. Since FET and bipolar devices are low-voltage devices, the power supplies that drive them must provide low voltage and high current. High-reliability connections must be guaranteed in the dc distribution. Since power output from a transistor varies as the square of the dc voltage, the output voltage must be tightly regulated as a function of ac line and temperature variations.

Efficiency of the power supply is important. The lost power results in heat as well as unnecessarily high utility costs. Any voltage or current transients at the ac input should be suppressed before reaching the transistor. Power supplies should pass the applicable portions of the ANSI/IEEE C62.41 (IEEE-587) transient testing standard.

POWER COMBINERS

There are several methods for combining RF power in solid-state transmitters, most of which make use of the impedance-transforming properties of quarter-wave transmission lines. A very effective combining method is the use of in-phase N-way combiners. Three specific combiner topologies of this type are described.

No matter the combining technique, the RF power amplifiers must be matched with respect to phase and gain to provide maximum power to the antenna and minimum power to the reject load. Low-level electronic phase shifters and attenuators internal to the module may be used to optimize system output. These devices should have memory so that they return to optimum phase and gain after a power failure.

WILKINSON COMBINER

Wilkinson power dividers and combiners are generally used for values of N in the range 2 to 6. A schematic drawing of a two-way Wilkinson[9] combiner is shown in Figure 4-9. The amplifier input ports are connected to the common output port by a pair of quarter-wavelength transmission lines. When both amplifiers are operating with identical output, equal voltages are present at both nodes of the reject load resistor so that no power is dissipated. When an amplifier failure occurs, the power is distributed equally between the load and the output port. The impedance of the transmission lines and the length of the lines is selected to achieve the desired impedance transformation. A balanced reject load is used to provide isolation between amplifiers. This combiner may be constructed using a variety of transmission line types, including coax and microstrip. The combiner may be housed in a shielded casing so that the balanced load configuration cannot be influenced by outside fields due to external components. The reject load is cooled by conducting heat through the flange to a heat sink where the heat is exchanged to a moving airstream.

RING COMBINER

A schematic drawing of a Gysel[10] or ring combiner is shown in Figure 4-10. Coaxial lines are used for the transmission lines in this example. The higher

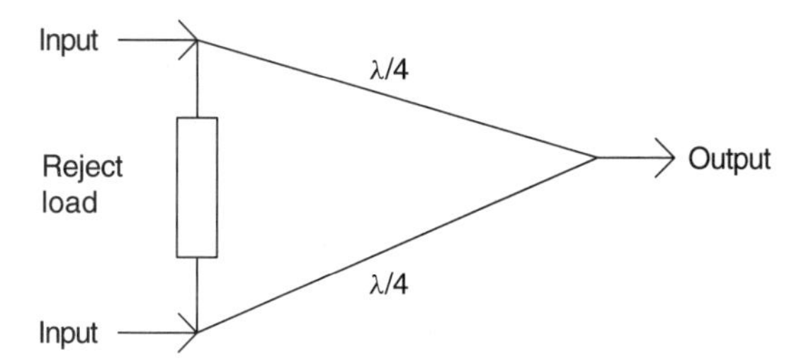

Figure 4-9. Schematic drawing of two-way Wilkinson combiner.

[9] E. J. Wilkinson, "An N-Way Hybrid Power Divider," *IRE Trans. Microwave Theory Tech.*, Vol. 8, January 1960, pp. 116–118.
[10] Ulrich H. Gysel, "A New N-Way Power Divider/Combiner Suitable for High Power Applications," *IEEE MTTS-5, International Symposium Digest*, 1975, p. 116.

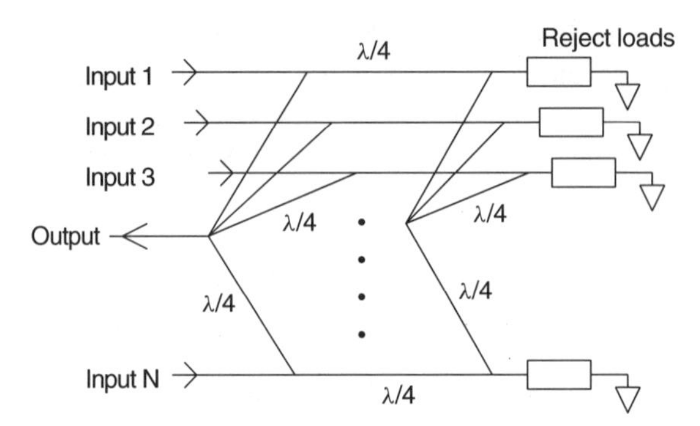

Figure 4-10. Schematic drawing of N-way Gysel combiner.

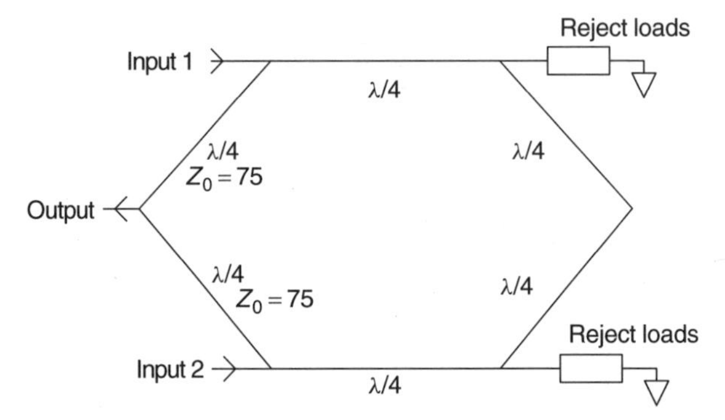

Figure 4-11. Schematic drawing of two-way Gysel combiner.

power-handling capability of the coax lines allows from 2 to 20 amplifiers to be combined. Isolation between amplifiers is provided using reject loads which are not in the direct RF path to the output. The operation of the ring combiner is easily understood by considering a two-way version shown in Figure 4-11. Each transmission line is a quarter-wavelength long. When identical voltages are applied to both input ports, the combined signals appear only at the output port. This occurs because (1) the distance from each input port to the output is electrically equal whether the signal follows the shorter or longer path, (2) the signal from one amplifier arrives at the load port out of phase with that from the other so that these signals cancel, and (3) the signals from input port 1 arriving at input port 2 via the short and long paths are out of phase, and vice versa. Under normal operating conditions, all power appears at the output, none is absorbed in the loads, and there is complete isolation between amplifiers.

Power is absorbed in the load resistors only when an unbalance occurs. Consider the case when only one amplifier is operating. The signal paths are electrically equal for the long and short paths to the output as well as for the right and left paths to either load. The power from the operating amplifier is therefore split equally between the output and the isolation loads. Due to the isolation inherent in the network, the input port remains matched even when one amplifier is removed.

Since the transmission line that connects the amplifier port to the output port is a quarter-wavelength long and 75-Ω transmission line is used, the 50-Ω amplifier impedance is transformed to $(75/50) \times 75 = 112.5\ \Omega$. When combining N amplifiers, the impedance at the output combiner junction is $112.5/N$. This impedance is readily matched to 50 Ω.

The reject loads may be mounted to a grounded heat sink or heat pipe. In case of amplifier removal, the reject load temperature rises. In the heat pipe, the fluid in the lower section warms until it vaporizes. The vapor rises to a finned heat exchanger, where the heat is transferred to a moving airstream. The vapor condenses as it releases the heat and returns to the bottom of the heat pipe to repeat the cycle.

STARPOINT COMBINER

The starpoint combiner is a simple low-loss circuit for summing the output of many amplifiers. For the sake of simplicity, consider the parallel operation of four isolated sources of equal phase and amplitude, each with a source impedance of 50 Ω. The resulting parallel impedance is 12.5 Ω. All that is needed to complete the combining process is to transform this impedance to 50 Ω. Obviously, this concept can be extended to any value of N. The Q of the impedance transformation determines the combiner bandwidth. For optimum bandwidth it may be necessary to combine and transform in several steps. Since the combined amplifiers have slightly differing phases and amplitudes, there is some nonresistive combining loss. A flat air-dielectric coaxial or stripline structure may be used to build a starpoint combiner with extremely low resistive loss.

The starpoint combining technique works best when N is large and all inputs are identical. For small values of N, the isolation may be insufficient to prevent interaction between modules when a failure occurs. Circulators may be used at the output of each module to provide isolation, producing a combiner with excellent isolation and bandwidth even for low values of N. Circulators also protect against high VSWR due to icing and other causes. However, the circulators contribute additional loss. The combiner band split should be consistent with RF module and circulator bandwidth.

COOLING

Proper cooling of a digital transmitter is important for high mean time between failures (MTBF). For example, the MTBF of a solid-state device approximately doubles for every 10°C drop in the junction temperature. Distributed cooling systems employing more than one fan offer good redundancy. Alternatively, a large direct-drive fan may be as reliable as many smaller fans. If transmitter output power is high, a large volume of air is needed to cool the heat sinks adequately. Low-pressure fans or blowers may be used if heat-sink fin density is not high. This aids in reducing audible noise. The heat is distributed over a large volume of air and the temperature rise is relatively low.

Solid-state transmitters are usually self-contained, except for the external air handler. Air that has already passed through the equipment and absorbed heat must be removed from the immediate vicinity to prevent the hot air from being recirculated. In addition, provisions must be made for sufficient intake air to replace the exhaust air.

It is common practice to set the transmitter into a sealed wall forming a plenum chamber that is supplied with outside air. The intake vent and blower should be sized to provide a slight positive room pressure. Installation of a manometer to sense pressure drop across the air prefilters helps determine the replacement interval of the filters. Equipment layouts usually provide for heated air to exit from the top of the cabinet. The size and location of this exhaust area is usually shown on a manufacturer-supplied outline drawing, similar to that shown in Figure 4-12. If space does not permit construction of the transmitter manufacturer's recommended air system, care must be taken to modify the design to fit the available space while cooling the transmitter properly.

The transmitter manufacturer's recommended air system is usually sized only for cooling the transmitter. Any additional cooling load in the building must be added. In areas with severely polluted air, it may be necessary to precondition the intake air to avoid bringing in corrosive salts or gaseous contaminants. Separate air conditioning is usually provided for the front side to cool personnel, program source, and test equipment. It is important to consult experienced professionals in the area of heating, ventilation, and air conditioning (HVAC) design for best results.

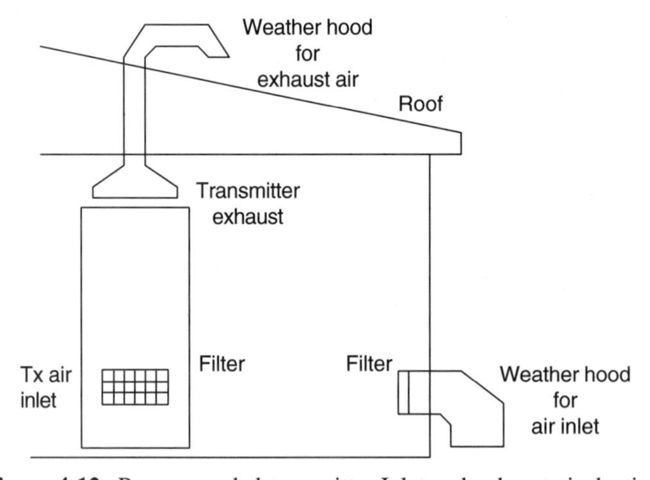

Figure 4-12. Recommended transmitter Inlet and exhaust air ducting.

AUTOMATIC GAIN OR LEVEL CONTROL

Ambient temperature changes cause gain changes in RF amplifiers. Automatic gain or level control (AGC or ALC) is used to maintain constant power output from the transmitter when this occurs. RF drive power must be boosted to maintain constant power output. A detected RF sample of the PA output is fed to an input of a comparator. With AGC, a voltage proportional to the exciter output is applied to the other input of the comparator. The dc output is then integrated and fed to an attenuator that varies the low-level RF drive. With ALC, the comparator reference is a fixed dc voltage; the output power level is maintained without regard for the exciter output level. Alternatively, the output RF sample may be taken from an intermediate stage. In this case, other means must be used to temperature-compensate the output stage, or some power reduction with increasing temperature must be tolerated.

AC DISTRIBUTION

A reliable method of ac distribution provides power to modular RF amplifier cabinets through a parallel arrangement. Each cabinet is protected by a separate ac breaker external to the transmitter. This approach allows a cabinet to be serviced safely while the remaining cabinets are operational. Phase monitors guard against low voltage, loss of phase, or phase reversal.

TRANSMITTER CONTROL

If individual amplifier modules and power supplies are self-protecting, control and monitoring functions can be simple and straightforward. One approach for the control system is to use a single controller to control and monitor all the functions of the transmitter as shown in Figure 4-13. Another approach is to

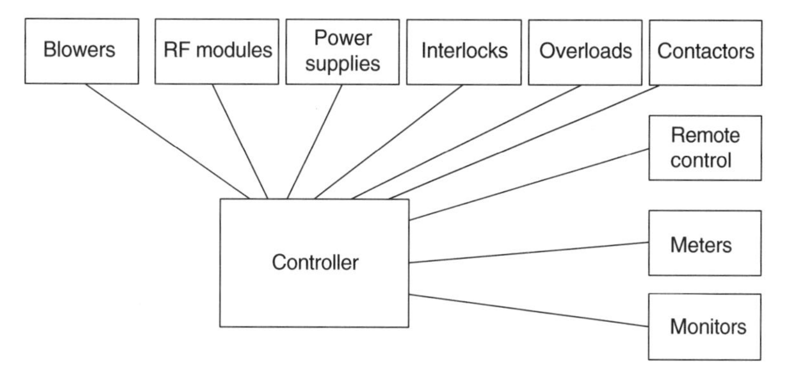

Figure 4-13. Block diagram of centralized transmitter control.

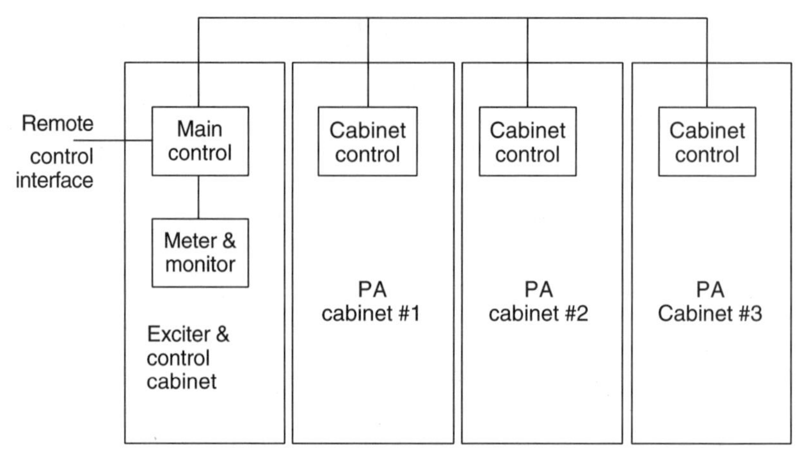

Figure 4-14. Block diagram of distributed transmitter control.

distribute the control system throughout the transmitter as shown in Figure 4-14. The distributed control system can be designed so that the failure of any individual controller does not affect the operation of the others. After ac power failure, the controller should have backup memory to restore the transmitter to the same operating condition as before.

Monitoring of transmitter output power, other operating parameters, and system status is essential to quick fault diagnosis. Monitoring should be independent of control so that a failure in the monitoring circuits will not affect the transmitter on-air status. Typical status conditions displayed include faults for exciter, VSWR, VSWR foldback, power supply, controller, air loss, door open, fail-safe interlock, phase loss, RF module, drive, and external interlock(s). VSWR foldback reduces power during high VSWR operation, such as antenna icing, and restores RF power to normal when the high VSWR is removed. A block diagram of a solid-state transmitter with VSWR foldback is shown in Figure 4-15.

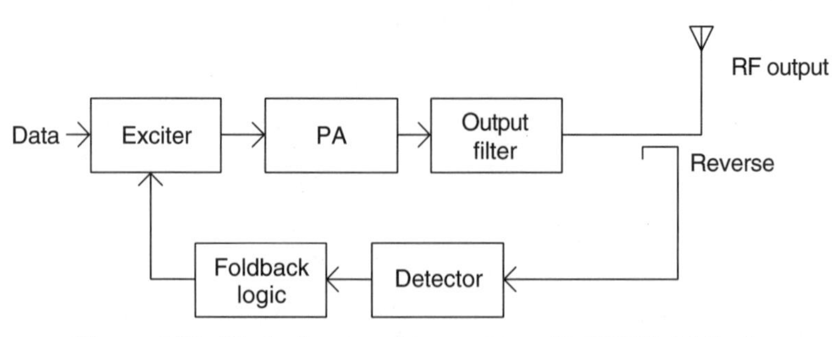

Figure 4-15. Block diagram of transmitter with VSWR foldback.

TUBE TRANSMITTERS

Due to the need for higher transmitter power, low power consumption and high efficiency are of utmost importance for UHF stations with high-AERP assignments. Although solid-state transmitters offer many advantages, many UHF stations find that the most cost-effective transmitter design is based on tube technology. Tube amplifiers have been used in transmitters since the dawn of television. They are field proven as linear amplifiers and can produce very high power at relatively low cost. The latest generation of tube amplifiers are reliable and efficient and provide the lowest cost per watt for high-power UHF transmitters.

The high beam voltage required by tubes tends to minimize power supply cost. Liquid cooling for the higher-power tubes results in efficient heat transfer and low acoustic noise within the transmitter plant. High-power-tube designers have shown great ingenuity in improving operating efficiency and bandwidth so that tubes remain a feasible alternative for UHF transmitters.[11]

Some of the best features of solid-state transmitters are also used in tube-based designs. For the highest powers, multiple output tubes, each with its own drive chain, are used to provide an acceptable level of redundancy and the benefits of a soft-fail architecture. Distributed control and monitoring architectures are also used. Some of the lower power and more efficient tubes [i.e., tetrodes and inductive output tubes (IOTs)] may be air-cooled.

A variety of tube technologies, including tetrodes, klystrons, multiple-stage depressed collector klystrons, and IOTs are available for UHF digital television requirements. Some of these are most suited for lower-power transmitter designs. Others are more appropriate for the highest power requirements. For example, the linear transfer and soft compression characteristic of the IOT may make this tube the technology of choice for power levels of 25 kW or higher.

In tetrode-based transmitters, the anode current is modulated by the RF drive applied between the cathode and control grid. The Diacrode[12] is a fairly recent development that operates on the same principles as the tetrode. It is, in effect, a dual tetrode and is capable of twice the power of conventional tetrodes with some improvement in gain and efficiency.

Klystrons are in the family of linear beam tubes and have the required bandwidth, linearity, and peak power capability to amplify a digital TV signal. They operate as class A devices and therefore are quite inefficient. As with other devices, the beam power is set to the level required for the highest output power without significant clipping.

Development of new depressed collector klystron transmitters for digital television has not been reported. Tests of 8 VSB on an EEV K3672BCD klystron

[11] Robert S. Symons, "Tubes: Still Vital After All These Years," *Spectrum*, April 1998, pp. 52–63; Robert Symons, Mike Boyle, John Cipolla, Holger Schult, Richard True, "The Constant Efficiency Amplifier: A Progress Report," *NAB Broadcasting Engineering Proceedings*, 1998, pp. 77–84.

[12] Diacrode is a registered trademark of Thomson Tubes Electroniques.

amplifier[13] indicate that good performance is possible with depressed collector klystrons but at the price of low efficiency, just as with conventional klystrons. The in-band and out-of-band performance was found to be satisfactory at an average power level of approximately 11 kW. However, the efficiency was less than 10%, compared to more than 30% measured for IOTs. This is a consequence of the operation of the klystron as a class A amplifier, whereas the IOT operates as a class AB. Since the efficiency and linearity of klystrons are inferior to the IOT, the focus of this discussion is on IOT transmitters. The principles discussed, however, apply to the other technologies.

The IOT has an electron gun and RF input section similar in design to that of a triode. The output section is similar to that of a klystron. The IOT is biased to operate as a class AB device and therefore is very efficient over much of the operating range of the digital television signal. Typical efficiency versus power output for a 60 kW IOT is shown in Figure 4-16. At an average power output of 15 kW, the efficiency is approximately 33%. Tube life is known to be in the range of 20,000 to 30,000 hours.

The IOT operates as a class AB linear amplifier, exhibiting a soft compression characteristic and low incidental phase distortion. Unlike the klystron, it does not have a hard-limiting characteristic. This reduces the amount of precorrection and makes setup and optimization easier.

Because all transmitter manufacturers must use the tubes produced by a limited number of tube suppliers, standard power levels for tube transmitters are common.

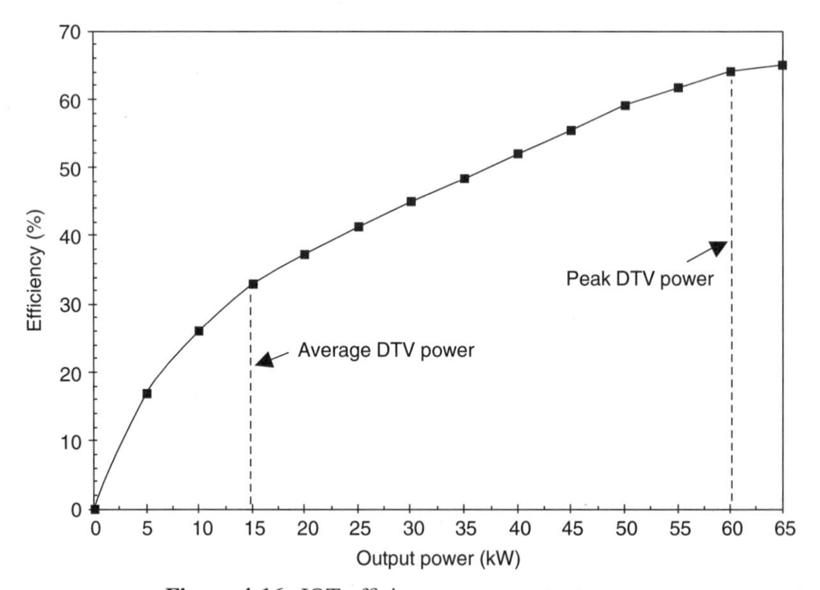

Figure 4-16. IOT efficiency versus output power.

[13] Roy Heppinstall, Alan Wheelhouse, and Geoffrey Clayworth, "The 8-VSB Performance Expected from Klystron Amplifier Systems," *NAB Broadcast Engineering Proceedings*, 1998, pp. 47–52.

The variations among manufacturers are often in the number of tube finals combined in a standard configuration and the effectiveness of precorrection and equalization circuitry. The gain of the IOT is in the range 20 to 21 dB. This means that at least 100 W is required to drive each tube. Since exciter output power is typically in the range of 250 mW, an intermediate power amplifier (IPA) is required. To minimize distortion and correction, the IPA must be very linear while producing the power required.

The type of IPA influences the transmitter efficiency. One approach is to use a class A amplifier. This is effective but expensive in terms of hardware and power consumption. Class A amplifiers exhibit efficiency in the range of 10%, so that a large number of solid-state amplifiers is required to generate the necessary drive power. This, in turn, leads to the need for more power supply capability. The added power supplies and class A operation require additional cooling. The result is a fairly large IPA rack and substantial increase in power consumption. A more efficient approach is to use a class AB amplifier for the IPA with a feedforward loop.

Uncorrected, the class AB amplifier has somewhat poorer linearity than the class A amplifier, due to crossover distortion. However, extremely good linearity can be obtained by adding a feedforward loop. In its simplest form, this technique subtracts a sample of the IPA output from a sample of the input. This difference signal is amplified in a low-power class A amplifier and reinserted at the proper phase and timing into the IPA output by means of a directional coupler. The phasing cable and coupler introduce some additional insertion loss in the IPA output which must be overcome by a corresponding increase in the output power of the class AB amplifier. Even with the increased amplifier size, the result is a linear IPA with substantially lower power consumption. Thus the linearity of the IPA can equal or exceed a class A amplifier while enjoying most of the efficiency advantage of class AB operation.[14] Assuming at least three times the channel bandwidth in the IPA, the presence of the feedforward loop does not affect the order of the baseband or IPA precorrection and equalization. Obviously, the feedforward loop must be set up correctly before adjusting the baseband or IF circuits.

Unlike solid-state transmitters, tube transmitters are usually not self-contained. In addition to the basic transmitter, other external components, such as beam power supplies, cooling pump modules, fan units, and ac line control cabinets are required. For TPO up to 25 kW, a single beam power supply, cooling pump module, fan unit, and ac line control cabinet are required. Since the beam supply is often unregulated, an ac line regulator may be desirable. Typical size and weight values for the electrical assemblies are listed in Table 4-3. For 35- and 50-kW transmitters, a pair of beam supplies and line control cabinets are required; three sets are needed for a 75-kW unit and four sets for a 100-kW unit. Each pump module is 36 in. wide by 55 in. deep and weighs 800 lb. A single unit is needed for power output through 75 kW; two are needed for an output of 100 kW. A single outside cooling fan unit is needed for TPO through 75-kW, but its

[14] Feedforward is not cost-effective for linearizing the final amplifier. The added loss of an injection coupler and phasing cable at the transmitter output would result in an unacceptable efficiency penalty.

TABLE 4-3. Size and Weight of IOT Electrical Assemblies

	Total Average Transmitter Output Power (kW)	Size (in.)	Approximate Weight (lb)
Beam Supply		52 W × 62 D	6000
Line Control		37 W × 12 D	365
Automatic	10–12.5	26 W × 18 D	600
voltage regulators	17.5–25	26 W × 18 D	900
(optional)	35	26 W × 18 D	1100
	50	57 W × 25 D	1600
	75	57 W × 25 D	2100
	100	57 W × 25 D	2300

Source: Data courtesy of Harris Communications.

TABLE 4-4. IOT Cooling Fan Size and Weight

Total Power Output (kW)	Size (in.)	Weight (lb)	Remarks
10–25	44 W × 92 D	700	
35–50	44 W × 132 D	950	
75	44 W × 172 D	1340	
100	44 W × 132 D	950	Two required

size increases with increasing power as shown in Table 4-4. Because of the number of assemblies involved and the complexity of installation, operation and maintenance of the IOT transmitters is greater than for solid-state.

TUBE OR SOLID-STATE TRANSMITTERS

For the very highest power output, tubes are clearly the most cost effective technology. For the lowest power levels, solid-state is usually selected. But what about the midpower range? As already seen, for moderate AERPs, both tube and solid-state transmitters are available for digital television. Having just discussed the number, size, and weight of the various IOT assemblies and their attendant installation, operation, and maintenance issues, many engineers might opt for solid-state transmitters. In fact, many stations have experienced dramatic reductions in maintenance cost when replacing their analog tube transmitter with a solid-state unit. But there are other considerations, including acquisition and operating cost, which in many cases, favor tube transmitters.

Figure 4-17 is a graph of power consumption versus TPO for several IOT and solid-state transmitters. Clearly, the tube transmitters consume less power. This is due, in part, to the absence of combining losses arising from the necessity of combining many low-power devices in solid-state transmitters. Both

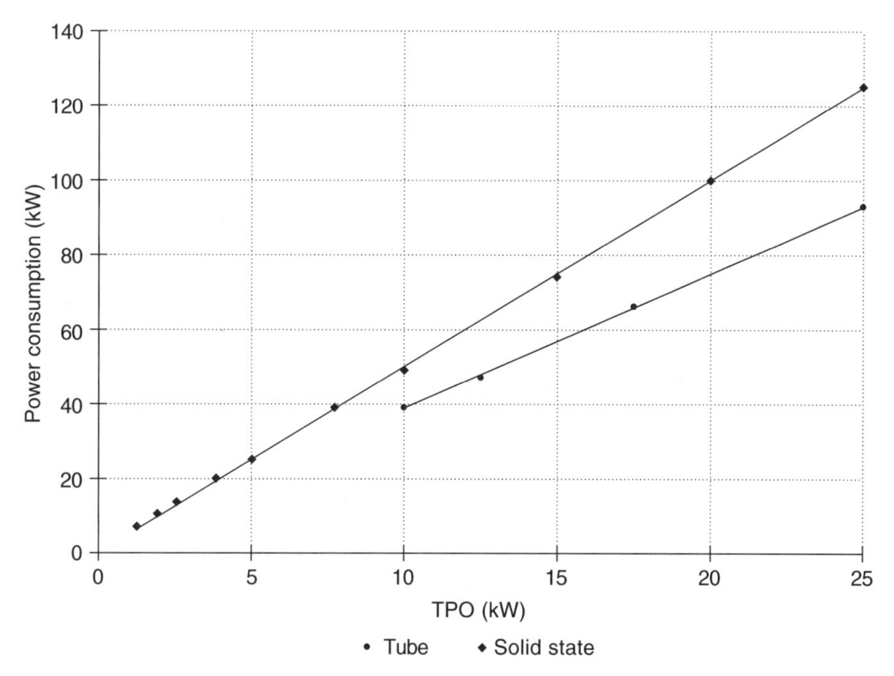

Figure 4-17. Transmitter power consumption.

amplifier types operate class AB, so there is not much difference in dc-to-RF conversion efficiency. However, solid-state transmitters use low voltage–high current supplies, which have lower ac-to-dc conversion efficiency than the high voltage–low current beam supplies used with tubes. Overall, the difference in power consumption is substantial — 11 kW at a TPO of 10 kW; at 25 kW, the difference is 32 kW. Depending on prime power cost and operating time, this could increase operating cost by as much as $25,000 per year.

Acquisition cost is an important factor in transmitter selection. This is illustrated in Figure 4-18, which shows the relative acquisition cost of IOT and solid-state UHF transmitters. The raw costs from which this graph was derived include dual exciters, RF systems, and surge suppressers in the price of the transmitter. For the IOT, an automatic voltage regulator (AVR) and calorimeter are included in the transmitter costs. Not surprisingly, the solid-state transmitters are more expensive than the IOT transmitters at the same power level. Comparative costs of the two technologies with increasing TPO also rise at different rates. As solid-state TPO doubles, the cost also doubles. In the TPO range where either solid-state or IOT transmitters may be used, the IOT transmitter cost rises only moderately with increasing TPO.

The effect of parallel redundancy on system reliability must also be considered. This factor generally favors solid-state transmitters. The formula for the reliability of a system, R_s, composed of several parallel amplifiers, each with

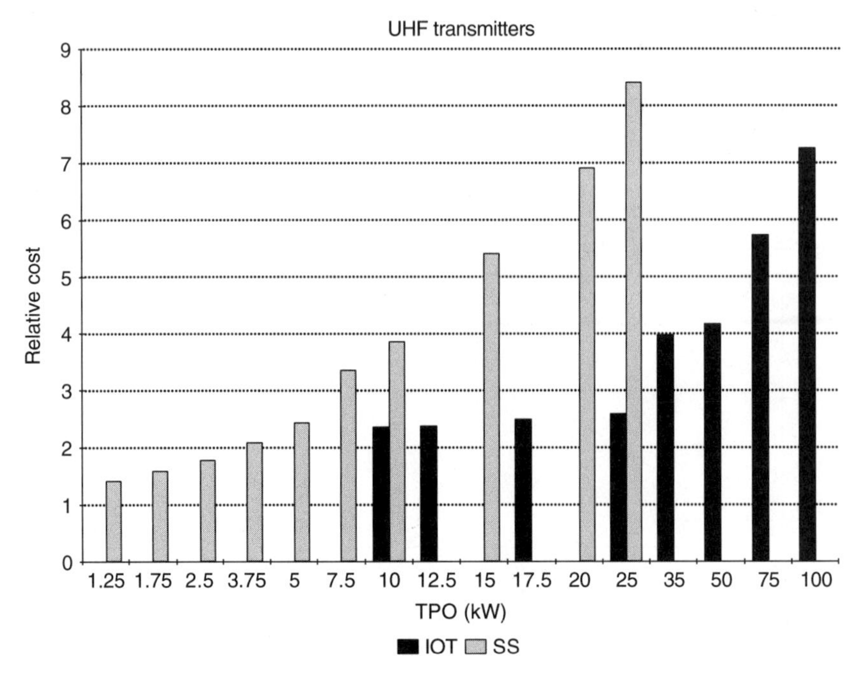

Figure 4-18. Acquisition cost versus TPO.

reliability of R_a, is[15]

$$R_s = 1 - (1 - R_a)(1 - R_a)(1 - R_a) \cdots (1 - R_a)$$

For example, a single amplifier with no parallel redundancy might have a reliability of 0.999 or lack of reliability of 10^{-3}. By paralleling two units, the system reliability increases to 0.999999 or a lack of reliability of 10^{-6}. To achieve high levels of reliability, it is assumed that the individual assemblies are reliable, that the system may continue to operate when one unit fails, that the system need not be turned off for maintenance and repair, and that spares are readily available. For solid-state transmitters, these criteria can be met. Large numbers of critical on-air items such as power supplies and RF modules are combined so that high levels of parallel redundancy can be achieved. When parallel redundancy in the PA is coupled with dual exciters and automatic switching and other fail-safe features, very reliable operation can be expected.

To a limited degree, this approach may be taken with multiple tube amplifiers. However, the number of parallel PAs is usually limited to four, so that the redundancy of solid-state amplifiers is not achieved. In addition, loss of one tube PA out of four means a greater loss of TPO than one out of a large number of solid-state PAs. Given the "cliff edge" effect characteristic of digital transmission

[15] *Reference Data for Radio Engineers*, op. cit., p. 43–26.

(see Chapter 8), this means a much greater loss of coverage with the failure of a single-tube PA than with a single solid-state PA. The need for quick repair of the failed unit is heightened. If continuous full-area coverage is required with a tube transmitter, consideration should be given to a complete backup transmitter system. If this is necessary, some of the cost advantage of tube transmitters is lost. If dual tube amplifiers are considered the norm, a single solid-state transmitter is cost competitive with a tube transmitter on an acquisition cost basis through a TPO of about 10 kW.

PERFORMANCE QUALITY

In the previous discussion it has been assumed that the available transmitters meet the required performance specifications. This is usually a good assumption; a TV transmitter manufacturer will not stay in business long if it does not meet published performance specs. However, the ease with which specifications are met, the effectiveness and ease of use of precorrection and equalization circuits, and system stability may vary with PA and exciter technology as well as manufacturer. The various manufacturers' offerings should be evaluated carefully.

Bandlimiting of the output is a fundamental difference between solid-state and tube technology. This issue is especially important since it relates to meeting emissions mask requirements. As a result of manufacturing and field service considerations, most solid-state transmitters use broadband PA modules. Transmitter manufacturing cost and spares inventory can be minimized if broadband RF modules rather than channelized designs are used. All power tube technologies, whether tetrode, klystron, or IOT, extract RF power from the tube by means of an output cavity or cavities. These are tuned circuits that are adjusted for a specific operating channel. Although the tunable nature of the cavity requires a certain level of maintenance, it has the advantage of bandlimiting the output and thereby aiding the ability to meet the requirements of the emissions mask.

Typical sideband response measured on a solid-state DTV transmitter is shown in Figure 4-19. As expected, the sideband level is a function of drive level. This is a result of driving the amplifier into compression, thereby limiting the less frequent peaks. Breakpoints occurs at the edges of the in-band signal. Outside the signal bandwidth the sideband level falls away gradually, following a nonlinear curve. However, there is no distinct bandlimiting beyond the breakpoints since there is no sharply tuned circuit in the output. No amount of external filtering can reduce the sideband level at the breakpoints. Although it might be effective with respect to amplitude response, even a brickwall filter would introduce a large amount of phase distortion or group delay. In practice, reduction in the breakpoints can only be achieved by reducing the drive to the PA and/or by precorrection of PA nonlinear distortion. This, in effect, establishes the maximum TPO of the transmitter. The transmitter drive is raised to the point that the sideband level

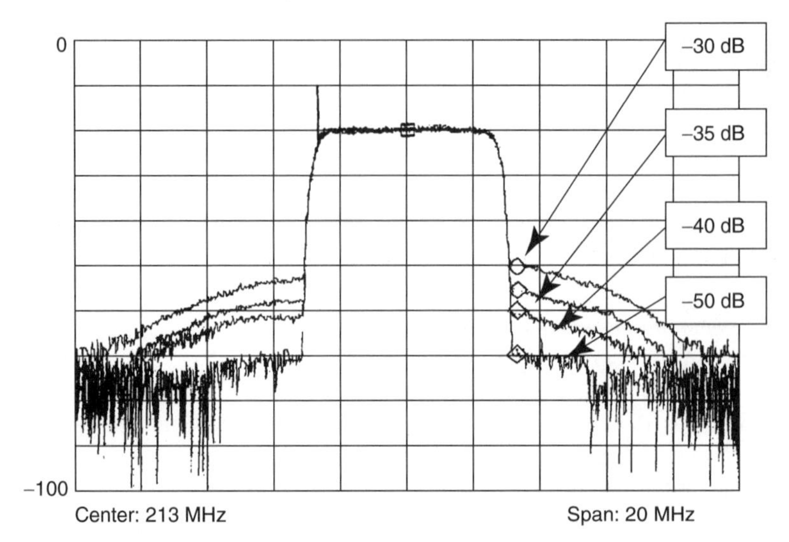

Figure 4-19. Sideband response of a solid-state transmitter for several drive levels.

at the breakpoints can be met (with some headroom) using the combination of precorrection available within the exciter and other low-level stages. The sideband level beyond 0.5 MHz from the breakpoints is controlled by a high-power bandpass filter.

A plot of sideband response for an IOT transmitter is compared to a similar plot for a UHF solid-state transmitter in Figure 4-20. Within a few megahertz of the channel, there is very little difference in the out-of-band emissions. (This comparison is not entirely valid since the IOT data are corrected for PA nonlinearity and the solid-state data are uncorrected). Nevertheless, at ±25 MHz, there is approximately a 20-dB difference in the respective levels. The bandlimiting of the output cavity accounts for most of the difference.

RETROFIT OF ANALOG TRANSMITTERS FOR DTV

In the United States, there is some interest in retrofitting NTSC transmitters for DTV during the transition from analog to digital. As has been seen, the fundamental characteristics of digital television transmitters hold many features in common with analog transmitters. Thus it is expected that most, if not all, NTSC power amplifiers could be retrofitted to transmit the DTV signal. For UHF, this would include tetrodes, klystrons, depressed collector klystrons, and IOTs. Even though the efficiency of klystron amplifiers is low, implementation costs may make this a feasible option. VHF retrofits would include older tube-model transmitters and newer solid-state PAs.

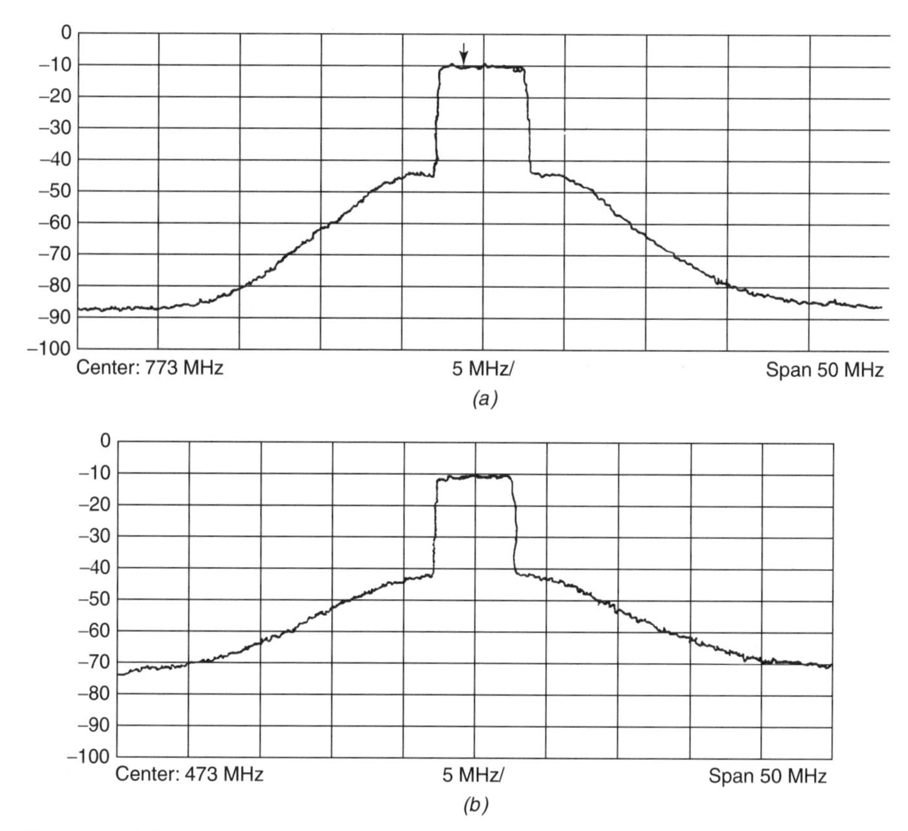

Figure 4-20. Unfiltered emissions from (*a*) IOT transmitter and (*b*) UHF solid-state transmitter.

No matter the amplifier technology, there are some common elements in retrofits. Obviously, there must be a new DTV modulator. This might include only the elements from the transport layer interface to the IF output. This would require use of an upconverter from an analog exciter. Because of the differences between NTSC and ATSC carrier frequencies and signal bandwidth, changes to frequency generation circuits and the output filter of the upconverter would generally be required. It may also be necessary to improve the precorrection circuits. A better solution might be replacement of the entire analog exciter with a new DTV exciter. Although this would mean greater expense, system integration is simpler. It should not be expected that either the DTV modulator or complete exciter would fit in the space previously occupied by the NTSC exciter. If not, it will be necessary to modify the exciter cabinet or mount the DTV exciter in a separate rack.

Modifications to the exciter/transmitter control and metering interface will be needed. It is certain that the time constant for power detection and metering must be modified to indicate average power rather than peak sync power. There will be no separate aural and visual power metering. Detection of reflected power will require modification. For automatic changeover of dual DTV exciters, a new exciter switcher may be required.

Tuning of output networks in tube PAs will require modification. Bandwidth must be increased to the full bandwidth of the digital channel. Aural PAs must be disabled if they are not used to transmit DTV. The diplexer must be replaced with a DTV bandpass filter. It may be necessary to increase the IPA output power since increasing the bandwidth of the PA will reduce the gain of the final stage. It may also be necessary to reoptimize the drive chain for best linearity.

Conversion of a typical NTSC transmitter to DTV is depicted in Figure 4-21. The existing transmitter is a two-tube 60-kW NTSC depressed collector klystron transmitter. This system could be converted to a main/standby DTV transmitter with each tube operating at 10 to 12 kW of average power. The cost of adding the DTV exciter and output filter plus that of making other transmitter modifications might be a fraction of the purchase price of a new transmitter of comparable power. The costs of making the required modifications and the costs of operation must be weighed carefully against the costs of a new DTV transmitter.

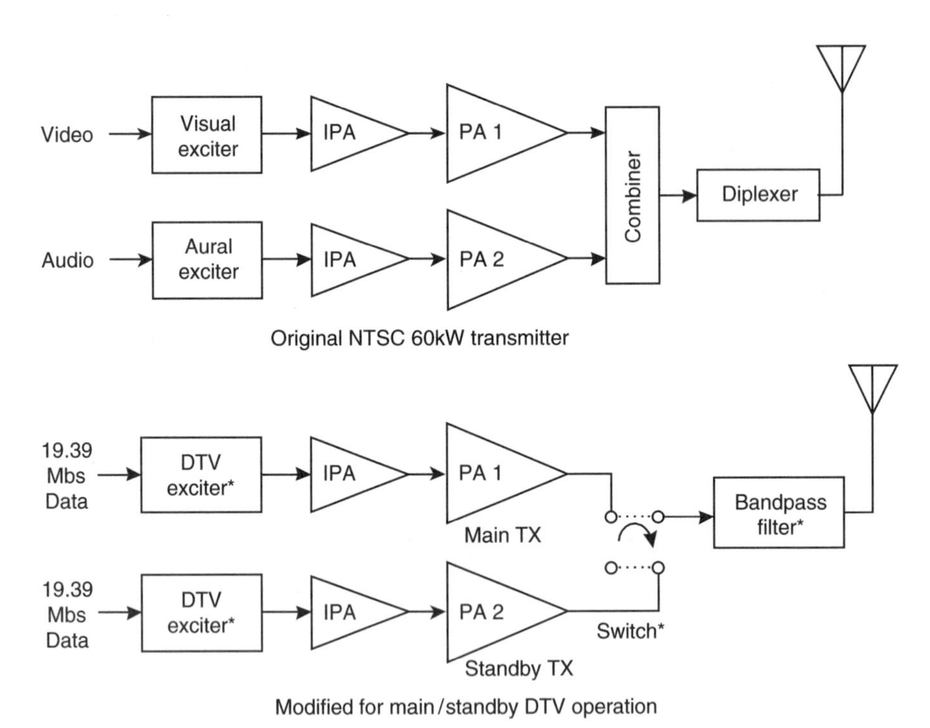

Figure 4-21. Conversion of NTSC transmitter for DTV operation.

Conversion of an NTSC klystron transmitter for simultaneous analog and digital TV transmission has been reported.[16] The klystron was a Varian (now CPI) VKP-7553S tube. The visual transmitter was reconfigured to transmit visual and aural in common amplification. The resulting visual power was 1 dB lower than the visual-only power. The aural amplifier was modified to transmit DTV at a level 12.5 dB below the peak visual level. Satisfactory spectral regrowth and C/N ratio were reported. However, the reported output power of 5500 W, beam voltage of 23.5 kV, and current of 2.6 A indicate an efficiency of only 9%. This is in substantial agreement with the efficiency reported by Heppenstal et al.[17]

[16] R. W. Zborowski and David Brooking, "Klystron Transmitter Conversion for Simultaneous Analog DTV Transmission," *NAB Broadcast Engineering Proceedings*, 1998, pp. 40–46.
[17] Op. cit.

5

RADIO-FREQUENCY SYSTEMS FOR DIGITAL TELEVISION

High-power output filters and channel combiners are usually included in the category of RF systems. These components are located at the output of the power amplifier. Output filters serve to limit the bandwidth of the radiated signal prior to the transmission line and antenna. Channel combiners are used to facilitate transmission of a pair of TV signals by a single transmission line and antenna. The paired TV stations may be analog and digital or a pair of digital stations. In many cases the paired stations have adjacent channel allocations.

In a sense, the term *RF system* is misleading in that every component in the transmission path onward of the modulator is a part of the RF system. This would include the RF portions of the exciter, the IPA, the PA, filters and combiners, transmission line, and antenna. However, this terminology is common for analog TV and has been carried forward to digital TV. Better practice is to call these devices by their names (i.e., output filters and channel combiners). This is the terminology adopted in the remainder of this book.

These components have an important role to play in overall system performance. Both in-band and out-of-band performance are affected. As related to the digital signal, the in-band amplitude and group delay variations of all components beyond the transmitter PA output must be flat with respect to frequency. The overall response of the filter must be stable under expected environmental conditions, especially temperature and humidity variations. There must also be a minimum of in-band insertion loss.

The requirement for low insertion loss relates to overall system efficiency. Any power dissipated beyond the final amplifier is expensive and must be minimized. If it is assumed that every increase in insertion loss of 0.1 dB raises the cost of generating TPO by 2.3%, the importance of minimizing insertion loss is obvious. Although transmitter output power tends to come in discrete increments, thinking of loss in these terms puts the issue in terms understood by all concerned. Waveguide construction of filters is preferred because waveguide losses are much lower than for rigid coaxial designs. In addition, waveguide provides extremely high power-handling capability.

Although substantial amounts of amplitude and group delay variations (up to 150 ns) can be equalized, low in-band linear distortions in the output components are important. In addition to filters and combiners, the output cavities of tube amplifiers contribute linear distortions. Use of waveguide for the main transmission line can also introduce group delay. The total linear distortions of output cavities, RF filter, channel combiner, and transmission line must be equalized for satisfactory in-band performance. The greater the distortions, the greater the difficulty in adjusting and maintaining equalization circuits. Fortunately, digital television exciters with adaptive equalization minimize the setup and maintenance of in-band response.

Performance of the output components must be closely integrated with the PA and exciter precorrection and equalization circuits. For example, it may be possible to exploit the lower out-of-band emissions of a tube transmitter to allow the use of a simpler and less expensive filter. If adaptive equalizers are used, appropriate sample points and signal detectors must be provided. For these reasons, it is preferable to purchase the output filter and channel combiner from the transmitter supplier. The output components and transmitter may then be properly integrated and tested to assure optimum overall performance. Output power should be set up and calibrated at the output of the filter. Precorrection and equalization circuits should be set up with the output components in place.

Filters may be designed as reflective or constant-impedance types. The reflective filter is by far the simpler and least costly of the two, consisting of only the filter proper. This type of filter presents a matched impedance to the in-band signals but a large reactance to the out-of-band signal. Because of this property, it is possible for the reflective filter to influence the frequency response of the PA. The in-band insertion loss of the reflective filter is somewhat lower than the constant impedance due to the absence of input and output hybrids. Several solid-state transmitters have gone on-air with reflective filters with satisfactory performance. IOT transmitters have generally been equipped with constant-impedance filters.

CONSTANT-IMPEDANCE FILTER

The constant-impedance filter is commonly used as a bandpass filter or a channel combiner, even though it is more complex than the reflective filter. It has the advantage of presenting a matched load to the transmitter output at all frequencies, in-band and out-of-band. Because of this property, the response of the PA is not affected by the impedance of the filter. This circuit topology, shown in Figure 5-1, takes advantage of the signal flow properties of quadrature

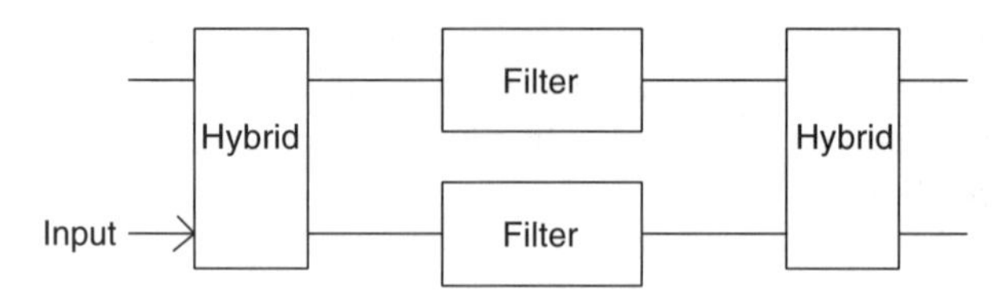

Figure 5-1. Constant-impedance filter topology.

hybrids and the frequency-response properties of RF filters. A pair of filters are sandwiched between input and output hybrids, much like many visual/aural diplexers used for analog TV. A reject load and a ballast load are also required. The transmitter output signal is applied at the input to the left-hand hybrid, where it is split equally into quadrature components. The passband signal passes through the pair of filters and is recombined into a single output by the right-hand hybrid. The stopband signals are reflected from the filters and recombined in the isolated port of the input hybrid.

It is well known that a quadrature hybrid divides the input signal at port 1 into two outputs of equal amplitude and quadrature phase at ports 2 and 3, as shown in Figure 5-2. By reciprocity, a pair of equal amplitude but quadrature signals at ports 2 and 3 will combine to a single signal at port 1; port 4 is isolated. By a similar process, shifting the phase of the signals at ports 2 and 3 by 180° results in the combined signal appearing at port 4; port 1 is then isolated. The phase shift can be accomplished by placing short circuits at ports 2 and 3. In this case, a signal may be applied at port 1 and the output will appear at port 4.

A pair of bandpass or band-reject filters may be used as frequency-selective short circuits, depending on the application. If bandpass filters are used, the result is a constant-impedance bandpass filter. If band-reject or stopband filters are used, a channel combiner is the result. No matter the filter type, in the stopband the frequency response at port 4 assumes the inverse characteristic of the filters. The signal at port 4 is absorbed by a suitable load, either an antenna or a reject load, depending on the application. Signals within the passband of the filters arrive at the input ports of the second hybrid, where they combine to a single output. The response at this port is identical to the response of the filters. This signal

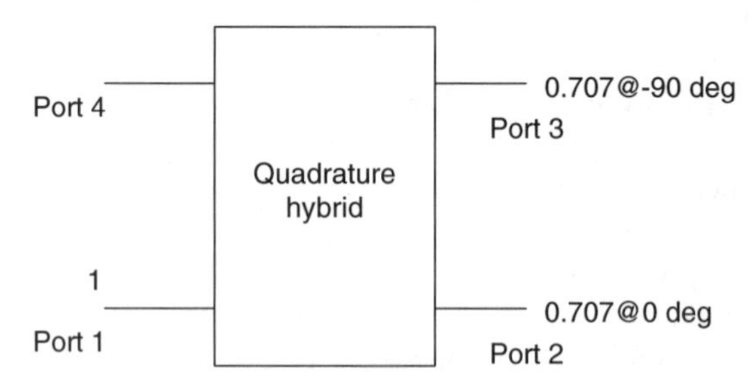

Figure 5-2. Quadrature hybrid operation.

is also absorbed by a suitable application-dependent load. Any power resulting from unbalanced splitting, filtering, or combining is absorbed in a ballast load.

Bandwidth is defined as the difference between upper and lower frequencies at a specified attenuation level, usually the peak ripple within the passband. The center frequency, f_0, is taken to be the geometric mean of the upper and lower band edges, f_2 and f_1.

$$f_0 = (f_1 f_2)^{1/2}$$

For digital television, these frequencies are, for all practical purposes, the same as the channel edges.

Although practical filters are constructed of coaxial and waveguide transmission lines, the operation of these devices is best understood in terms of their lumped-element prototypes or equivalent circuits. The elements of the prototypes include capacitors, self- and mutual inductance, and incidental resistance. Coupled cavities are used for the practical realization of these elements.

The transmission properties or attenuation and group delay as a function of frequency are the operating characteristics of greatest interest. For the prototype designs, the transmission function is determined by the choice of lumped reactive elements plus the impedance of the generator (the transmitter) and the load (the antenna). In mathematical terms, the shape of the attenuation versus frequency curve is the ratio of two polynomials. The roots of the numerator are called transfer function zeros or attenuation poles. The roots of the denominator are called transfer function poles or attenuation zeros. The roots are complex numbers, meaning that both amplitude and phase are important. The values of the prototype elements and the resulting locations of poles and zeros in the complex plane may be determined using modern filter synthesis techniques.

OUTPUT FILTERS

Given the requirements imposed by out-of-band emissions masks, there is little doubt that virtually every digital TV transmitter will require an output filter. Tube transmitters with their output cavity may require somewhat less filtering than do solid-state transmitters; however, it is not feasible to reduce nonlinear distortions sufficiently to meet out-of-band requirements without a filter. In addition to providing the filtering required to meet the emissions masks, the output filter must attenuate harmonics to acceptable levels. There is no upper-frequency limit on the FCC mask. The DVB-T mask definition is confined to ± 12 MHz from the channel center frequency.

The ideal filter would have an in-band response that is perfectly matched with zero insertion loss, a flat in-band amplitude and phase response, and a discontinuous increase in attenuation at the band edge to the required out-of-band rejection level. Although a function of frequency, the rejection level depends on the particular mask requirements and the out-of-band characteristics of the transmitter. In addition, the filter must handle the transmitter output power. Output filters may be mounted on the floor or ceiling, often with welded frames for support. In any case, performance requirements must be met while minimizing size, weight, and cost.

In practice, some compromise must be made to the ideal. The transition from passband to stopband is gradual in practical filters. Thus, perfectly flat in-band amplitude response cannot be achieved. Steep transitions from passband to stopband are associated with rapid changes in phase with respect to frequency. Thus perfectly flat in-band phase response is not feasible. In fact, a small amount of amplitude and phase ripple must be tolerated throughout the passband. The quality factor, $\boldsymbol{Q}$, of practical cavities is finite, so that a small amount of ohmic loss must be accepted. Power rating is also related to losses. Out-of-band attenuation is also limited. In the transition between passband and stopbands, the filter cannot provide the ideal attenuation curve. The transmitter must be sufficiently linear to provide adequate IP suppression in this region.

Since the purpose of the filter is to reduce out-of-band emissions to acceptable levels, this requirement must be defined first. This is done by subtracting the transmitter output emissions from the applicable emissions mask (see Chapter 4). From these data the required attenuation versus frequency may be plotted in the form of a filter response mask. The interdependence of the filter and transmitter reinforces the need to procure both items from the same source to assure good system performance.

A typical response mask for an ATSC UHF DTV output filter is shown in Figure 5-3. In-band ripple is specified to be less than ±0.05 dB over a

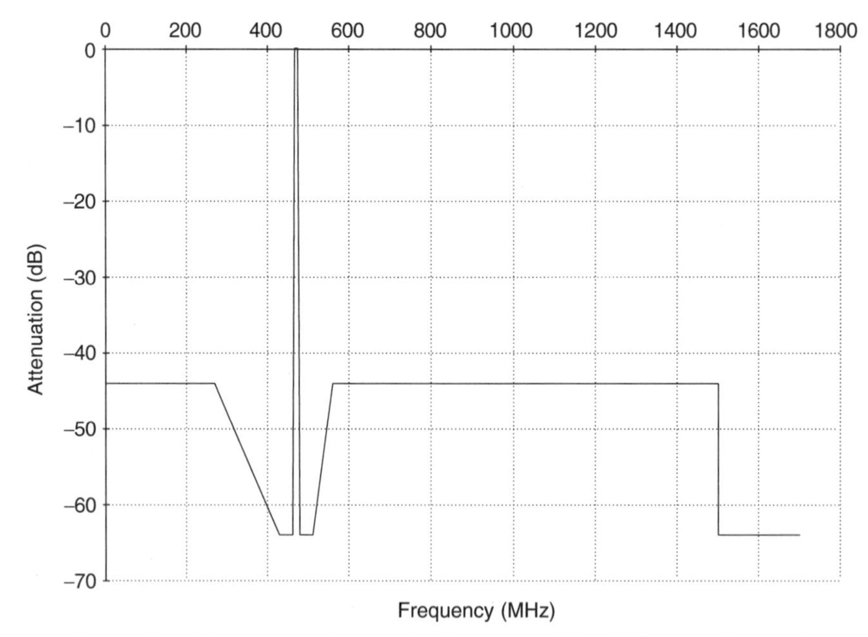

Figure 5-3. Filter attenuation mask. (Courtesy of Harris Communications.)

minimum bandwidth of 6 MHz. The transition from passband to stopband extends from ±3 to ±9 MHz. The maximum stopband attenuation of 64 dB extends to ±40 MHz. Beyond these frequencies the attenuation varies in accordance with FCC requirements, which includes attenuation of harmonics to required levels and protection of other services.

To illustrate the adequacy of the stopband response, the unfiltered and uncorrected IP output of a typical solid-state transmitter of −40 dB (see Chapter 4) may be added to the filter response at ±9 MHz. This yields total out-of-band suppression of −104 dB. At 90 MHz, the filter response is −44 dB; the transmitter's unfiltered response is down more than 60 dB. Again, the total out-of-band suppression is −104 dB.

The in-band amplitude response is specified to be flat enough that no additional equalization is required. Substantial amounts of group delay may be tolerated, however, with the assumption that sufficient equalization is available in the transmitter. Typical measured group delay response for a filter of this type is shown in Figure 5-4. There is nearly a 120-ns delay variation at ±3 MHz from band center.

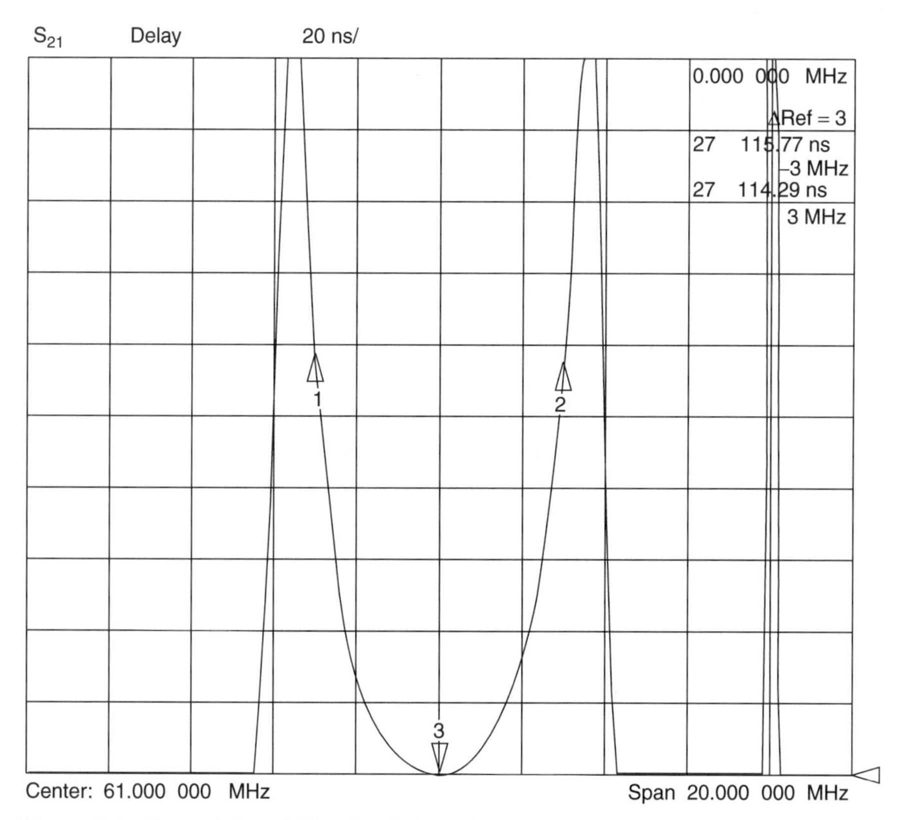

Figure 5-4. Group delay of filter for digital television. (Data courtesy of Scott Durgin of Passive Power Products, Gray, Marine.)

The cutoff slope is a key design parameter and is defined as

$$S_{\text{dB}} = \frac{A_{\text{sb}} - A_{\text{pb}}}{f_{\text{sb}}/f_{\text{pb}} - 1} \qquad \text{dB/MHz}$$

where A_{sb} and A_{pb} are the attenuation at the stopband and passband edge frequencies, f_{sb} and f_{pb}, respectively. For the mask shown in Figure 5-3 at US channel 14,

$$S_{\text{dB}} = \frac{64 - 0.05}{479/473 - 1} = 7568 \text{ dB/MHz}$$

The number of filter sections is related directly to the cutoff slope as well as the ripple in the passband and attenuation in the stopband. For a specified passband ripple and stopband attenuation, the greater the cutoff slope, the more sections are required.

ELLIPTIC FUNCTION FILTERS

To achieve the required level of performance demands advanced, complex filter designs. Minimum in-band ripple, steep skirts in the passband-to-stopband transition region and high stopband attenuation, high power-handling capability, and minimum cost necessitate all the filter designer's skills. This has led to the nearly universal use of designs based on lumped-element prototypes using the early work of Cauer and Darlington on elliptic functions and modern network filter theory. These functions provide poles of attenuation near the cutoff frequencies so that the slope in the transition region may be extremely large with a reasonable number of filter sections.

Elliptic function filters are characterized by equiripple response in both the passband and stopbands. This means that the peak-to-peak ripple in the passband is of low magnitude and constant; similarly, the peak-to-peak attenuation in the stopband is constant, although very high. These filters are optimum in the sense that they provide the maximum slope between the passband and stopbands for specified ripple in the passband and stopbands and for a given number of filter sections. This is in contrast to Butterworth or even Chebyshev designs, in which a large number of sections would be required for similar performance. For example, an elliptic function design may be less than half the length of a corresponding Chebyshev design.[1] An elliptic function design may also have less insertion loss and group delay variation than the Chebyshev design with equivalent rejection.

The normalized response or transmission power function, t_f^2, of a filter is defined in terms of the ratio of the power delivered by the transmitter, P_t, to the

[1] William A. Decormier, "Filter Technology for Advanced Television Requirements," *IEEE Broadcast Technology Society Symposium Proceedings*, September 21, 1995.

power delivered to the load, P_l; that is,

$$t_f^2 = \frac{P_t}{P_l}$$

The filter attenuation is simply 10 $\log(t_f)^2$. Since in the ideal or lossless case, the filter consists only of reactive elements, any power not delivered to the load is reflected. Thus the output power must be the difference between the power delivered by the transmitter and the reflected power. The lossless filter may therefore be fully characterized by the transmission and reflection coefficient functions, that is,

$$t_f^2 = 1 + \Gamma^2$$

where Γ is the reflection coefficient function. To achieve attenuation of less than 0.05 dB in an ideal filter, the reflection coefficient must be less than about 0.1. In practice, resistive losses are always present. This requires that the reflection coefficient be reduced to make allowance for internal circuit losses.

For elliptical function filters, t_f^2 is given by

$$t_f^2 = \frac{1}{1 + \varepsilon^2 R_n^2}$$

where ε is the passband ripple ($A_{\text{pb}} = 20 \log \varepsilon$), R_n is the ratio of a pair of polynomials defining the filter poles and zeros, and n is the number of poles or filter order.[2] Transmission zeros occurring when the frequency is on the imaginary axis of the complex frequency plane result in high attenuation; transmission zeros occurring when the frequency is on the real axis result in group delay self-equalization. By combining transmission zeros on the real and imaginary axes, filters with the desired rejection and acceptable group delay may be designed.

It is has not been possible to apply the necessary degree of phase correction to high-power elliptical function filters.[3] This has led to the use of a similar class of filters with cross couplings between nonadjacent resonators. These filters are referred to as cross-coupled or pseudoelliptic filters. These may be implemented in a variety of ways, including interdigital structures for low-power applications or in-line or single-mode TE101 or TE102 resonators in rectangular waveguide. Either of the latter are suitable for high-power applications.

In-line single-mode resonators can provide the levels of performance approaching those required. However, overall filter size can become an issue due to the extreme amount of rejection required by the emissions masks. Each resonator contributes only one resonance, so that the minimum filter length must

[2] Albert E. Williams, "A Four-Cavity Elliptic Waveguide Filter," *IEEE Trans. Microwave Theory Tech.*, Vol. 18, No. 12, December 1970, pp. 1109–1114.

[3] Graham Broad and Robin Blair, "Adjacent Channel Combining in Digital TV," *NAB Broadcast Engineering Conference Proceedings*, 1998, p. 13.

equal the number of resonators times one-half the waveguide wavelength. For a 10-resonator filter operating at 470 MHz, this may amount to a length exceeding 20 ft. If the TE102 mode is required to achieve sufficient Q, the resonator is a full wavelength long and the filter length is double.

Use of in-line, dual-mode resonators or cavities in a square or circular waveguide permit construction of filters with approximately half the size of the single-mode filters. In this structure, illustrated in Figure 5-5, each resonator supports a pair of orthogonal modes or polarizations. These modes are depicted by mutually perpendicular vectors. Since there are two electrical resonances, each resonator functions as the equivalent of a pair of resonators.[4] The equivalent circuit of a waveguide pseudoelliptic function filter is shown in Figure 5-6. A

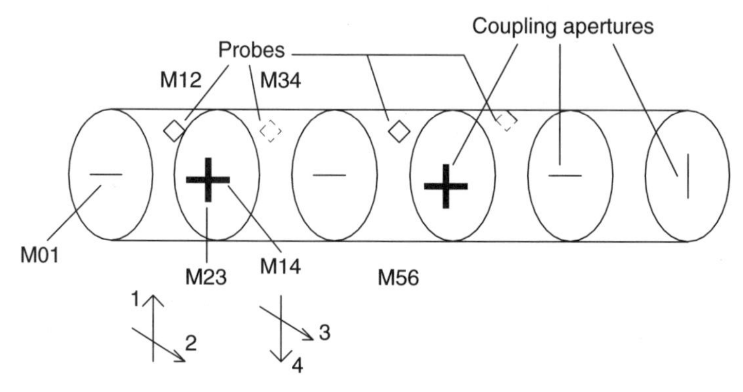

Figure 5-5. In-line dual-code filter. (From Ref. 6; used with permission.)

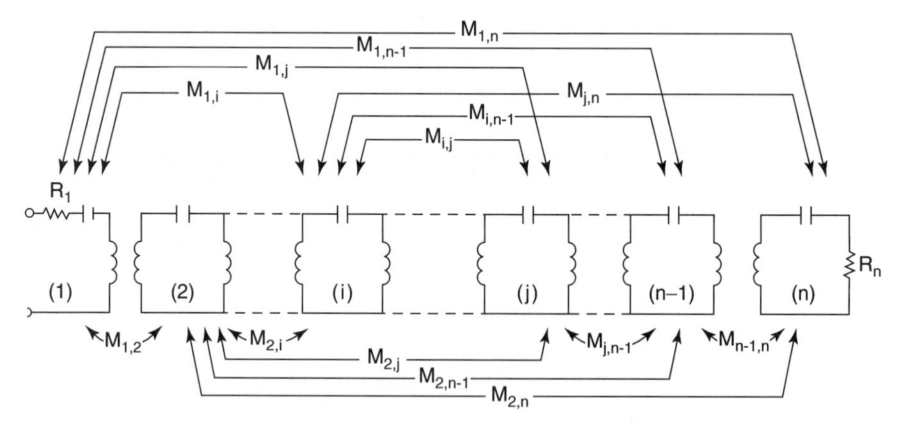

Figure 5-6. Equivalent circuit of n coupled cavities. (From Ref. 6 © 1972 IEEE; used with permission.)

[4] D. J. Small, "High Power Multimode Filters for ATV... Systems," available on the World Wide Web at ppp.com.

total of n coupled resonators are employed to produce the desired transmission zeros at the desired frequencies. Each resonator is a single resonant circuit with multiple couplings to the other resonators.[5] The value of the coupling factors, M_{mn}, determine the degree to which the cavities are coupled. The resonators produce transmission zeros at the edges of the stopband and at $f = \infty$. In practice, $R_1 = R_n$, so that the filter is matched to the system characteristic impedance.

CAVITIES

An ideal cavity is a lossless dielectric region completely enclosed by perfectly conducting walls. The operation of a cavity is based on the properties of a short-circuited transmission line. At certain frequencies, the cavity is resonant just like a shorted line. The input impedance, Z_{sc}, of a short-circuited lossless transmission line as a function of frequency is

$$Z_{sc} = jZ_0 \tan \frac{\pi f}{2f_0}$$

where f_0 is the frequency at which the transmission line is $\frac{1}{4}$ wavelength long. This is just the product of the characteristic impedance, Z_0, and a complex frequency variable, S, given by

$$S = j \tan \frac{\pi f}{2f_0}$$

so that Z_{sc} is directly proportional to this complex frequency, that is,

$$Z_{sc} = Z_0 S$$

When used as a series element, a shorted stub produces a transmission zero when $f = f_0$. Since S is periodic in $2f_0$, the response of the line section repeats at this interval.

A cavity may be visualized as a pair of short-circuited transmission lines connected at their inputs as shown in Figure 5-7. It supports the appropriate transmission line mode and is an integer number of half-wavelengths long at the resonant frequency. Key design parameters include the resonant frequency and quality factor. A cavity may be constructed of either waveguide or coax, depending primarily on the frequency of operation, allowable losses, and power-handling requirements.

To minimize insertion loss, the cavities used in filters for digital television operating at UHF are constructed of air-dielectric circular waveguide operating

[5] A.E. Atia and A.E. Williams, "Narrow-Bandpass Waveguide Filters," IEEE *Trans. Microwave Theory Tech.*, Vol. 20, No. 4, April 1972, pp. 258–265.

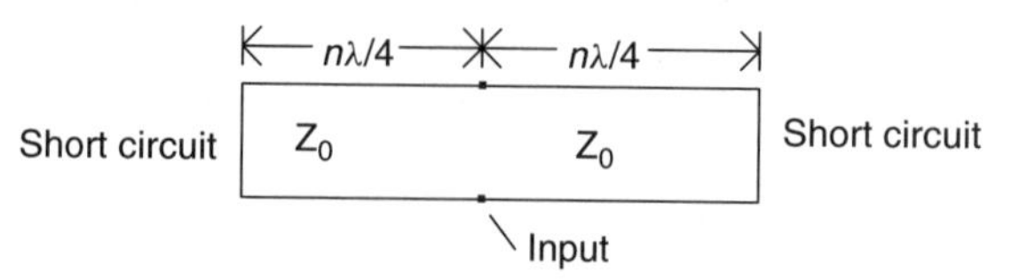

Figure 5-7. Cavity equivalent circuit.

in the TE11 mode. For this, the lowest-order mode, resonance occurs when the total length of the cavity, $2h_c$, is equal to one-half the guide wavelength, λ_g. The guide wavelength is

$$\lambda_g = \frac{\lambda}{[\varepsilon_r - (\lambda/\lambda_c)^2]^{1/2}}$$

where λ_c is the cutoff wavelength of the guide. In an air-dielectric cavity, ε_r, the relative dielectric constant, is approximately unity. From these relationships it can be shown that the resonant wavelength of a circular cavity operating in the dominant mode[6] is given by

$$\lambda = \frac{4}{[(1/h_c)^2 + (1.17/a^2)]^{1/2}}$$

The mode designation, TE111, indicates that the cylindrical waveguide is operating in the TE11 mode and the cavity length is one-half guide wavelength.

Means must be provided for coupling the input cavity to the transmitter output, the cavities to each other, and the output cavity to the transmission line and antenna. This involves removal of sections of the cavity walls and the introduction of coupling apertures, such as inductive slots or irises. These apertures, illustrated in Figure 5-5, must be shaped, located, and oriented to excite the proper mode and in such a way as to minimize the perturbation of the field configuration and resonant frequency of the cavity. By proper selection of the point and degree of coupling, the cavity input impedance at resonance and the loaded Q are determined.

The pair of modes within each cavity are coupled to each other by a tuning plunger or probe oriented at 45° with respect to the desired mode polarization. The probe introduces asymmetry to the cavity, giving rise to two identical but orthogonal modes which are polarized parallel to one coupling iris and perpendicular to the other. The degree of coupling between the orthogonal modes is determined by the probe depth. This type of coupling is represented in Figure 5-5 by M12, M34, and M56.

Coupling between successive cavities and nonadjacent resonances is inductive and frequency dependent. It is achieved by apertures or irises in the end wall of each resonator. For example, M14 provides coupling between nonadjacent

[6] *Reference Data for Radio Engineers*, 6th ed., Howard W. Sams, Indianapolis, Ind., 1977, p. 25–19.

resonances 1 and 4. The sign of the coupling factors between adjacent modes, M01, M12, and M23 must be positive; the cross or nonadjacent coupling factors must be negative. The cross or reverse coupling produces a pseudoelliptical response with two poles of attenuation per cross-coupled cavity. Group delay compensation may be designed in by adding cavities with positive couplings between cross-coupled modes. The effect of the reactance of the probes and irises is to increase the electrical length of the cavities; this requires the cavity to be shortened to compensate.

The cavity Q is defined as 2π times the ratio of the energy stored to the energy dissipated per cycle and is closely related to the bandwidth and loss of the cavity. Unloaded Q, which accounts only for losses internal to the cavity, is designated Q_u. Loaded Q accounts for the added effects of coupling to external circuits and is designated Q_l. The effects of all sources of dissipation are thus included.

The relationship between loaded and unloaded Q may be derived by reference to Figure 5-8, which shows the equivalent circuit of a cavity with single input and output couplings to external circuits. The cavity is modeled as a shunt resonant circuit with shunt conductance G_c. Similarly, the coupling to input and output circuits are modeled as shunt conductances, G_{in} and G_{out}, plus shunt suceptance. At resonance, the combination of all susceptances appears as an open circuit; all that remains is the shunt conductances. In the absence of coupling, the energy dissipated is proportional to $V^2 G_c$. The coupling results in additional dissipation, $V^2(G_{\text{in}} + G_{\text{out}})$. In both cases, the stored energy is the same. Thus the ratio of the unloaded Q to the loaded Q is

$$\frac{Q_u}{Q_l} = \frac{G_c}{G_c + G_{\text{in}} + G_{\text{out}}}$$

The coupling factors, M_{in} and M_{out}, quantify the efficiency with which energy stored in the cavity is coupled to the external circuits[7] and are equal to the

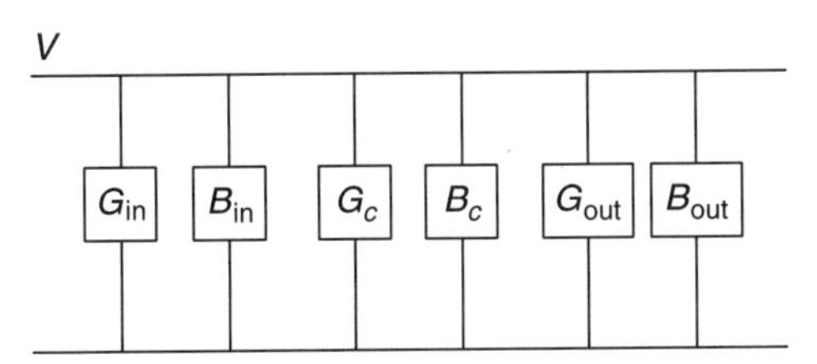

Figure 5-8. Equivalent circuit of cavity with input and output coupling.

[7] Carol G. Montgomery, *Techniques of Microwave Measurements*, Boston Technical Publishers, Lexington, Mass., 1963, p. 290.

corresponding conductances normalized to the cavity conductance[8]:

$$M_{\text{in}} = \frac{G_{\text{in}}}{G_c}$$

and

$$M_{\text{out}} = \frac{G_{\text{out}}}{G_c}$$

so that

$$\frac{Q_u}{Q_l} = \frac{1}{1 + M_{\text{in}} + M_{\text{out}}}$$

This relationship can be extended to include any number of coupling factors. The bandwidth of the cavity is related to Q_l by

$$\Delta\omega = \omega_0 / Q_l$$

where $\Delta\omega$ is the radian frequency difference between half-power points and ω_0 is the radian resonant frequency.

The unloaded ***Q*** is related to the size of the cavity; the larger the cavity, the higher the value of Q_u and the lower the insertion loss. In theory, unloaded Q_u of 35,000 to over 40,000 can be achieved with half-wavelength circular cavities operating in the TE111 mode, depending on cavity dimensions, material, and frequency. The theoretical Q_u value of aluminum and copper cavities operating at 800 MHz as a function of the length-to-radius ratio, h_c/a, is shown in Figure 5-9. Maximum Q_u occurs for h_c/a of approximately 0.76. The Q_u of copper cavities is approximately 23% greater than aluminum cavities.

The variation of Q_u with frequency is shown in Figure 5-10. The surface resistance of the metal walls increases with increasing frequency due to the skin effect. Consequently, Q_u is highest at the lower frequencies. In practice, Q_u is limited to about 75% of these values, due to limitations in fabrication and assembly.[9]

Insertion loss is inversely proportional to Q_u,[10] that is,

$$Q_u = \frac{\pi}{\alpha_c \lambda_g}$$

where α_c is the cavity attenuation in nepers per unit length. For half-wave cavities with Q_u of 35,000, this expression implies that attenuation is on the order of

[8] Williams, op. cit.; Darko Kaifez, "*Q*-Factor Measurement Techniques," *RF Design*, August 1999, p. 60.

[9] Small, op. cit., p. 1.

[10] William Sinnema, *Electronic Transmission Technology*, Prentice Hall, Upper Saddle River, NJ, p. 75, 1988, 2nd Edition.

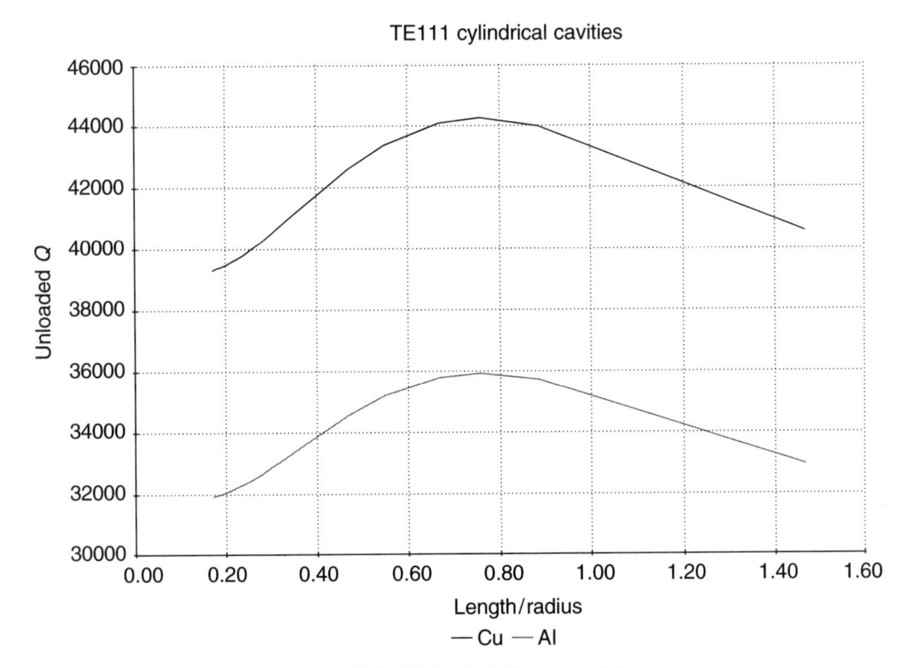

Figure 5-9. Unloaded Q versus h/a.

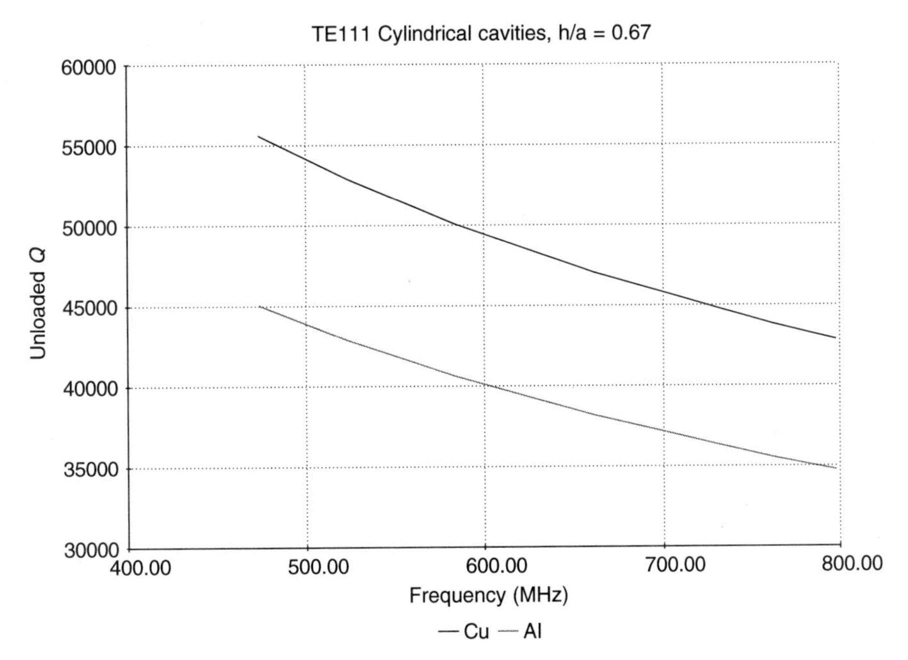

Figure 5-10. Unloaded Q versus frequency.

0.002 dB per cavity. Since the size of the cavities is related to the insertion loss, it follows that average power handling is also determined by the cavity dimensions. Insertion loss of 0.002 dB represents dissipation of about 0.5 W per kilowatt of input. It is estimated that up to 10 sections will be required to meet the rejection specification of the FCC mask. This would imply ohmic losses in the filter of about 0.02 dB.

At the input and output, the slots must properly couple to the main transmission line. A long thin slot is used for this coupler. The coupling coefficient is determined by the magnetic polarizability of the slot, which is related to the length of the slot with appropriate correction for slot thickness.

The rejection specification determines the minimum number of resonators to achieve a given passband ripple. The number of cavities and the size of each determine the overall size of the filter. Thus, even with the use of dual-mode cavities, the space required for the filter is related directly to the key electrical specifications.

The resonant frequency of a cavity changes because of expansion and contraction of the cavity due to temperature changes. Thus, it is important to select cavity materials to minimize losses while minimizing the effects of temperature variations. If the cavity is made of a single type of metal, the change in resonant frequency will be very nearly directly proportional to the linear coefficient of expansion of the metal and the absolute temperature. This is because the resonant frequency is inversely proportional to the linear dimensions of the cavity. For copper, the linear coefficient of expansion at a temperature of 25°C is $16.8 \times 10^{-6}\ {}^{\circ}\mathrm{C}^{-1}$. A 25°C change in temperature will produce a 0.044% change in dimensions and a corresponding resonant frequency change. At 800 MHz, this amounts to 0.35 MHz, a significant change. For aluminum cavities, the coefficient of expansion and change in resonant frequency is 38% greater. In practice, combinations of materials may be used. Aluminum waveguide may be used for the body of the filter with either aluminum or copper irises.

The resonant frequency also changes as a function of temperature and humidity due to changes in dielectric constant of the atmosphere. The relative dielectric constant of standard atmospheric air at sea level is approximated by

$$\varepsilon_r = 1 + 207 \times 10^{-6}\frac{P_a}{T_a} + 169.2 \times 10^{-6}\left(1 + \frac{5880}{T_a}\right)\frac{P_w}{T_a}$$

where P_a and P_w are the partial pressures of dry air and water vapor in millimeters of mercury, respectively, and T_a is the absolute temperature in Kelvins. For dry air, P_a is the same as atmospheric pressure (~760 mmHg) and $P_w = 0$. For saturated air, P_a ranges from ~755 to 667 mmHg and P_w ranges from 5 to 93 mmHg, as temperature ranges from 0 to 50°C. The relative dielectric constant of both dry and saturated air is plotted in Figure 5-11. Even when the air is dry, the dielectric constant is slightly greater than unity. If a cavity is tuned at a temperature of 25°C and a humidity of 60%, a change in temperature to 50°C with a relative humidity increase to 100% results in a change in the

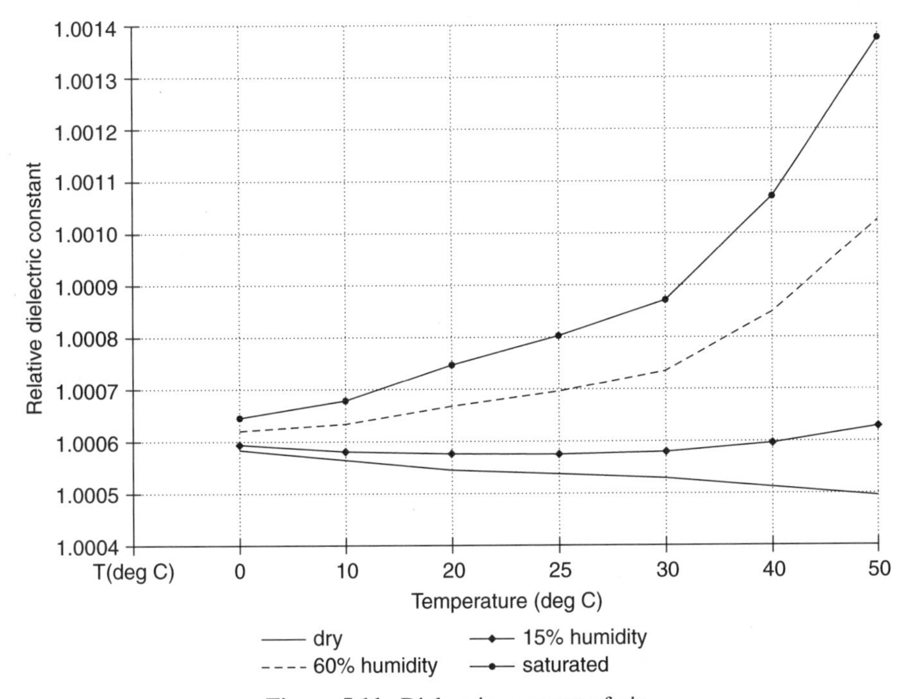

Figure 5.11. Dielectric constant of air.

dielectric constant of air from 1.0007 to 1.0014. This results in a change in resonant frequency of 0.035%. At 800 MHz, this amounts to 0.28 MHz. The combination of frequency shifts due to cavity expansion and changes in the dielectric constant of air impose additional constraints on the trade-off between passband and stopband characteristics.

These considerations also indicate possible strategies for mitigating resonant frequency changes due to temperature and humidity. The maximum use of copper may be worth the extra weight and cost. The use of air conditioning in the space occupied by the filter would prevent large temperature variations. Air conditioning will also serve to reduce humidity. Evidently, when the relative humidity is in the neighborhood of 20%, the dielectric constant of air is nearly constant over a wide temperature range. The importance of testing the transmitter, cooling system, and filter as a system is also reinforced by these considerations; this assures that the filter performance is known at the typical ambient temperature along with the effects of heating due to ohmic losses.

The resonant frequency may also depend on the load and transmitter impedances, especially if those impedances are reactive. Fortunately, digital broadcast systems are normally well matched to maintain maximum power transfer. For convenience of design and measurement, system impedances are also resistive. Unless severe mismatches occur due to antenna icing or other emergency condition, filter response should not be affected.

CHANNEL COMBINERS

Finding suitable tower space for a new digital television antenna and transmission line while continuing to operate the analog system is one of the major hurdles that a broadcaster must overcome during the transition period. Co-location of the digital transmitter with the analog and using a common antenna and transmission line for both is a possible solution. This give rise to the need for suitable channel combining techniques.

To provide sufficient spectrum to accommodate the required number of DTV allocations in the United States, upper $(N + 1)$ and lower $(N - 1)$ adjacent channel assignments were necessary, especially in major markets. This has given rise to the potential for severe adjacent channel interference from the signal transmitted by the NTSC station to the transmitted DTV signal. Even if the paired signals are not in adjacent channels, there is potential for interference, although to a lesser degree. For these reasons, it is important to provide adequate isolation when combining the two signals.

The average power of a DTV signal at the transmitter is nominally 12 dB below the NTSC peak of sync. At these relative levels, any signal outside the NTSC channel has the potential for creating interference to the DTV signal. This is especially true if the DTV channel is below the NTSC. The relationship between the desired and undesired signals is illustrated in Figure 5-12. The potential for interference is apparent. The specification for the NTSC lower sideband is only 20 dB below the peak sync level. Although this specification is usually met with some margin in well-maintained transmitters, significant interference is still possible. For pulsed UHF systems, the level of the reinserted lower sideband may be up to 10 dB higher than shown, further increasing the potential for interference. Even without the effects of unequal antenna patterns and propagation, the lower sideband might be only 8 dB below the average DTV power in the absence of additional filtering.

Several factors must be considered when determining the level and effect of the interference as well as strategies to minimize its effect. Obviously, combining to a common antenna is feasible only if the stations are co-located.[11] The transmission line loss and antenna gain will generally be approximately equal for both channels. The coverage for both stations should also be nearly equivalent, assuming comparable AERP. Assuming that the antenna and transmission line have sufficient pattern and impedance bandwidth to accommodate both channels, the signals may be combined without the use of a separate channel combiner. This may involve the use of a hybrid combiner with a turnstile antenna, similar to the method used for combining visual and aural signals in VHF batwing antennas. Isolation between inputs is obtained by virtue of the isolation inherent in the hybrid less the effect of the return loss of the antenna. This approach may be

[11] In this context, co-located means to be physically co-sited. The FCC rules define co-location as being located within a 10-mile separation. Obviously, this definition does not apply when considering channel combining of any type.

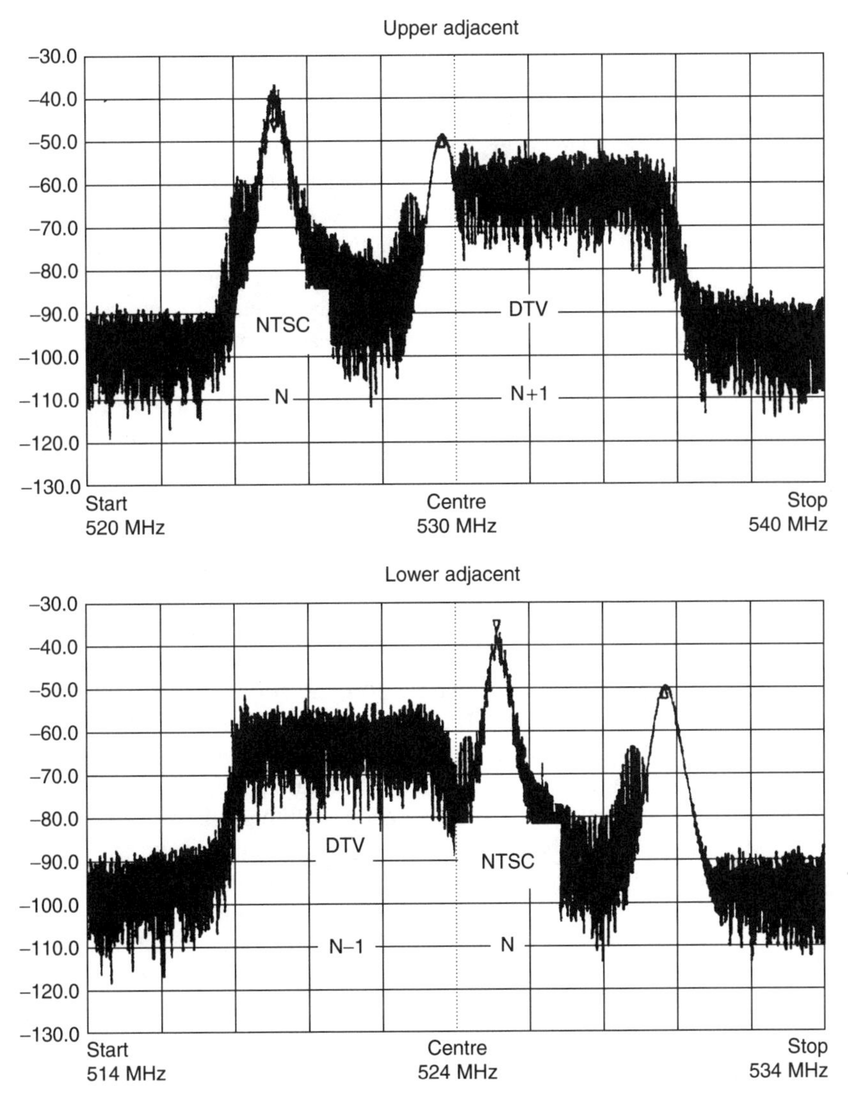

Figure 5-12. Adjacent channel signals. (From R.J. Plonka, "Planning Your Digital Television Transmission System," *NAB Broadcast Engineering*, 1997; used with permission.)

used for either $N + 1$ or $N - 1$ combining. The power rating of the antenna and transmission line must be adequate to support both signals simultaneously.

Channel combining can also be done with separate collinear antennas. For example, the antenna for analog could be mounted above the antenna for digital TV. In this case the antennas and transmission lines must be capable of providing the bandwidth and power-handling capability for only one channel each. Isolation is achieved by virtue of element phasing if radiation of both antennas toward the zenith and nadir is minimized. If equivalent coverage is desired for both signals, the azimuth and elevation radiation patterns should be matched to the degree possible. Pattern matching to within ±2 dB is desirable. Matching to this degree is difficult with side-mounted antennas; for this reason, top-mounted antennas are preferred if possible. However, no matter how well the patterns are matched, local reflections will change the analog-to-digital power ratio at the receiver. This is discussed further in Chapter 8. It should be noted that combining with either turnstile or collinear antennas does not necessarily provide any filtering. Thus filtering of the respective signal should be done prior to the combining functions. It is assumed that this will be necessary to meet the emissions mask requirements. It may also be necessary to provide additional filtering to attenuate the lower sideband of the analog signal.

In the event that combining in the antenna is not feasible, a separate channel combiner is necessary. The ideal channel combiner accepts signals from each transmitter at the respective input ports and combines them at the output port without loss to either signal. In practice, this ideal cannot be achieved for adjacent channels. Loss of at least 3 dB must occur at the common band edge.[12]

The most common form of channel combiner is the constant-impedance type. This system provides a good impedance match for both signals. This configuration is similar to the constant-impedance bandpass filter, except that the bandpass filter provides a different response. In the case of a lower adjacent digital assignment, the analog signal is fed into the narrowband port, then through high-Q analog TV bandpass filters to the antenna port. The bandpass filter removes the lower 300- to 400-kHz portion of the vestigial sideband. This is necessary to provide sufficient bandwidth for and to minimize distortion to the digital signal. The digital signal enters the broadband port, reflects off the bandpass filters, and recombines at the output port of the hybrid, where it enters the transmission line and antenna. The effect on the analog TV signal is limited by the ability of the transmitter circuitry to equalize the response to acceptable performance. The linear distortions to the digital TV signal will also be affected and must also be equalized. It is important that the insertion loss be low in both bands. The antenna and transmission line must have sufficient bandwidth and power-handling capability to accommodate both signals.

[12] Broad and Blair, op. cit., pp. 11–15.

6

TRANSMISSION LINE FOR DIGITAL TELEVISION[1]

The performance of the main transmission line is key to a successful digital television transmission facility. The ideal line would handle the transmitter power with adequate margin, be lossless, would add no linear distortions to the modulated signal or wind loading to the transmission tower, and would be inexpensive. Depending on the line selected, some of these characteristics can be achieved. For very short lines or lines operating at low frequencies, these characteristics can be approximated. For tall towers and allocations at high channels, careful trade-offs must be made to minimize loss, wind load, linear distortions, and cost while providing adequate power-handling capability.

In this chapter the essential criteria for transmission line selection for a digital television system are presented. The relationship between transmission line attenuation, dissipation, and power-handling capability is examined. These relationships are applied to the transmission line options available to the broadcaster, including rigid coax, corrugated cables, and waveguides. The purpose of this discussion is to show the effect of line selection on the ability to achieve a specified system output and the effect of selection decisions on other system performance parameters. The importance of maintaining low-voltage standing wave ratio (VSWR) on system performance, the extent of linear distortions for the various line types, frequency and bandwidth limitations, wind loading, and the benefits of pressurization is also reviewed. A transmission line figure of merit that should be useful as an aid to transmission line selection is also described.

[1] This chapter was originally written for Harris Corporation for use in the 1998 DTV Express Handbook, *Transmission Line for Digital Television*, and is used here with permission.

FUNDAMENTAL PARAMETERS

The purpose of the transmission line is to transfer RF power efficiently from the transmitter to the transmitting antenna. The key parameters by which transmission line is defined are the characteristic impedance; the velocity of propagation, v_p; attenuation constant, α, and power-handling capability. The characteristic impedance is the terminating impedance, in ohms, which results in maximum power transfer from the transmission line to the antenna. In an ideal lossless transmission line, this impedance is a resistor whose value may be stated in terms of the inductance and capacitance per unit length of the line. The characteristic impedance of the ideal uniform transmission line may be written as

$$Z_0 = \left(\frac{L}{C}\right)^{1/2}$$

where L is the inductance per unit length and C is the capacitance per unit length. The velocity of propagation may also be determined by these quantities, that is,

$$v_p = (LC)^{1/2}$$

For air-dielectric lines, the velocity of propagation is the same as the speed of light in air, 300×10^6 m/s. Using these equations it is an easy matter to compute the inductance and capacitance per unit length of any transmission line, given the characteristic impedance. Fortunately, many practical, air-filled transmission lines exhibit sufficiently low attenuation so that they may be considered to be "ideal" for the purpose of determining characteristic impedance and velocity of propagation. However, when attenuation and power rating are determined, the nonideal aspects of a transmission line must be considered.

EFFICIENCY

The efficiency of the transmission line is a very important factor in overall system efficiency and a major consideration when designing a digital television facility. For high-AERP stations, transmission line attenuation occurs in the system where it greatly affects the amount spent to purchase and operate a high-power transmitter. Losses occurring between the transmitter and antenna are very expensive, making it essential to select the transmission line with great care. For high-power UHF installations, it is important to make the optimum choice among large rigid coaxial lines, corrugated coaxial cables, and waveguides as a means of achieving high system efficiency and minimum cost.

Transmission line efficiency is dependent on the attenuation per unit length and the total length of line. Standard values for attenuation per unit length are published by the various transmission line manufacturers. The larger the line, the lower the attenuation per unit length. However, some variation in the published loss tables will be found, depending on the assumptions made by the manufacturers, such as estimates of dielectric losses, conductor conductivity, inner conductor temperature, and other derating factors. When comparing line efficiency for the different line types and suppliers, care should be taken to understand the underlying assumptions.

The attenuation per unit length of a matched coaxial transmission line may be expressed as

$$\alpha = Af^{1/2} + Bf$$

where A is the conductor loss factor, B is the dielectric loss factor, and f is the frequency in megahertz. From this equation it is apparent that conductor losses are proportional to the square root of frequency. For air-dielectric lines, this factor predominates. For copper lines, the conductor loss factor is approximated by[2]

$$A = \frac{0.433}{Z_0}\left(\frac{1}{D_i} + \frac{1}{d_o}\right)$$

where D_i is the inside diameter of the outer conductor and d_o is the outside diameter of the inner conductor, both dimensions in inches. Two conclusions may be reached by detailed inspection of this expression. First, note that 75-ohm line is preferred over 50-ohm for highest efficiency. For 75-ohm line, $D_i/d = 3.49$ and $A = 0.0259/D_i$; for 50-ohm line $D_i/d = 2.3$ and $A = 0.0286/D_i$. From these ratios we see that the conductor loss for 50-ohm line of a given outer conductor diameter and material is greater than for the same size and type 75-ohm line. Second, for a specific characteristic impedance, conductor loss decreases as line size increases.

Dielectric loss increases directly with increasing frequency. Although this factor is much smaller than the conductor loss for air-dielectric lines, it cannot be ignored. It becomes more important as line size increases because of the decreasing importance of conductor loss.

The transmission line efficiency in percent, η_l, is related to total line attenuation by the formula

$$\eta_l = 100\% \times 10^{-(N_l\alpha/10)}$$

where N_l is the length of the line in standard units of length. Typically, efficiency is expressed in decibels per 100 ft. In this case, N_l represents the number of 100-ft lengths in the transmission line run. Line efficiency may also be expressed in terms of input power, P_i, and output power, P_o:

$$\eta_l = \frac{P_o}{P_i} \times 100\%$$

[2] Kerry W. Cozad, "A Technical Review of Transmission Line Designs and Specifications," *NAB Broadcast Engineering Proceedings*, 1998, pp. 16–24.

From conservation of energy, the power dissipated, P_d, is

$$P_d = P_i - P_o$$

Now the efficiency may be written as

$$\eta_l = \frac{P_i - P_d}{P_i} \times 100\%$$

If P_i is defined to be the maximum power-handling capability, P_d represents the maximum dissipation per unit length of the line. This should be independent of frequency. The heating of the inner conductor is solely dependent on the maximum allowed dissipation per unit length and determines the maximum average power-handling capability. Thus the maximum average input power-handling capability is closely related to the line efficiency, that is,

$$10^{-(N_l\alpha/10)} = \frac{P_i - P_d}{P_i}$$

or

$$P_d = P_i(1 - 10^{-(N_l\alpha/10)})$$

EFFECT OF VSWR

In the foregoing discussion it is assumed that the transmission line is well matched. This is generally true under normal operating conditions. Typical VSWR of a well-designed broadcast antenna should be less than 1.1:1 over the full channel bandwidth. However, under icing conditions or in the event of an antenna malfunction, it is possible that the line will be mismatched. In this case the dissipation and attenuation will be higher and power ratings will be lower. This is a consequence of the periodic current peaks associated with the standing wave. As the current rises at these peaks, the power dissipated in the line increases. This increase is in proportion to the square of the reflection coefficient[3].

To understand this effect, consider a wave propagating along a transmission line in the z direction, terminated in its characteristic impedance. The variation of all field components in the z direction is expressed as

$$e^{-\gamma z}$$

where γ is the complex propagation constant or $\gamma = \alpha + j\beta$. The attenuation constant, α, has already been defined; β is the phase constant of the line. In

[3] The magnitude of the reflection coefficient, $|\Gamma|$, is related to VSWR by the expression

$$\text{VSWR} = \frac{1 + |\Gamma|}{1 - |\Gamma|}$$

general, waves travel in both directions on the line since a reflection occurs when the line is not terminated in its characteristic impedance. In this case, the total voltage, V_l, and current, I_l, on the line are each the sum of two waves, one traveling in the positive z direction, the other in the negative z direction. That is,

$$V_l = V'e^{-\gamma z} + V''e^{+\gamma z}$$

and

$$I_l = I'e^{-\gamma z} + I''e^{+\gamma z}$$

where V' and I' represent the direct wave and V'' and I'' represent the reflected wave.

To determine the effect of mismatch on transmission line dissipation due to series conductor losses,[4] it is necessary to know the total current relative to the direct wave. Dividing the expression for total current by the direct wave current, we have the current relative to the direct wave:

$$\frac{I_l}{I'} = e^{-\gamma z} + \frac{I''}{I'}e^{+\gamma z}$$

The current reflection coefficient is defined as $\Gamma_i = I''/I'$, so that

$$\begin{aligned} \frac{I_l}{I'} &= e^{-\gamma z} + \Gamma_i e^{+\gamma z} \\ &= (1 + \Gamma_i)\cos\gamma z - j(1 - \Gamma_i)\sin\gamma z \end{aligned}$$

This expression represents a standing wave with a period of one-half wavelength.

The next step is to compute the square of the current, that is,

$$\left(\frac{I_l}{I'}\right)^2 = (1 + \Gamma_i)^2\cos^2\gamma z + (1 - \Gamma_i)^2\sin^2\gamma z$$

Applying a little calculus, it can be shown that the average value of the square of the current for a half-wave section of line is

$$\begin{aligned} \left(\frac{I_l}{I'}\right)^2_{\text{av}} &= \frac{1}{\pi}\frac{\pi}{2}[(1 + \Gamma_i)^2 + (1 - \Gamma_i)^2] \\ &= 1 + \Gamma_i^2 \end{aligned}$$

For good conductors such as copper and aluminum, the heat generated over a half-wave section should be spread uniformly so that this ratio represents the increase in dissipation due to impedance mismatch. The ideal case, $\Gamma_i = 0$, represents a matched transmission line. Under this condition, the square of the current wave is unity.

[4] A similar derivation could be written for losses due to the shunt conductance using the total voltage on the line. The result would be identical.

Referring to the expression relating dissipation, power rating, and attenuation, we may write

$$P_d = (1 + \Gamma_i^2)P_i(1 - 10^{-(N_l\alpha/10)})$$

For example, the increase in conductor dissipation due to a VSWR of 1.05:1 ($\Gamma = 0.025$) is a factor of only 1.000625, a negligible amount. For a VSWR of 2:1 ($\Gamma = 0.333$), the increase is a factor of 1.111. For a line operating near its maximum rating, a sudden change in antenna impedance could result in transmission line failure. This highlights the importance of maintaining low antenna VSWR as well as the need for automatic VSWR foldback in the transmitter design.

SYSTEM AERP

The transmission line output power may also be expressed in terms of dissipation. After a bit of algebra, it can be shown that

$$P_o = P_d \frac{\eta_l/100}{1 - \eta_l/100}$$

Ordinarily, the average effective radiated power of a digital television station is written as

$$\text{AERP} = \text{TPO} \times g_a \times \eta_l$$

where g_a is the antenna gain. Now the product of TPO and line efficiency is the power at the output of the line, P_o.

Thus AERP may be written as

$$\text{AERP} = g_a P_o$$

Therefore,

$$\text{AERP} = g_a P_d \frac{\eta_l/100}{1 - \eta_l/100}$$

and the maximum AERP that can be accommodated by a transmission line may be determined without explicit reference to the transmitter output power provided that the antenna gain, line efficiency, and maximum line dissipation are known.

RIGID COAXIAL TRANSMISSION LINES

To illustrate the foregoing ideas, consider the performance of rigid coaxial transmission line. Although there are several specific offerings among the various manufacturers, these lines share some common characteristics. They are generally made using copper inner and outer conductors. The inner conductors are supported at intervals by Teflon pins or disks. Under normal conditions, the dielectric material may be considered to be equivalent to dry air. The velocity of propagation is very nearly equal to the speed of light in free space.

DISSIPATION, ATTENUATION, AND POWER HANDLING

Consider a matched 100-ft length of unpressurized $6\frac{1}{8}$-in. rigid coaxial line operating at U.S. channel 69. Since 100 ft is a standard length for published data, $N_l = 1$. From a manufacturer's table of attenuation and power handling,[5] $\alpha = 0.154$ dB per 100 ft and $P_i = 49.54$ kW. The maximum dissipation is, therefore, 1.72 kW per 100 ft. (A slightly higher value would be obtained for shorter lengths, since the dissipation is not uniform along a 100-ft length of line; e.g., using a 1-ft length for the calculation yields a dissipation of 0.0179 kW/ft.)

It is interesting to note that the published tabular attenuation and power rating data do not always yield the same dissipation. Calculating dissipation using the foregoing method yields higher values at lower frequencies. For example, at U.S. channel 41, the maximum dissipation of matched $6\frac{1}{8}$-in. line is 1.76 kW. At high-band channels, 1.81 kW is computed; at low-band channels, values as high as 1.86 kW are calculated. From a physical point of view, the temperature rise of the line should depend only on the total dissipation. This is a consequence of Newton's law of cooling, which states that the rate at which a body loses or dissipates heat to its surroundings, whether by convection or radiation, is proportional to the difference in temperature[6]; that is,

$$P_d = c(T_2 - T_1)$$

For a fixed ambient temperature, T_1, and maximum allowable inner conductor temperature, T_2, the maximum dissipation should be constant with no frequency dependence.

For the purpose of this analysis, the lowest calculated value of dissipation (1.72 kW) will be used to compute derated power ratings. This should result in conservative estimates of power rating and assure reliable performance. With this value of dissipation and the attenuation at any other frequency, maximum power-handling capability may be computed at any other frequency. The data presented should be considered representative and not used for line supplied by all manufacturers. The reader may apply this technique to the lines being considered for a specific installation.

In practice, published attenuation and power-handling specifications include a derating of 15 to 19%. Up to 4% of this derating is due to loss at the flange interfaces and oxidation of the copper, which causes some reduction in conductivity. The remaining derating accounts for the inner conductor temperature when operating

[5] Andrew Corporation, Catalog 36, p. 287.

[6] Harvey E. White, *Modern College Physics*, 3rd ed., D. Van Nostrand, New York, 1957, p. 288.

at maximum power as well as the effects of higher ambient temperatures. The increase in loss is

$$M_\alpha = [1 + 0.0039(T_2 - 20)]^{1/2}$$

The published attenuation is usually calculated for an inner temperature of 20 to 24°C. When operating at rated power, the inner conductor temperature is limited to a temperature of 100°C. At this temperature the derating factor is in the range 1.139 to 1.145.

TABLE 6-1. Power Rating and Attenuation of $6\frac{1}{8}$-in. 75-Ω Rigid Coaxial Transmission Line

Channel	F (MHz)	P_i (kW)	Attenuation (dB/100 ft)	Channel	F (MHz)	P_i (kW)	Attenuation (dB/100 ft)
2	57	182.37	0.041	36	605	48.91	0.156
3	63	172.89	0.043	37	611	48.62	0.157
4	69	164.67	0.046	38	617	48.34	0.158
5	79	153.13	0.049	39	623	48.07	0.158
6	85	147.21	0.051	40	629	47.80	0.159
7	177	98.60	0.077	41	635	47.53	0.160
8	183	96.79	0.078	42	641	47.27	0.161
9	189	95.07	0.079	43	647	47.01	0.162
10	195	93.43	0.081	44	653	46.75	0.163
11	201	91.87	0.082	45	659	46.50	0.164
12	207	90.38	0.084	46	665	46.25	0.165
13	213	88.94	0.085	47	671	46.01	0.166
14	473	56.48	0.134	48	677	45.76	0.167
15	479	56.06	0.135	49	683	45.52	0.167
16	485	55.66	0.136	50	689	45.29	0.168
17	491	55.26	0.137	51	695	45.06	0.169
18	497	54.87	0.138	52	701	44.83	0.170
19	503	54.49	0.139	53	707	44.60	0.171
20	509	54.12	0.140	54	713	44.38	0.172
21	515	53.75	0.141	55	719	44.16	0.173
22	521	53.38	0.142	56	725	43.94	0.174
23	527	53.03	0.143	57	731	43.73	0.174
24	533	52.68	0.144	58	737	43.51	0.175
25	539	52.34	0.145	59	743	43.30	0.176
26	545	52.00	0.146	60	749	43.10	0.177
27	551	51.67	0.147	61	755	42.89	0.178
28	557	51.34	0.148	62	761	42.69	0.179
29	563	51.02	0.149	63	767	42.49	0.180
30	569	50.70	0.150	64	773	42.29	0.181
31	575	50.39	0.151	65	779	42.10	0.181
32	581	50.08	0.152	66	785	41.91	0.182
33	587	49.78	0.153	67	791	41.72	0.183
34	593	49.49	0.154	68	797	41.53	0.184
35	599	49.19	0.155	69	803	41.34	0.185

TABLE 6-2. Power Rating and Attenuation of $8\frac{3}{16}$-in. 75-Ω Rigid Coaxial Transmission Line

Channel	F (MHz)	P_i (kW)	Attenuation (dB/100 ft)	Channel	F (MHz)	P_i (kW)	Attenuation (dB/100 ft)
2	57	269.79	0.032	27	551	73.34	0.117
3	63	255.46	0.033	28	557	72.85	0.118
4	69	243.06	0.035	29	563	72.38	0.119
5	79	225.64	0.038	30	569	71.91	0.120
6	85	216.70	0.039	31	575	71.45	0.120
7	177	143.49	0.059	32	581	70.99	0.121
8	183	140.78	0.061	33	587	70.55	0.122
9	189	138.19	0.062	34	593	70.11	0.123
10	195	135.73	0.063	35	599	69.67	0.123
11	201	133.39	0.064	36	605	69.25	0.124
12	207	131.15	0.065	37	611	68.83	0.125
13	213	129.00	0.066	38	617	68.41	0.126
14	473	80.48	0.107	39	623	68.01	0.126
15	479	79.87	0.107	40	629	67.61	0.127
16	485	79.27	0.108	41	635	67.21	0.128
17	491	78.68	0.109	42	641	66.82	0.129
18	497	78.10	0.110	43	647	66.44	0.130
19	503	77.53	0.111	44	653	66.06	0.130
20	509	76.97	0.112	45	659	65.69	0.131
21	515	76.43	0.112	46	665	65.32	0.132
22	521	75.89	0.113	47	671	64.96	0.133
23	527	75.36	0.114	48	677	64.60	0.133
24	533	74.84	0.115	49	683	64.25	0.134
25	539	74.33	0.116	50	689	63.90	0.135
26	545	73.83	0.116	51	695	63.56	0.135

Charts and graphs of attenuation and maximum average power for matched rigid coaxial lines are shown in Tables 6-1, 6-2, and 6-3 and Figures 6-1 and 6-2 for $6\frac{1}{8}$-, $8\frac{3}{16}$-, and $9\frac{3}{16}$-in. lines, respectively. These are based on the formulas above and a 17% derating factor. The average power rating is determined using maximum dissipations of 1.72, 1.95, and 2.18 kW per 100 ft for each respective line. Because of the derating factors used and the assumptions with regard to dissipation, these charts and curves are considered reasonably conservative (except for lines exposed to direct solar radiation[7]) and may be used as presented to estimate the operating specifications for most digital television installations. Using the data of Figure 6-1 and Tables 6-1 and 6-2, graphs of maximum AERP that can be supported by matched rigid $6\frac{1}{8}$- and $8\frac{3}{16}$-in. coaxial lines versus frequency for typical line lengths and antenna gains are shown in Figures 6-3

[7] Consult manufacturer's data for derating factors for solar radiation. Additional derating at temperate latitudes of 15% may be need for rigid coaxial lines; derating up 35% may be needed at tropical latitudes.

TABLE 6-3. Power Rating and Attenuation of $9\frac{3}{16}$-in. 75-Ω Rigid Coaxial Transmission Line

Channel	F (MHz)	P_i (kW)	Attenuation (dB/100 ft)
2	57	335.32	0.028
3	63	317.34	0.030
4	69	301.77	0.032
5	79	279.91	0.034
6	85	268.71	0.035
7	177	176.98	0.054
8	183	173.59	0.055
9	189	170.36	0.056
10	195	167.28	0.057
11	201	164.35	0.058
12	207	161.54	0.059
13	213	158.86	0.060
14	473	98.33	0.097
15	479	97.56	0.098
16	485	96.82	0.099
17	491	96.08	0.100
18	497	95.36	0.101
19	503	94.66	0.101
20	509	93.96	0.102
21	515	93.28	0.103
22	521	92.61	0.104
23	527	91.96	0.104
24	533	91.31	0.105
25	539	90.68	0.106
26	545	90.06	0.107
27	551	89.44	0.107
28	557	88.84	0.108
29	563	88.25	0.109
30	569	87.67	0.109
31	575	87.09	0.110
32	581	86.53	0.111
33	587	85.98	0.112
34	593	85.43	0.112
35	599	84.89	0.113
36	605	84.36	0.114
37	611	83.84	0.115
38	617	83.33	0.115

and 6-4. For example, a 2000-ft run of $6\frac{1}{8}$-in. line will not support an AERP of 1000 kW for any UHF channel unless antenna gain is greater than 30. In general, an antenna with horizontal directivity must be used if greater gain is desired. If the line length is 1000 ft, AERP of 1000 kW can be achieved with a gain of 30 up to U.S. channel 43. Alternatively, AERP of 1000 kW may be supported with $8\frac{3}{16}$-in. line and antenna gain of 25 for U.S. channels through 35.

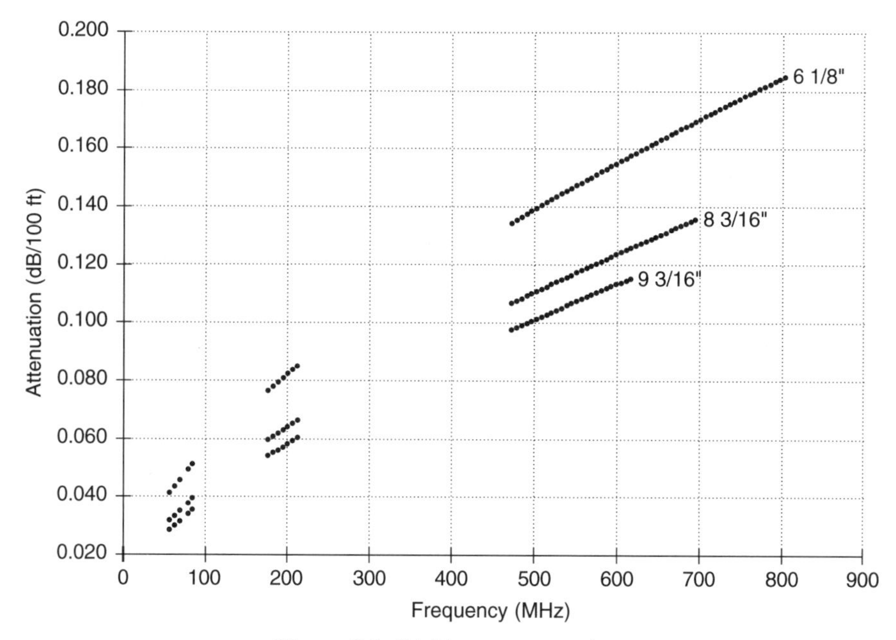

Figure 6-1. Rigid coax attenuation.

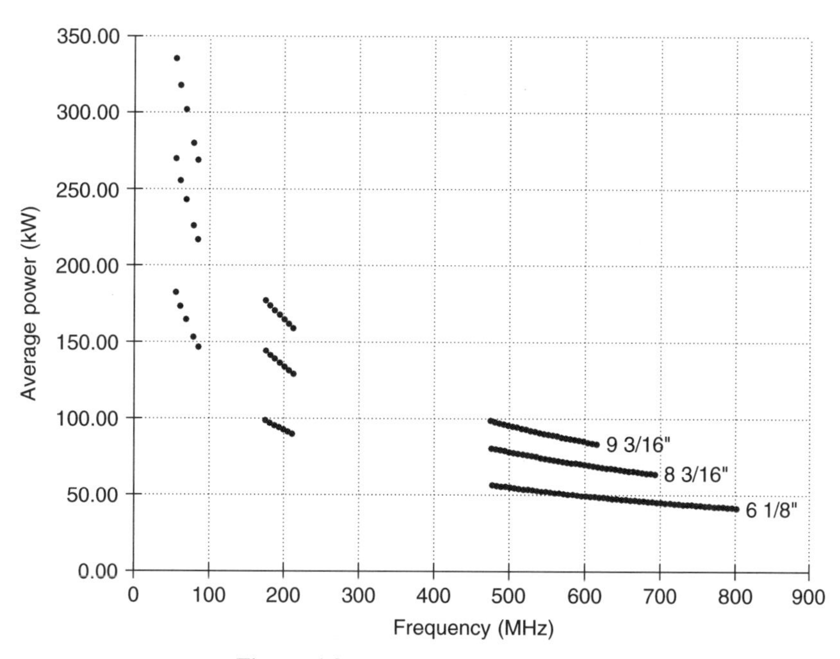

Figure 6-2. Rigid coax power rating.

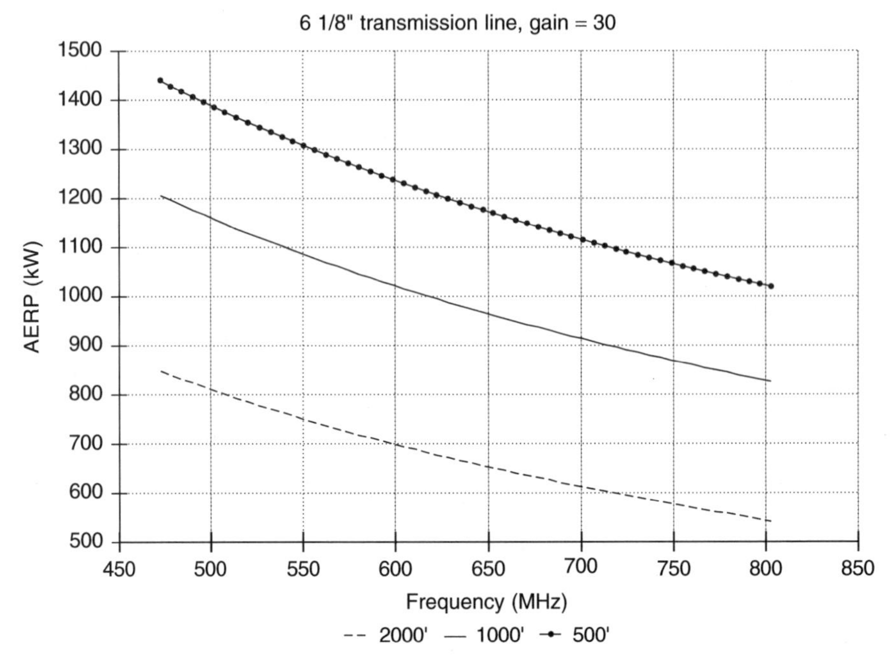

Figure 6-3. Maximum AERP.

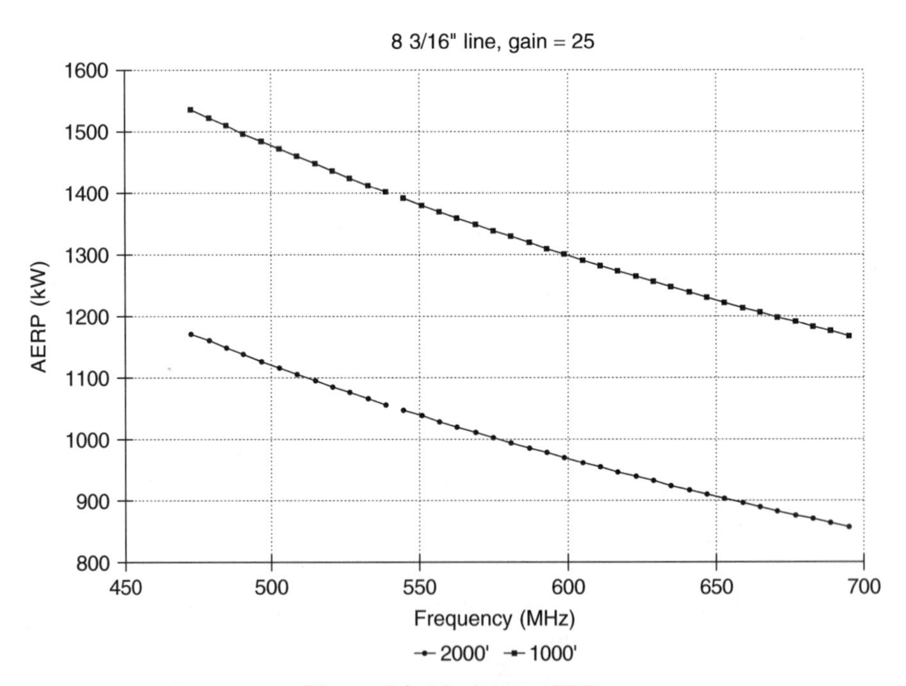

Figure 6-4. Maximum AERP.

HIGHER-ORDER MODES

In addition to an upper-frequency limit based on AERP, attenuation, and average power rating, the possibility of the first higher-order mode must also be considered. For coaxial lines, the TEM mode is the dominant or fundamental mode. The cutoff frequency in megahertz, f_c, for the first higher-order mode in air-dielectric coaxial lines is

$$f_{\text{co}} = \frac{7514}{D_i + d_o}$$

The calculated cutoff frequency is usually reduced by 5 to 10% to allow for the effects of manufacturing tolerances, elbows, transitions, and connections at flanges. Using a 5% criterion, $8\frac{3}{16}$-in. rigid line is usable through U.S. channel 51; $9\frac{3}{16}$-in. line is usable through U.S. channel 38. Smaller rigid lines are usable throughout the UHF broadcast band.

PEAK POWER RATING

The peak power rating of a transmission line does not vary with frequency, since it is determined by voltage breakdown instead of ohmic heating of the inner conductor. In general, peak ratings are quite high. For example, the peak ratings of $3\frac{1}{8}$- and $6\frac{1}{8}$-in. rigid lines published by at least one manufacturer[8] are 900 and 2000 kW, respectively. Thus the peak rating is generally not a factor in selecting a transmission line for a single digital television station.

If multiple digital television signals are combined into a single transmission line, peak power-handling limits may be important. The instantaneous peak voltage is the sum of all voltages. For two or more signals of equal power, the peak power increases by the square of the number of signals in the worst case. For two signals, the increase can be up to a factor of 4; for three signals, there can be a ninefold increase in peak power. Peak-to-average ratios up to 11 dB have been reported in some highly linear systems.[9] The possibility of this high value should be considered when specifying transmission line breakdown voltage.

FREQUENCY RESPONSE

Linear distortions include non-constant-amplitude response and nonlinear phase within the channel. These distortions result in intersymbol interference, which is evidenced by closure of the eye pattern, decrease in the signal-to-noise ratio, and increase in error vector magnitude. It is therefore important to minimize linear distortions wherever they occur in the transmission path.

[8] Andrew Corporation, Catalog 36, p. 274.

[9] Robert J. Plonka, "A Fresh Look at 8 VSB Peak to Average Ratios," *NAB Broadcast Engineering Proceedings*, 1998, pp. 53–62.

The frequency response of coaxial lines is simply the slope of the attenuation versus frequency curve. Reference to Figures 6-1 and 6-2 shows that the response tilt of coaxial lines is dependent on frequency and the length of line. In general, the response tilt is small. In mathematical terms, the slope of the frequency response is the first derivative of the formula for attenuation with respect to frequency, or

$$\frac{d\alpha}{df} = \frac{1}{2}Af^{-1/2} + B \qquad \text{dB/100 ft per megahertz}$$

From the foregoing graphs and this formula, it is apparent that the response tilt is greatest at the lowest frequencies. For example, $3\frac{1}{8}$-in. rigid line operating at U.S. channel 2, the response tilt is 0.0007 dB per 100 feet per megahertz. Even a 2000-ft run would exhibit only 0.09 dB over 6 MHz. Thus for all practical purposes, frequency-response tilt may be ignored in rigid coaxial lines. For those who wish to do so, this amount of tilt could be preequalized. This could be accomplished using either analog IF equalizers or programmable digital equalizers using the equation above and the appropriate frequency, line size, and line length.

Phase nonlinearity and group delay variations do not occur in matched coaxial lines. Since the operating mode is TEM, the phase is linear with frequency and there is uniform group delay. If the line is mismatched, however, phase nonlinearity and group delay is present, depending on the antenna reflection coefficient and the line length. Eilers has published an analysis of this effect.[10] The group delay for a constant antenna VSWR of 1.05:1 ($\Gamma = 0.025$) and lossless transmission line are shown in Figure 6-5. The group delay is periodic with ripple frequency and magnitude directly proportional to line length. For this example, a maximum group delay of 100 ns is computed for a 2000-ft line length. The ripple frequency increases to 24 ripples across a 6-MHz band for a line length of 2000 ft.

The group delay is a consequence of phase ripple due to the mismatch. For this example, the phase ripple is constant for all transmission line lengths with a value of $\pm 1.43°$. A response ripple of ± 0.2 dB is also present. Like the magnitude of the phase ripple, the magnitude of the amplitude ripple is independent of line length. For higher VSWR, the phase ripple, group delay, and amplitude ripple are proportionately greater. The ripple frequency is independent of the magnitude of the reflection.

In the practical case, antenna VSWR is not constant with frequency, and neither the phase ripple, group delay, or amplitude ripple can easily be predicted as a function of frequency. To some extent, the effect of antenna VSWR will be reduced due to the transmission line losses. Thus, these linear distortions are difficult to determine without measurement and thus are difficult to preequalize. The best approach is to specify and maintain antenna VSWR as low as possible.

[10] Carl G. Eilers, "The In-Band Characteristics of the VSB Signal for ATV," *IEEE Trans. Broadcast.*, Vol. 42, No. 4, December 1996, p. 298.

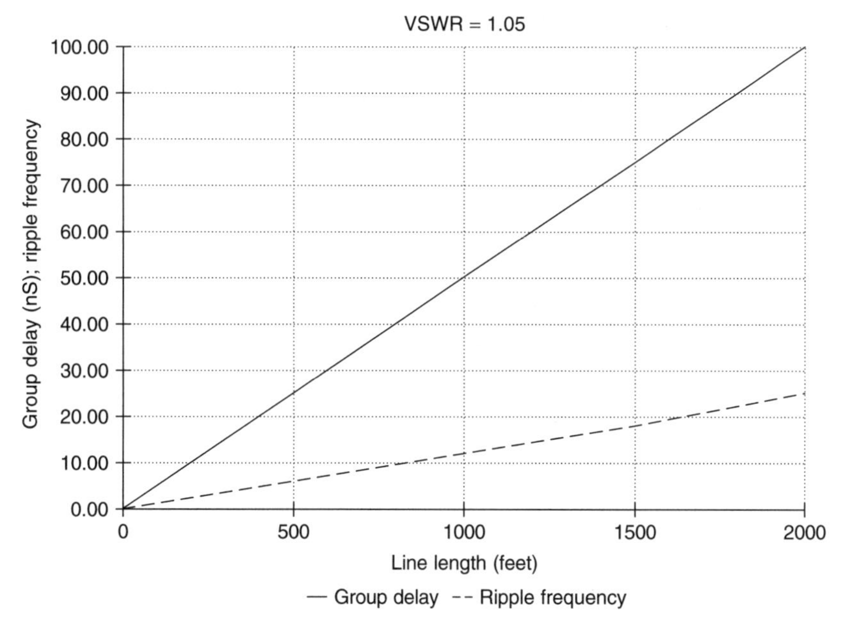

Figure 6-5. Group delay and ripple frequency.

STANDARD LENGTHS

Rigid coaxial lines are manufactured in standard section lengths for television of 19.5, 19.75, and 20 ft. The optimum section length for a particular channel is based on the need to minimize the accumulated reflections from the flange connections for a long length of line. It is well known that identical reflections spaced an odd number of quarter-wavelengths apart will cancel, so that the total reflection coefficient is zero. Since the bandwidth of the digital television signal is at least 6 MHz, this condition cannot be achieved precisely for all frequencies within the band. However, by proper selection of the section length, the effect of accumulated reflections can be minimized within the channel bandwidth. The proper section length for each channel is available in tabular form from transmission line suppliers.

CORRUGATED COAXIAL CABLES

Installation time and cost are yet other major factors in transmission line selection. Continuous runs of semiflexible air-dielectric corrugated coaxial cables are available to reduce these costs. As with rigid lines, there are variations among the various manufacturers' offerings. However, these cables share some characteristics. Inner and outer conductors are usually corrugated copper, although at least one manufacturer[11] offers a 9-in. diameter line with a corrugated aluminum outer conductor. Inner conductors are supported by spirals of polyethylene, polypropylene, or Teflon. Under normal conditions, the dielectric material may be considered to be almost equivalent to dry air. The velocity of propagation is more than 90% that of free space, somewhat lower than rigid coaxial lines. These cables are also covered with a black polyethylene or fluoropolymer jacket.

Acquisition cost of corrugated cables is usually lower than for rigid lines. Corrugated cables are often easier to install than rigid lines since the continuous sections are longer. A longer continuous run reduces the concern for performance at a specific channel, since there are fewer flanges. Cable sizes of 3-, 4-, 5-, $6\frac{1}{8}$-, 8-, and 9-in. diameter are available. The standard characteristic impedance is 50 Ω.

In general, efficiency is somewhat lower than the corresponding-size rigid line. This a consequence of the 50-Ω characteristic impedance and the requirement for more dielectric material to support the inner conductor properly. Although the expression for attenuation is of the same form as for rigid line, the constants, A and B, are different. For example, consider $6\frac{1}{8}$-in. corrugated cable. At 800 MHz, one manufacturer's graph shows this cable to have attenuation of 0.18 dB per 100 ft[12] compared to 0.154 dB per 100 ft for rigid $6\frac{1}{8}$-in. line. The manufacturer's graph fits closely if A and B are chosen as 0.00472 and 0.0000632, respectively.

Charts and graphs of attenuation and maximum average power for corrugated coaxial cables are shown in Tables 6-4, 6-5, 6-6, and 6-7 and Figures 6-6 and 6-7 for 5-, $6\frac{1}{8}$-, 8-, and 9-in. lines, respectively. These are based on the formulas above using the same derating factor for attenuation as for rigid lines. The average power rating is determined using maximum dissipations of 1.58, 1.82, 2.4, and 3.23 kW per 100 ft for each respective cable. The dissipation ratings for these cables are higher than for the rigid lines, as a consequence of the larger inner conductor needed for the 50-Ω characteristic impedance and higher heat transfer coefficient.[13] Except for lines exposed to direct solar radiation,[14] the derating factors used and the assumptions made with regard to dissipation are considered adequate. Consequently, these charts and curves are considered to be conservative and may be used to estimate the operating specifications for most digital television system designs. However, the data presented should be considered representative and not used for cables supplied by all manufacturers. The reader may apply the computational technique to the cables being considered for a specific installation.

[11] Radio Frequency Systems, Inc., Catalog 720C, pp. 44–45.

[12] Ibid., p. 40.

[13] Cozad, op. cit.

[14] Consult manufacturers' data for derating factors for solar radiation. Additional derating at temperate latitudes of 8% may be need for cables using Teflon supports; 15% for those using polyethylene. Even higher derating may be needed at tropical latitudes.

TABLE 6-4. Power Rating and Attenuation of 5-in. 50-Ω Corrugated Coaxial Cable

Channel	F (MHz)	P_i (kW)	Attenuation (dB/100 ft)	Channel	F (MHz)	P_i (kW)	Attenuation (dB/100 ft)
2	57	132.02	0.052	36	605	34.88	0.202
3	63	125.10	0.055	37	611	34.68	0.203
4	69	119.11	0.058	38	617	34.47	0.204
5	79	110.69	0.063	39	623	34.27	0.205
6	85	106.38	0.065	40	629	34.08	0.206
7	177	70.96	0.098	41	635	33.89	0.208
8	183	69.64	0.100	42	641	33.70	0.209
9	189	68.39	0.102	43	647	33.51	0.210
10	195	67.20	0.103	44	653	33.32	0.211
11	201	66.06	0.105	45	659	33.14	0.212
12	207	64.98	0.107	46	665	32.96	0.214
13	213	63.94	0.109	47	671	32.78	0.215
14	473	40.36	0.174	48	677	32.61	0.216
15	479	40.06	0.175	49	683	32.44	0.217
16	485	39.77	0.176	50	689	32.27	0.218
17	491	39.48	0.178	51	695	32.10	0.219
18	497	39.20	0.179	52	701	31.93	0.221
19	503	38.92	0.180	53	707	31.77	0.222
20	509	38.65	0.181	54	713	31.61	0.223
21	515	38.38	0.183	55	719	31.45	0.224
22	521	38.12	0.184	56	725	31.29	0.225
23	527	37.86	0.185	57	731	31.14	0.226
24	533	37.61	0.187	58	737	30.98	0.228
25	538	37.36	0.188	59	743	30.83	0.229
26	545	37.12	0.189	60	749	30.68	0.230
27	551	36.88	0.190	61	755	30.53	0.231
28	557	36.64	0.192	62	761	30.39	0.232
29	563	36.41	0.193	63	767	30.25	0.233
30	569	36.18	0.194	64	773	30.10	0.234
31	575	35.95	0.195	65	779	29.96	0.236
32	581	35.73	0.197	66	785	29.82	0.237
33	587	35.51	0.198	67	791	29.69	0.238
34	593	35.30	0.199	68	797	29.55	0.239
35	599	35.09	0.200	69	803	29.42	0.240

Using the data of Figures 6-6 and 6-7 and Tables 6-4, 6-5, 6-6, and 6-7, graphs of maximum AERP that can be supported by 5-, $6\frac{1}{8}$-, and 8-in. corrugated cable versus frequency for typical line lengths and antenna gains are shown in Figures 6-8, 6-9, and 6-10. For example, a 1000- or 2000-ft run of 5-in. cable will not support an AERP of 1000 kW for any UHF channel unless antenna gain is greater than 30. If the line length is 500 ft, AERP of 900 kW can be achieved with a gain of 30 up to U.S. channel 25. Alternatively, AERP of 1000 kW may

TABLE 6-5. Power Rating and Attenuation of $6\frac{1}{8}$-in. 50-Ω Corrugated Coaxial Cable

Channel	F (MHz)	P_i (kW)	Attenuation (dB/100 ft)	Channel	F (MHz)	P_i (kW)	Attenuation (dB/100 ft)
2	57	173.15	0.046	36	605	44.71	0.181
3	63	163.98	0.049	37	611	44.44	0.182
4	69	156.03	0.051	38	617	44.18	0.183
5	79	144.88	0.055	39	623	43.92	0.184
6	85	139.15	0.057	40	629	43.66	0.185
7	177	92.27	0.087	41	635	43.41	0.186
8	183	90.53	0.088	42	641	43.16	0.187
9	189	88.88	0.090	43	647	42.91	0.188
10	195	87.30	0.092	44	653	42.67	0.190
11	201	85.80	0.093	45	659	42.43	0.191
12	207	84.36	0.095	46	665	42.20	0.192
13	213	82.99	0.096	47	671	41.96	0.193
14	473	51.91	0.155	48	677	41.73	0.194
15	479	51.52	0.156	49	683	41.51	0.195
16	485	51.13	0.158	50	689	41.29	0.196
17	491	50.75	0.159	51	695	41.07	0.197
18	497	50.38	0.160	52	701	40.85	0.198
19	503	50.02	0.161	53	707	40.63	0.199
20	509	49.66	0.162	54	713	40.42	0.200
21	515	49.31	0.164	55	719	40.21	0.201
22	521	48.97	0.165	56	725	40.01	0.202
23	527	48.63	0.166	57	731	39.81	0.203
24	533	48.30	0.167	58	737	39.60	0.205
25	539	47.97	0.168	59	743	39.41	0.206
26	545	47.65	0.169	60	749	39.21	0.207
27	551	47.33	0.170	61	755	39.02	0.208
28	557	47.02	0.172	62	761	38.83	0.209
29	563	46.72	0.173	63	767	38.64	0.210
30	569	46.42	0.174	64	773	38.45	0.211
31	575	46.12	0.175	65	779	38.27	0.212
32	581	45.83	0.176	66	785	38.09	0.213
33	587	45.54	0.177	67	791	37.91	0.214
34	593	45.26	0.178	68	797	37.73	0.215
35	599	44.99	0.180	69	803	37.55	0.216

be supported with 8-in. cable and antenna gain of 25 for U.S. channels through 39 if the line length is 1000 ft or less.

The cutoff frequency of corrugated cables is generally lower than that of rigid lines by virtue of their lower characteristic impedance. The larger inner conductor coupled with the usual reduction by 5% to allow for the effects of manufacturing tolerances, transitions, and connections at flanges results in 8-in. cable being usable through U.S. channel 39; 9-in. cable is usable through U.S. channel 27. Smaller corrugated cables are usable throughout the UHF broadcast band.

TABLE 6-6. Power Rating and Attenuation of 8-in. 50-Ω Corrugated Coaxial Cable

Channel	F (MHz)	P_i (kW)	Attenuation (dB/100 ft)
2	57	291.60	0.036
3	63	275.78	0.038
4	69	262.08	0.040
5	79	242.85	0.043
6	85	233.00	0.045
7	177	152.48	0.069
8	183	149.50	0.070
9	189	146.68	0.072
10	195	143.98	0.073
11	201	141.41	0.074
12	207	138.76	0.076
13	213	136.61	0.077
14	473	83.80	0.126
15	479	83.13	0.127
16	485	82.49	0.128
17	491	81.85	0.129
18	497	81.22	0.130
19	503	80.61	0.131
20	509	80.01	0.132
21	515	79.42	0.133
22	521	78.84	0.134
23	527	78.27	0.135
24	533	77.71	0.136
25	539	77.16	0.137
26	545	76.62	0.138
27	551	76.08	0.139
28	557	75.56	0.140
29	563	75.05	0.141
30	569	74.54	0.142
31	575	74.05	0.143
32	581	73.56	0.144
33	587	73.08	0.145
34	593	72.60	0.146
35	599	72.14	0.147
36	605	71.68	0.148
37	611	71.23	0.149
38	617	70.78	0.150
39	623	70.35	0.151

WIND LOAD

For comparable line sizes, there is obviously no significant advantage with respect to windload to use corrugated cables. For the smaller cross sections, however, the wind load can sometimes be reduced by "hiding" the line behind or within a tower leg.

TABLE 6-7. Power Rating and Attenuation of $9\frac{3}{16}$-in. 50-Ω Corrugated Coaxial Cable

Channel	F (MHz)	P_i (kW)	Attenuation (dB/100 ft)
2	57	435.68	0.032
3	63	411.75	0.034
4	69	391.05	0.036
5	79	362.00	0.039
6	85	347.11	0.041
7	177	225.69	0.063
8	183	221.22	0.064
9	189	216.96	0.065
10	195	212.91	0.066
11	201	209.05	0.068
12	207	205.36	0.069
13	213	201.83	0.070
14	473	122.65	0.116
15	479	121.66	0.117
16	485	120.69	0.118
17	491	119.74	0.119
18	497	118.81	0.120
19	503	117.89	0.121
20	509	116.99	0.122
21	515	116.11	0.123
22	521	115.25	0.124
23	527	114.40	0.125
24	533	113.56	0.125
25	539	112.74	0.126
26	545	111.93	0.127
27	551	111.14	0.128

WAVEGUIDE

For some UHF installations, rectangular or circular waveguide may be desirable. The attenuation per 100 ft and efficiency for a 2000-ft run of WR1800 and WC1750 waveguides at U.S. Channel 14 are listed in Table 6-8 together with the data for rigid coaxial lines. Obviously, there is much to be gained with respect to line efficiency by using waveguide.

The effect of line efficiency on TPO and the choice of final amplifier is illustrated in Figure 6-11. Recognizing that UHF power tubes come in average DTV power ratings of 10, 12.5, 17.5, and 25 kW, it is seen that a variety of system design options are available. For a transmitter power output between 25 and 50 kW, a possible configuration could be a pair of 17.5- or 25-kW final amplifier. Either of these might be a good choice, assuming the use of any one of the rigid coaxial lines. For a TPO below 25 kW, a pair of 12.5-kW finals could be used with any one of the waveguide types. Because the line efficiency has an impact on

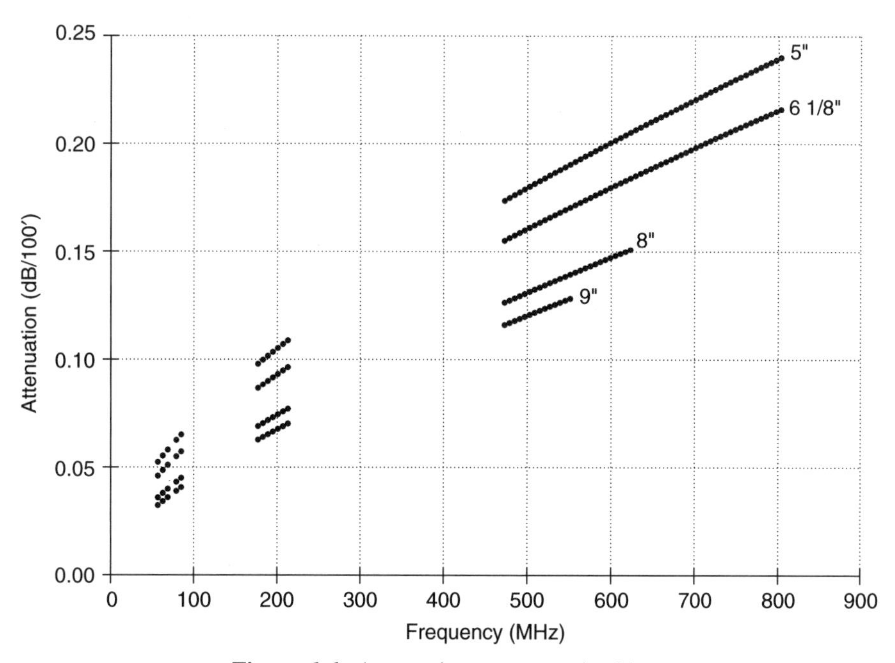

Figure 6-6. Attenuation, corrugated cables.

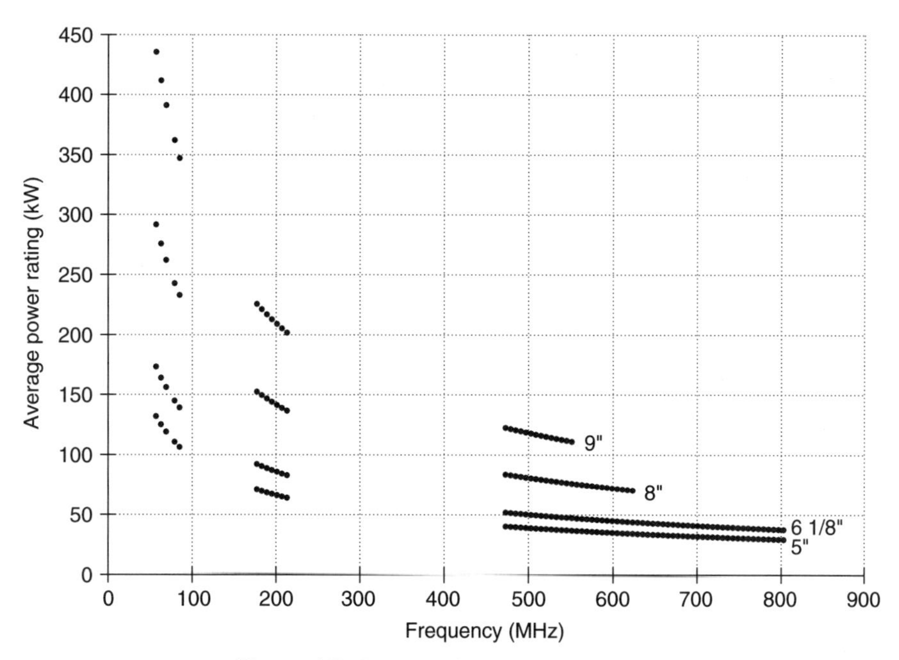

Figure 6-7. Power rating, corrugated cable.

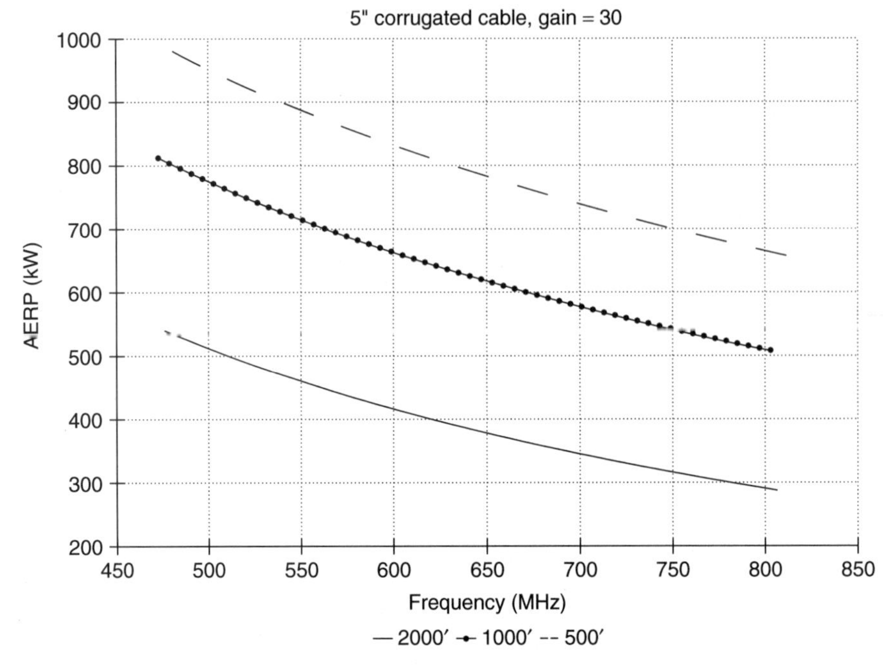

Figure 6-8. Maximum AERP.

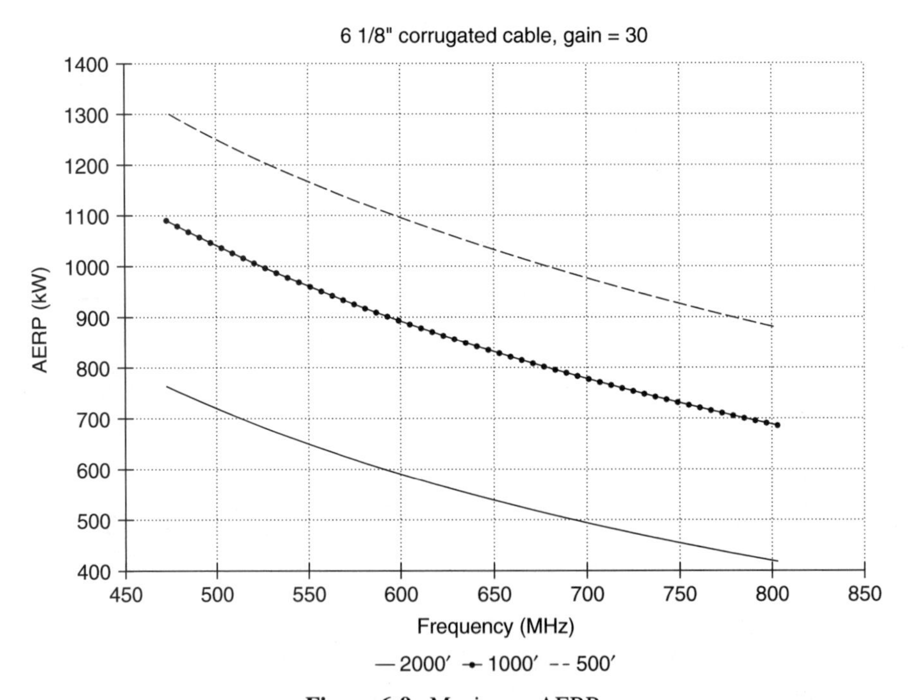

Figure 6-9. Maximum AERP.

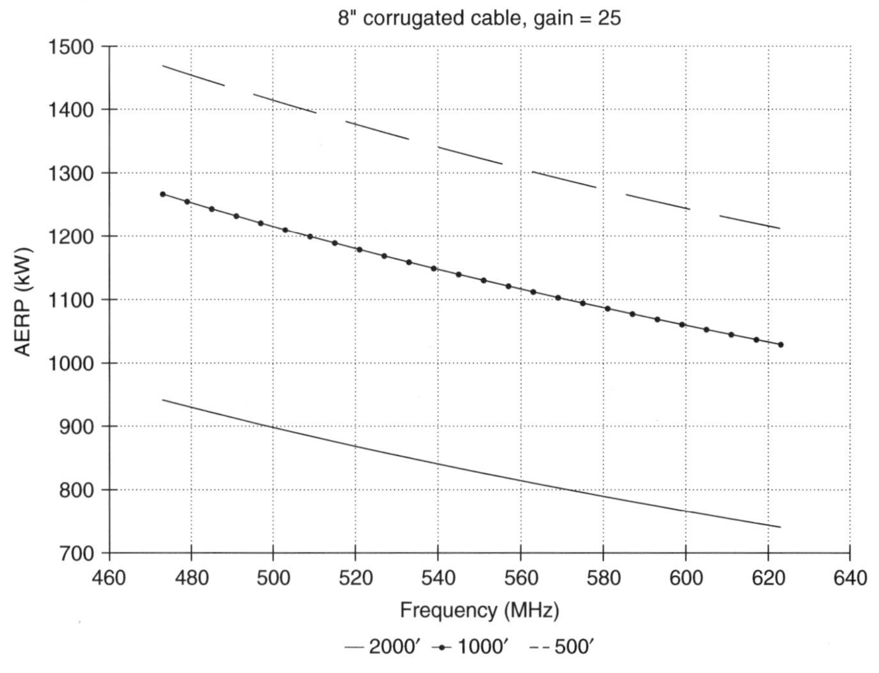

Figure 6-10. Maximum AERP.

TABLE 6-8. Attenuation of UHF Transmission Lines (U.S. Channel 14)

Type	Size (in.)	Attenuation (dB/100 ft)	Efficiency (%) (2000 ft)
Coaxial	6 1/8 75Ω	0.134	54.7
Coaxial	8 3/16	0.107	61.9
Coaxial	9 3/16	0.097	64.6
Rectangular waveguide	WR1800	0.057	77.0
Circular waveguide	WC1750	0.052	78.7

the power output of each final amplifier, it consequently affects cooling subsystem design, transmitter cost, and system operating costs.

BANDWIDTH

Waveguides are offered only for UHF channels and are band limited at both the lower and upper frequencies of operation. The lower frequency of operation is determined by the cutoff frequency of the dominant mode. For rectangular guide, this is the TE01 mode, which has a cutoff wavelength, λ_c, given by

$$\lambda_c = 2a_i$$

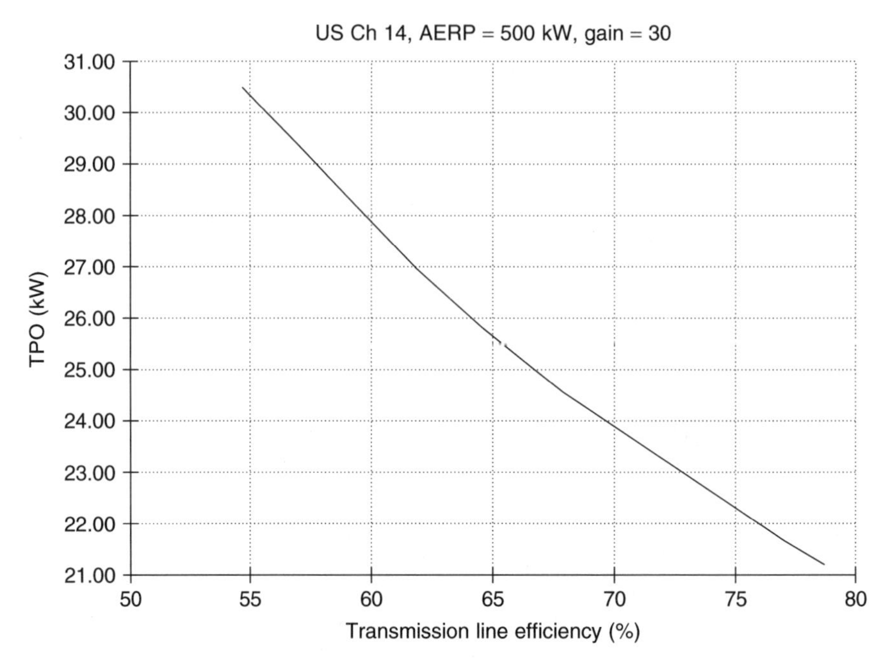

Figure 6-11. TPO versus line efficiency.

where a_i is the wide inside dimension of the waveguide. In practice, the actual operating frequency is set approximately 25% above the cutoff frequency to assure minimum attenuation and phase nonlinearity.

Like coaxial lines, the upper-frequency limit of rectangular waveguide is determined by the cutoff frequency of the first higher-order mode. There are two modes with the same cutoff frequency, the TE11 and TM11 modes. For standard waveguides for which the height, b_i, is half the width, the cutoff wavelength is

$$\lambda_c = 0.894a_i$$

The upper frequency of operation is usually reduced below the first higher-order-mode cutoff frequency by 17 to 19% to allow for the effects of manufacturing tolerances and moding at elbows, transitions, and connections at flanges.

The maximum bandwidth ratio, BW, of rectangular waveguide is the ratio of the cutoff wavelength for the TE10 and TE11 modes, or

$$\text{BW} = \frac{2}{0.894} = 2.237$$

Accounting for the allowances made for operation above the dominant mode and below the first higher-order-mode cutoff frequencies, the actual operating bandwidth ratio of a rectangular waveguide is approximately 1.5:1.

Due to the rectangular cross section, there is a preferred direction for the field orientation in rectangular waveguide. There is therefore no need for special design techniques to maintain polarization, as in circular waveguide.

For circular waveguide, the dominant or fundamental mode is the TE11, for which the cutoff wavelength is

$$\lambda_c = 1.706D_i$$

where D_i is the inside diameter of the guide. In practice, the actual operating frequency is set at least 19% above the cutoff frequency to assure minimum attenuation and phase nonlinearity.

The first higher-order mode for unmodified circular waveguide is the TM01 mode, for which the cutoff wavelength is given by

$$\lambda_c = 1.307D_i$$

The ideal bandwidth ratio, BW, of unmodified circular waveguide is, therefore, the ratio of the cutoff wavelength for the TE11 and TM01 modes, or

$$BW = \frac{1.706}{1.307} = 1.305$$

This is considerably less than the bandwidth ratio of rectangular waveguide. Accounting for the allowances made to operate above the dominant mode and below the first higher-order-mode cutoff frequencies, the actual operating bandwidth ratio of unmodified circular waveguide would be less than 1.09:1. If some means were not devised to increase the bandwidth ratio, circular waveguide would find very little application.

When care is taken to avoid discontinuities that can generate higher-order modes, circular waveguide may be manufactured with greater operating bandwidth. This involves precise manufacturing and installation techniques to minimize deformations from true circular cross section and discontinuities at flange junctions, as well as avoidance of elbows. Because of circular symmetry, there is no preferred field direction in an unmodified circular waveguide. To assure that the TE11 mode is properly oriented at the output to couple the RF energy properly to the antenna, it is necessary to control the polarization of the dominant mode. Recognizing that the electric field must always go to zero in the presence of a tangential conducting boundary, it is apparent that a preferred orientation of the electric field may be established by inserting a conducting boundary in the guide in a direction orthogonal to the TE11 field lines. Such a boundary would have no appreciable effect on propagation of the TE11 mode. This boundary can be constructed with metallic pins at appropriate intervals, introducing only minimal mismatch to the transmission line. These pins also serve to stabilize the precise circular shape of the guide, thereby helping to suppress higher-order modes.

If the TM01 mode is suppressed, the next-higher-order mode in circular waveguide is the TE21. This mode has a cutoff wavelength given by

$$\lambda_c = 1.0285 D_i$$

The bandwidth ratio for the cutoff frequencies of the TE11 and TE21 modes is

$$\mathrm{BW} = \frac{1.706}{1.0285} = 1.659$$

Now accounting for the allowances made to operate above the dominant mode and below the second-higher-order mode cutoff frequencies, the actual operating bandwidth ratio of modified circular waveguide may be increased to as much as 1.36:1. Circular waveguide manufacturers provide charts indicating the recommended operating frequencies for each waveguide size. To operate over this much bandwidth requires that input and output connections and elbows be constructed in rectangular waveguide. If dimensions are measured in inches, the cutoff frequency of the modes in both rectangular and circular guide is given by

$$f_{\mathrm{co}} = \frac{11{,}803}{\lambda_c}$$

WAVEGUIDE ATTENUATION

The attenuation constant of standard aluminum rectangular waveguide is given by

$$\alpha = 1.75a^{-3/2}\left\{\left(\frac{\lambda_c}{\lambda}\right)^{3/2} + \frac{(\lambda_c/\lambda)^{-1/2}}{[(\lambda_c/\lambda)^2 - 1]^{1/2}}\right\}$$

where α is in decibels per 100 ft of line. A similarly complicated formula for aluminum circular waveguide is

$$\alpha = 1.72D^{-3/2}\left\{\frac{0.4185(\lambda_c/\lambda)^{3/2} + (\lambda_c/\lambda)^{-1/2}}{[(\lambda_c/\lambda)^2 - 1]^{1/2}}\right\}$$

The results of calculations based on these formulas is shown graphically for rectangular waveguide in Figure 6-12 and for circular waveguide in Figure 6-13. Aside from having lower attenuation than coaxial lines, waveguides are unique in that the attenuation for any specific-sized waveguide actually decreases with increasing frequency. The attenuation of circular waveguide is generally lower than the corresponding rectangular waveguide. For multiple stations using a common waveguide, the operating bandwidth must cover all channels of interest.

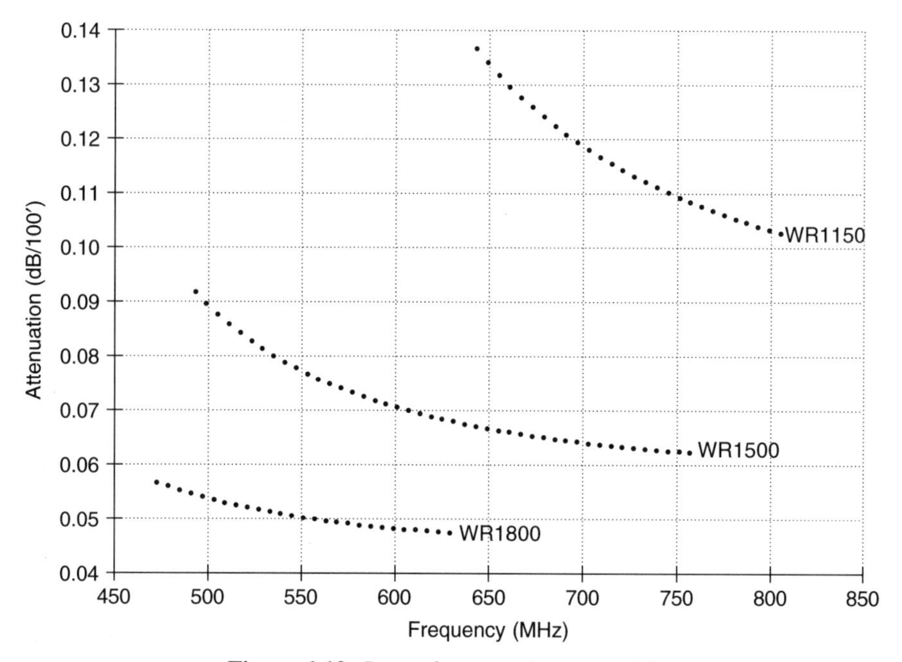

Figure 6-12. Loss of rectangular waveguide.

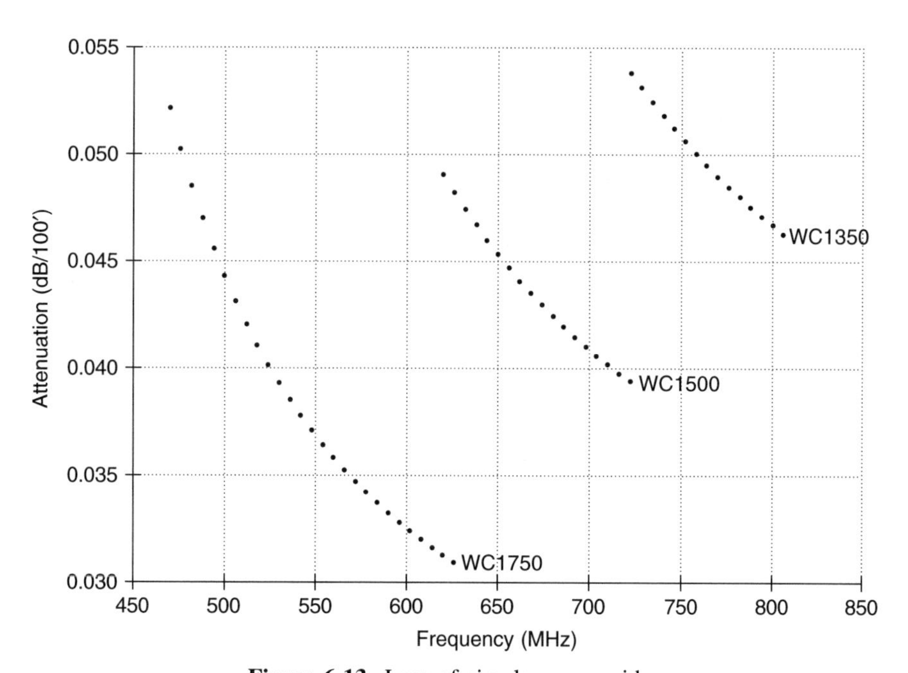

Figure 6-13. Loss of circular waveguide.

POWER RATING

For all practical purposes, the average power rating of waveguide may be considered "unlimited" for digital television applications. Obviously, this is not literally true. But at least one manufacturer gives the average power rating as 360 kW or more for all waveguides, independent of frequency.[15] This is above the TPO of all currently available digital television transmitters. Thus, waveguide may be selected on the basis of channel and desired transmission line efficiency without regard for power-handling limitations.

FREQUENCY RESPONSE

Unlike coaxial lines, waveguides exhibit inherent-phase nonlinearity and group delay. This is a consequence of the lower-frequency bandlimiting. This gives rise to a phase constant that is nonlinear with frequency. For any transmission line,

$$\beta = \frac{2\pi}{\lambda_g}$$

where λ_g is the guide wavelength of the line.

For air-dielectric coaxial lines, the free-space wavelength is equal to the waveguide wavelength. However, for hollow waveguides, the guide wavelength is given by

$$\lambda_g = \frac{\lambda}{[1 - (\lambda/\lambda_c)^2]^{1/2}}$$

The phase constant for WR1800 waveguide is plotted in Figure 6-14 along with the linear-phase constant of free-space and air-dielectric coaxial lines. The difference between these phase constants and the group delay for a 300-m length of guide is also plotted. For long lines, the group delay due to the waveguide cannot be neglected.

There is also small but usually negligible amplitude tilt in the frequency response of waveguide. The worst case is for WR1150 operating at U.S. channel 42, for which the tilt is 0.00259 dB per 6 MHz per 100 ft. Even for a 2000-ft run, this amounts to a response tilt of only 0.05 dB. Since both the phase nonlinearity and the amplitude tilt are predictable, they may be preequalized in the exciter in a manner similar to that indicated for coaxial lines.

Antenna VSWR produces phase and amplitude ripple in the transmission line response just as it does for coaxial lines. In practice, this effect is somewhat more noticeable for waveguide because the line efficiency is higher. As indicated before, the need to minimize the antenna VSWR is extremely important.

[15] Andrew Corporation, Catalog 36, p. 288.

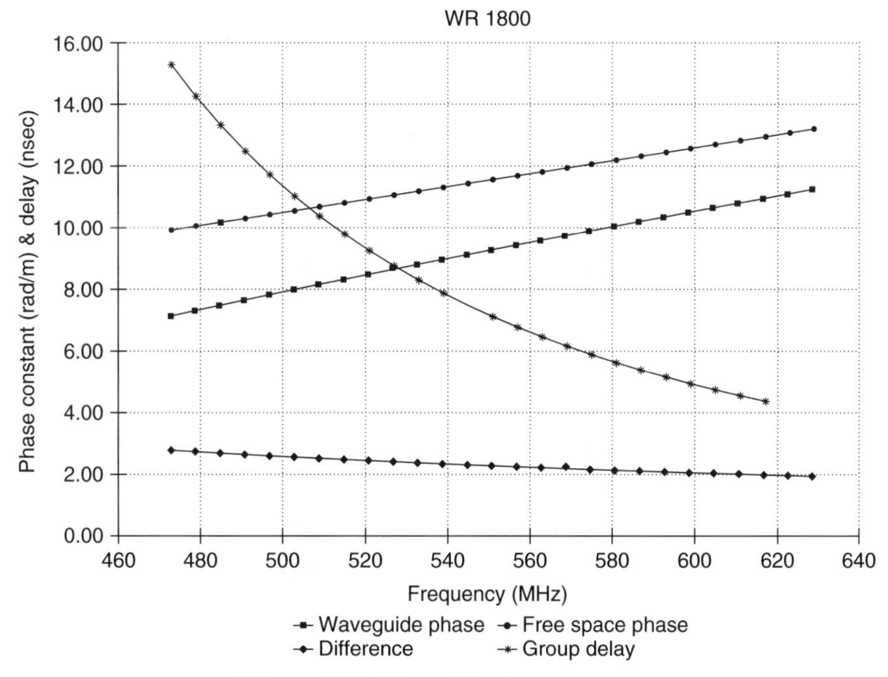

Figure 6-14. Waveguide phase response.

SIZE TRADE-OFFS

Standard rectangular waveguide sizes are such that there is considerable overlap in frequency coverage. For example, the recommended frequency range for WR1800 is from U.S. channels 14 through 39, and the range for WR1500 is from U.S. channels 18 through 60. Thus, for U.S. channels 18 through 39, the choice can be made for either size of waveguide. From a RF performance point of view, the larger guide is the obvious choice, since both attenuation and group delay are less as a consequence of the larger size and operation farther from the cutoff frequency of the TE10 mode. From a wind load and cost viewpoint, WR1500 is preferred, since this waveguide is smaller in cross section. A similar trade-off between WR1500 and WR1150 is evident for stations operating on U.S. channels 42 through 60. For commercially available circular waveguides, there is far less overlap and far less trade-off to be made. As is the case of rigid coax, waveguides are manufactured in standard section lengths. The correct section length must be selected to assure minimum buildup of reflections on long runs.

WHICH LINE? WAVEGUIDE OR COAX?

As is usually the case, improving line efficiency with larger rigid coaxial lines or waveguides comes with a price. The trade-off is in acquisition cost of the line and

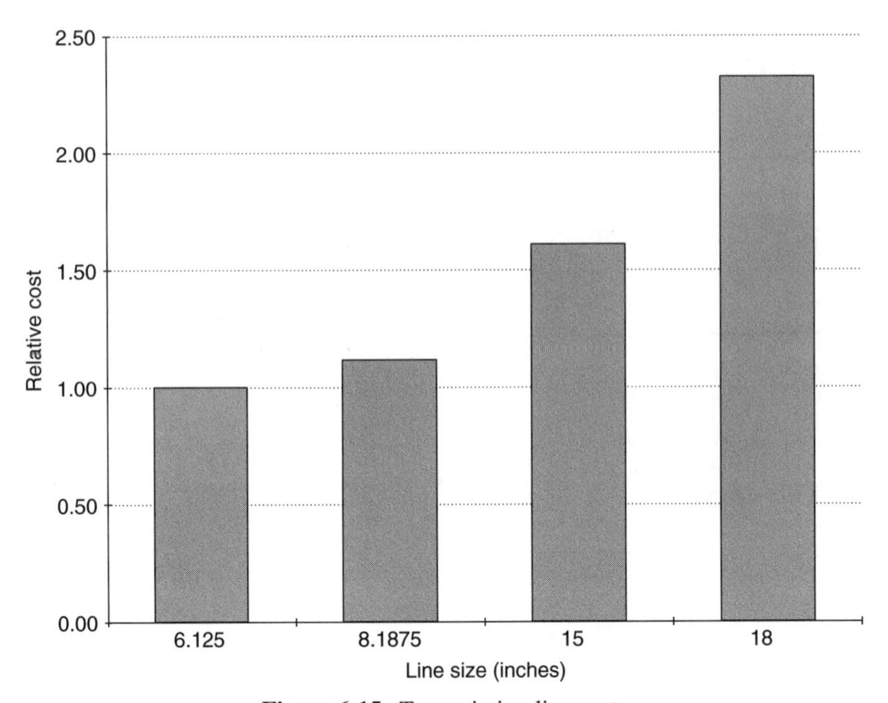

Figure 6-15. Transmission line cost.

wind loading. As would be expected, the relative acquisition cost of transmission line generally increases with increasing line size as shown in Figure 6-15. For rigid coaxial lines, the slope of the increase is at a rate greater than the square of the line size. Interestingly, the cost of rectangular waveguide may be somewhat less than $9\frac{3}{16}$-in. coax.

Line selection should consider the present value of the annual savings due to improved efficiency versus the initial purchase cost of transmitter and line. When installing a full-service digital television facility, payback may be expected over the full-service lifetime of the equipment. When installing an interim facility with a low-power transmitter and line, opting for the higher-efficiency line may not make much sense.

To help make the trade-off of line cost and efficiency, it is useful to define a transmission line figure of merit, FOM, where

$$\text{FOM} = \frac{\text{acquisition cost}}{\text{efficiency}}$$

With this FOM it is possible to quantify the cost/benefit ratio of using large line sizes. Line FOM is a function of the line length and channel because efficiency

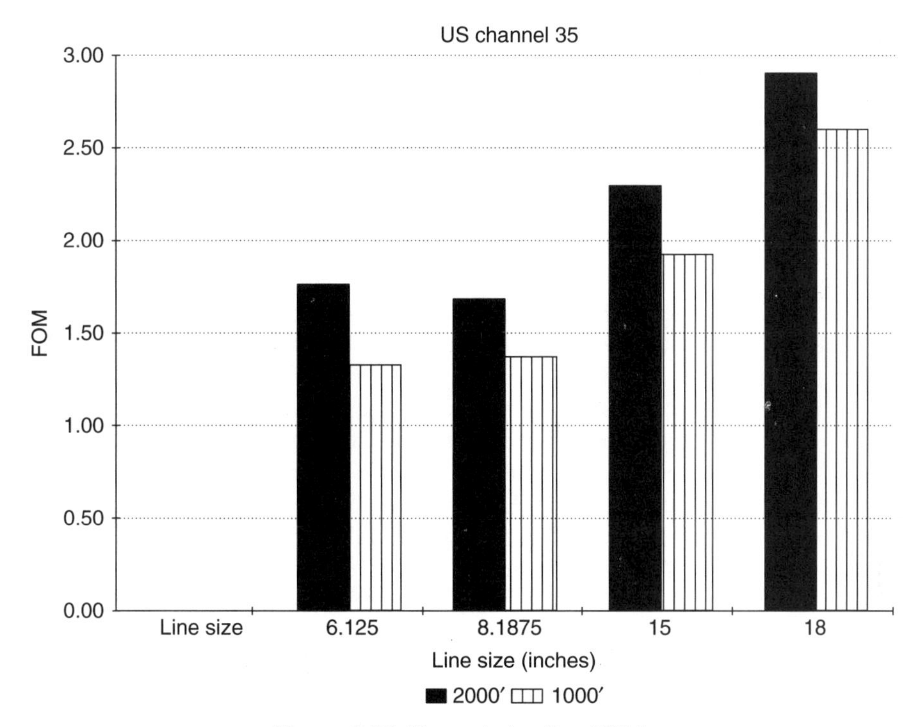

Figure 6-16. Transmission line FOM.

is dependent on these factors. FOM is illustrated for some representative line lengths at U.S. channel 38 in Figure 6-16. For the 1000′ case shown, a minimum in the cost-to-efficiency ratio occurs for a line size of $6\frac{1}{8}$-in. Although this may be the best choice for low-power installations, the low efficiency of $6\frac{1}{8}$-in. line may require an uneconomical transmitter choice. In some cases, $6\frac{1}{8}$-in. line will not handle the required power. In this event, the choice will be between 8- and 9-in. coax or 15- and 18-in. coax. On the other hand, if a large expenditure of money is required to upgrade the tower capability to support the larger lines, the $6\frac{1}{8}$-in. line might be preferred, assuming that power handling is not an issue. In general, the shorter the line and the lower the channel, the less the advantage of using larger line sizes. Transmission line suppliers should be consulted for current pricing and other specifications when making these trade-offs.

The maximum wind-load capability of the tower may also limit the line size. The larger the line, the greater the wind load. Obviously, the coaxial lines present less wind load than the waveguides. If the decision is made to use waveguide but the tower will not support rectangular waveguide, adding a circular shroud to

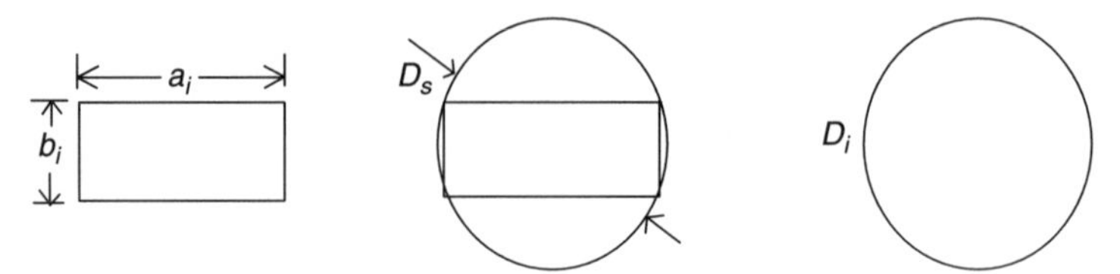

Figure 6-17. Waveguide cross sections.

the rectangular guide as shown in Figure 6-17 might be considered as a means of reducing wind load. As is well known, the wind load due to circular cross section is two-thirds that of flat surfaces. Using a circular shroud on rectangular waveguide obviously hides the flat surfaces of the waveguide, thereby presenting a more favorable cross section to the wind. The improvement is not the full 33%, however. Since the shroud must be at least as large as the diagonal of the waveguide, the shroud diameter, D_s, is greater than the largest flat dimension of the rectangular waveguide. For standard waveguide with $b = \frac{1}{2}a$, the diagonal is 11.8% greater than the wide dimension of the guide, a. Thus, the actual wind load reduction is $(1 - 1.118/1.5) \times 100\%$, or 25%.

For further reduction in waveguide wind load, circular waveguide should be considered. For example, from U.S. channels 14 through 19, either WR1800 or WC1750 way be used. The wind load of WC1750 is actually 37% less than WR1800.

PRESSURIZATION

Ordinarily, all transmission lines should be pressurized to a low positive gauge pressure up to 5 pounds per square inch (psig) with dry air or nitrogen. This prevents ingress of moisture and subsequent collection of water, thereby minimizing corrosion of conductors, deterioration of dielectric materials, and increases in attenuation over time. Temperature variations and humidity will cause water to condense inside an unpressurized transmission line. The presence of water and humidity can cause voltage breakdown. If coaxial lines are to be operated near their average power rating, higher pressures up to 30 psig should be considered to provide operating margin. For example, for standard atmospheric pressure, overpressure of 1 atmosphere (14.7 psig) will increase the average power rating by approximately 20%. For high-altitude sites, proper account must be taken of the reduced atmospheric pressure and temperature to determine the appropriate gauge pressure.

SUMMARY

There are many considerations in selecting the transmission line for a digital television installation. Some of these affect the electrical performance of the system. These factors include line efficiency, power-handling capability, and linear distortions. Nonelectrical factors include wind loading, installation time and cost, and the need to pressurize the line. Virtually all factors affect acquisition, installation, and operating cost. The choice of transmission line is a critical decision, the impact of which is felt over the life of the system and therefore should be made with much care.

7

TRANSMITTING ANTENNAS FOR DIGITAL TELEVISION

As with the transmitter and transmission line, the antenna and its performance characteristics plays an important role with regard to digital television system performance. Together with AERP, the shape and orientation of the antenna pattern determines a station's coverage. The shape of the antenna pattern determines the directivity, which with the transmission line and antenna efficiency, determines the transmitter output power required to achieve the AERP. Antenna power rating must be consistent with these requirements. The antenna wind load contributes significantly to tower loading and cost. Antenna impedance, pattern bandwidth, and associated frequency response are an important component of overall system response. Other antenna characteristics affecting system performance include tower mounting and channel combining strategies. These factors are examined in detail in this chapter.

The purpose of the transmitting antenna is twofold. First, the antenna directs RF energy into desired directions and suppressly energy in other, undesirable directions. Second, the antenna is an impedance-matching device that matches the impedance of the transmission line to that of free space. These are known as the directional and impedance properties of an antenna. The degree to which the antenna efficiently performs these functions determines, in large measure, the effectiveness of a digital television system.

ANTENNA PATTERNS

The antenna radiation pattern is a graphical representation of the energy radiated by an antenna as a function of direction. The complete radiation pattern is determined, whether by computation or measurement, by recording the field intensity at a fixed distance from the antenna in all directions. The result is a rectangular or polar plot that indicates the intensity of the radiated field as a function of direction. In practice, measurement of the complete radiation pattern is not desirable or necessary. Usually, cross sections of the complete pattern are shown in certain planes of interest called the principal planes. In a spherical coordinate system as shown in Figure 7-1, cross sections of the pattern for the elevation ($\phi = 0°$) and azimuth planes ($\theta' = 90°$) are commonly called the vertical and horizontal patterns, respectively. In practice, the direction of the peak of the elevation or vertical pattern of broadcast antennas is referenced to $\theta' = 90°$. The depression angle, θ, is measured from this plane. The shape of the radiation pattern is independent of radial distance from the antenna to the observer as long as the distance is sufficiently large.

For convenience of computation and measurement, the coordinate system is selected so that the elevation and azimuth patterns are independent. This requires that the axis of the antenna be oriented along the z-axis. With this orientation, the elevation pattern is measured in the θ direction. The azimuth pattern is measured in the ϕ direction.

Antenna patterns may be presented in terms of field strength, in which case the pattern is expressed as volts per meter. However, it is more likely that the pattern will be normalized to the peak field strength, so that the units will be simply proportional to field strength. Patterns may also be presented in terms of relative power. In this case the pattern represents power per unit solid angle. The power pattern is proportional to the square of the field strength pattern. Alternatively, the pattern may be presented in terms of decibels below a fixed reference, usually the peak radiation intensity or field strength.

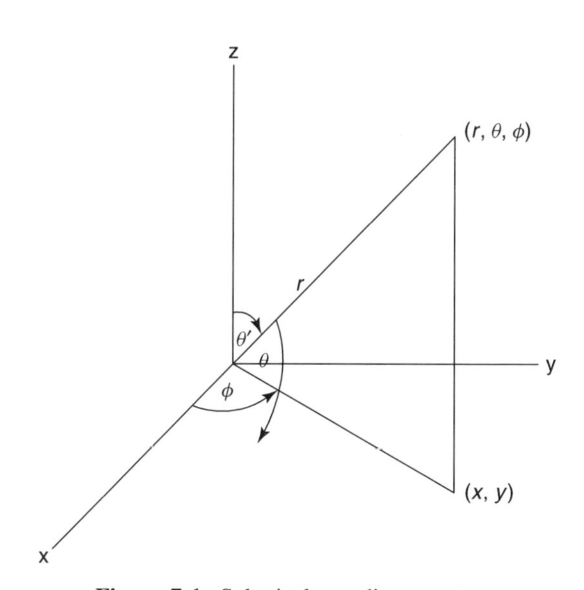

Figure 7-1. Spherical coordinate system.

In the far field, the direction of the electric field is tangent to the spherical surface of the coordinate system. Since polarization is defined in terms of the direction of the electric field, the antenna is said to be horizontally polarized when the electric field is in the azimuth direction or parallel to the earth's surface. The antenna is vertically polarized when the electric field is in the elevation direction or perpendicular to the earth's surface. If the antenna is elliptically polarized, both vertical and horizontal field components are present. The ellipticity is determined by the ratio of the vertical and horizontal components and their relative phase. Ellipticity is often specified in terms of axial ratio, which is the logarithmic ratio of the maximum to minimum field. Circular polarization is a special case of elliptical polarization for which the vertical and horizontal components are equal and in phase quadrature. In this case the axial ratio is unity or 0 dB. Only right-hand elliptical or circular polarization is permitted for television broadcast applications in the United States. This means that the electric vector rotates in a clockwise direction as viewed in the direction of propagation.

Implementation of digital television offers an excellent opportunity to take advantage of the reflection canceling benefits of circular polarization. As discussed in Chapter 8, reflection of a right-hand circularly polarized wave from a plane surface produces a wave rotating in the left-hand direction. Right-handed receiving antennas respond primarily to the right-hand circularly polarized wave. Thus the reflected wave is largely rejected by the receiving antenna, thereby reducing the linear distortions due to multipath. Since digital television stations are assigned primarily to the UHF band, physically small, aesthetically pleasing, easy to install, and inexpensive circularly polarized receiving antennas could be manufactured. With reduced multipath, the receiver equalization required for satisfactory reception would be reduced. If circular polarization were universally adopted for the UHF band, it would be possible to obtain the same, or in some cases, improved coverage without increasing system AERP. The need to orient the antenna properly to match the transmitted polarization would also be reduced.

ELEVATION PATTERN

The geometry of a typical terrestrial digital television broadcast station with respect to the earth's surface permits the use of antennas with reasonably directional elevation patterns. Consider a transmitting antenna at height, h_t, over the surface of the earth with radius, R_e, as shown in Figure 7-2. Obviously, there is no need to radiate a signal above the horizon. Thus, the ideal antenna would direct all transmitter power into a cone centered on the transmitting tower toward all angles at and below the radio horizon. Television towers are most often located 1 to 10 miles from the city of license, and consequently, some distance from population concentrations. For a tower height of 2000 ft at a distance of 1 mile, the maximum depression angle is approximately 20°. Thus it is seen that antennas with relatively narrow elevation patterns may be used.

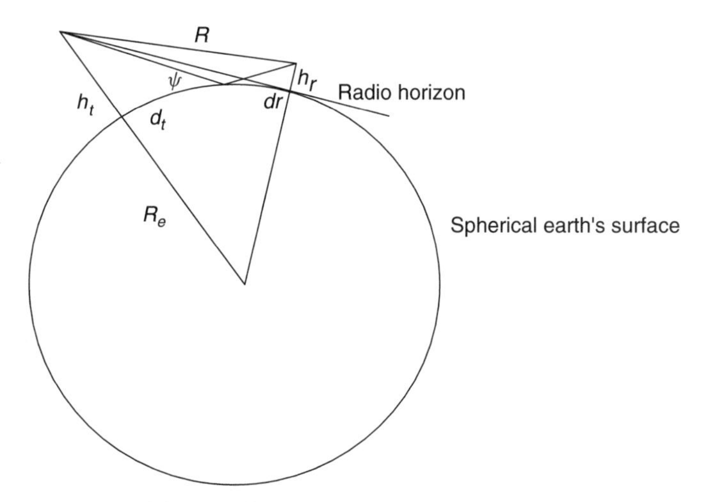

Figure 7-2. Broadcast antenna geometry.

This can readily be accomplished by using antennas with extended aperture in the vertical direction.

In practice, elevation patterns of digital TV broadcast antennas may be as broad as 30° to as narrow as 1.5°. The width of the beam is defined as the angular distance between the half-power points. These are the elevation angles at which the field pattern is 0.707 times the maximum field intensity. Figure 7-3 is a plot of an elevation pattern of a typical UHF antenna. The beamwidth is approximately 3.75°. (It is customary to plot the elevation angle on the horizontal axis; for proper orientation, the pattern should be rotated by 90°.) Note that the angular distance between the peak and first nulls is also about equal to the half-power beamwidth.

The beamwidth of the elevation pattern is affected most by the length of the antenna, L_a, and the distribution of current amplitude and phase on the radiating elements. To a first approximation, the half-power beamwidth, Θ_3, of an antenna with a nearly constant aperture distribution may be estimated by

$$\Theta_3 = \frac{57.3\lambda}{L_a}$$

where λ is the wavelength and Θ_3 is in degrees. For the pattern shown in Figure 7-3, the antenna length is approximately 18 wavelengths. This implies an antenna physical length of 22 to 38 ft for this example, depending on the specific channel in the UHF band.

Note that the peak of the beam of the elevation pattern shown in Figure 7-3 is at 0.75° below the 0° reference or the horizontal direction. Referring to Figure 7-2, it is apparent that the radio horizon for the spherical earth is depressed a small angle below the horizontal. To assure maximum field strength on the

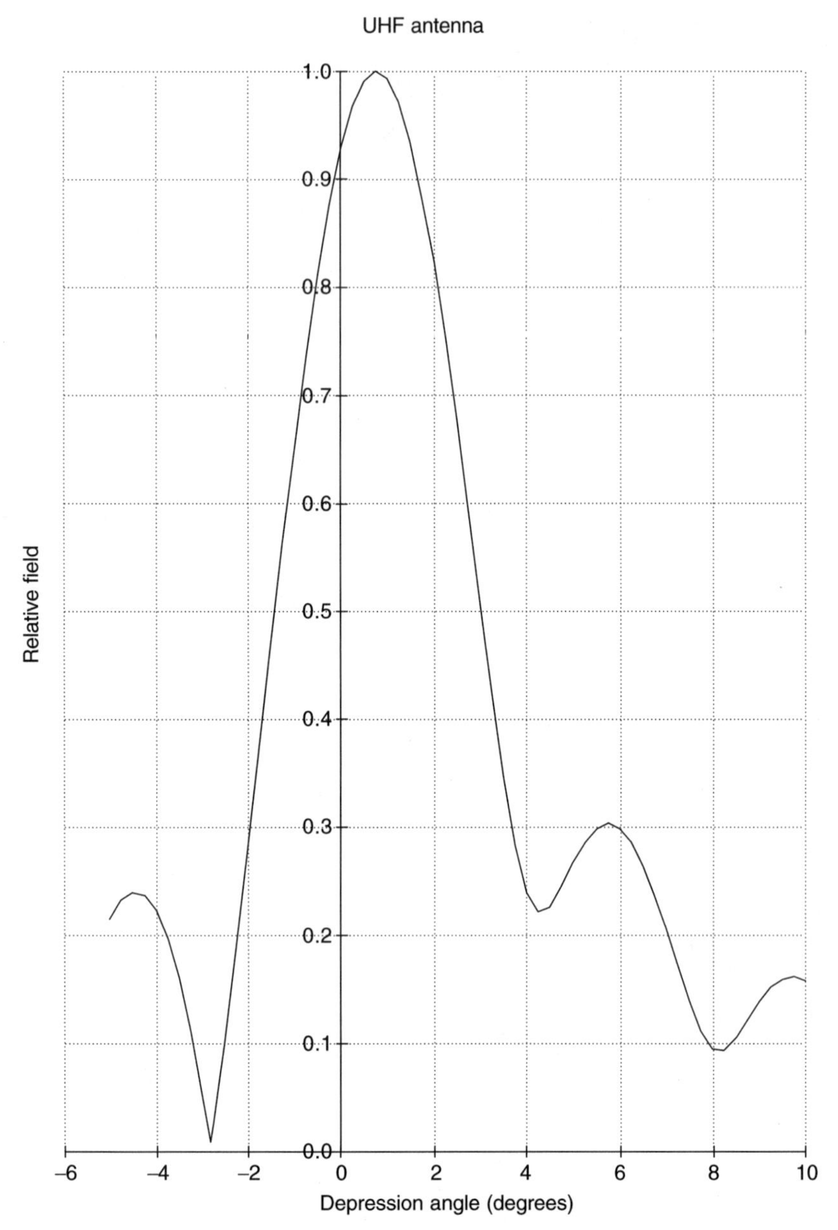

Figure 7-3. Typical elevation pattern.

surface of the earth, it is necessary to tilt the beam downward as shown. The minimum beam tilt angle, Θ_t, is proportional to the square root of the tower height and is given by

$$\Theta_t = 0.0153(h_t)^{1/2}$$

where the tower height is measured in feet. For example, the depression angle as viewed from the top of a 2000-ft tower is approximately 0.7°. It is common practice to tilt the beam downward slightly more than the angle to the radio horizon, say 0.2°, to assure maximum signal level within the coverage area.

Beam tilt may be implemented by either electrical or mechanical means.[1] Mechanical beam tilt results in a directional azimuth pattern. Electrical beam tilt usually produces the same beam direction in all azimuth directions. The peak of the beam is normal to the antenna axis ($\Theta_t = 0$ degrees) if all elements are excited at the same relative phase. This condition is referred to as a uniform phase distribution. Electrical beam tilt can be included as a part of the antenna pattern only by adjusting the phase distribution to the antenna elements. The physical implementation of this adjustment varies depending on the type of antenna design. In any case, the phase must be advanced for antenna elements above the center of the aperture; for elements below the center, the phase must be retarded. The approximate amount of phase advance or retardation from element to element in radians, α_e, is

$$\alpha_e = \frac{2\pi d}{\lambda} \tan \Theta_t$$

where d is the distance between elements. For the example shown in Figure 7-3, the element-to-element phase adjustment is approximately 0.0877 rad. or 5° for an element spacing of one wavelength. Inclusion of beam tilt in the pattern need not appreciably reduce the antenna directivity or gain.

The beam stability is a parameter closely related to beamwidth and beam tilt. This parameter is especially important for UHF antennas, where narrow beam antennas are commonly used. However, it should not be ignored for the wider beam antennas used for VHF. To ensure beam stability, both electrical and mechanical factors must be considered. The antenna must be stiff enough to minimize wind-induced deflections that would change the beam direction as a function of time. In addition, the radiating elements and feed system must produce stable phase as a function of frequency, thereby assuring stable beamwidth over the full 6- to 8-MHz frequency band.

The effect of electrically and mechanically induced phase shifts on beam stability versus frequency and time may be illustrated by considering the radiation pattern of an N-element linear array with uniform amplitude distribution and spacing. Such an end-fed array of isotropic radiating elements is illustrated in Figure 7-4. It can be shown that the array factor,[2] AF, for this configuration may

[1] Y.T. Lo and S.W. Lee, eds., *Antenna Handbook*, Van Nostrand Reinhold, New York, 1988, p. 27–7.

[2] The complete antenna pattern is the product of the array factor and the element pattern. Since the element pattern for broadcast antennas is generally quite broad, it has very little impact on the shape of the elevation pattern for angles of interest. It is, therefore, ignored in this discussion.

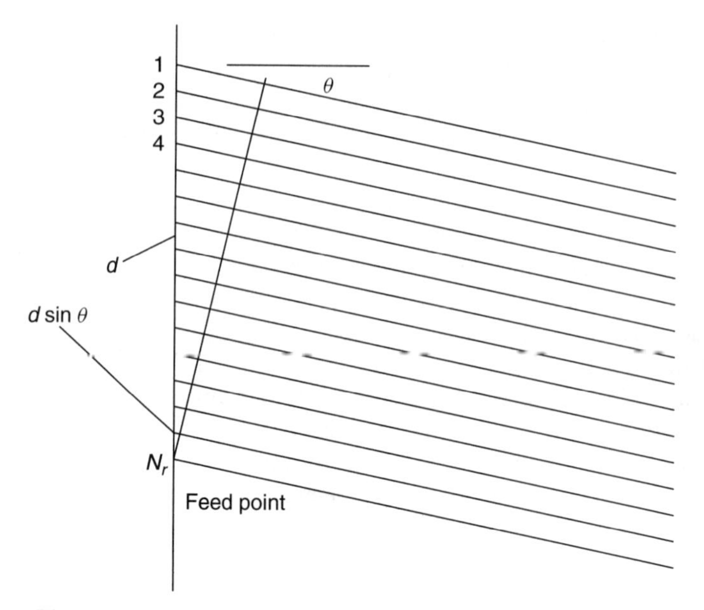

Figure 7-4. Geometry of N_r-element array of isotropic sources.

be written approximately as the familiar sinc function[3]

$$\mathrm{AF} \sim \frac{\sin(N_r \psi/2)}{N_r \psi/2}$$

where N_r is the number of radiating elements, and

$$\psi = \frac{2\pi d}{\lambda} \sin\theta + \alpha_e$$

The radiation pattern for a 30-element array is plotted in Figure 7-5 for the upper and lower frequencies of U.S. channel 14. The variation in beam tilt is evident. This is a consequence of the change in spacing with respect to wavelength between elements and the resulting change in phase shift from element to element. The frequency-response tilt produced as a function of elevation angle is shown in Figure 7-6. Although the response tilt is quite acceptable near the peak of the beam, it increases to unacceptable levels at angles below the main beam, in the null regions, and in much of the sidelobe region.

In practice, the beam direction may be stabilized by using the resonant and reactive properties of slotted elements, thereby compensating for changes in element spacing and associated phase shift. These techniques, although effective, are not perfect, and some residual beam shift versus frequency will be present.

[3] Constantine A. Balanis, *Antenna Theory*, Harper & Row, New York, 1982, p. 214.

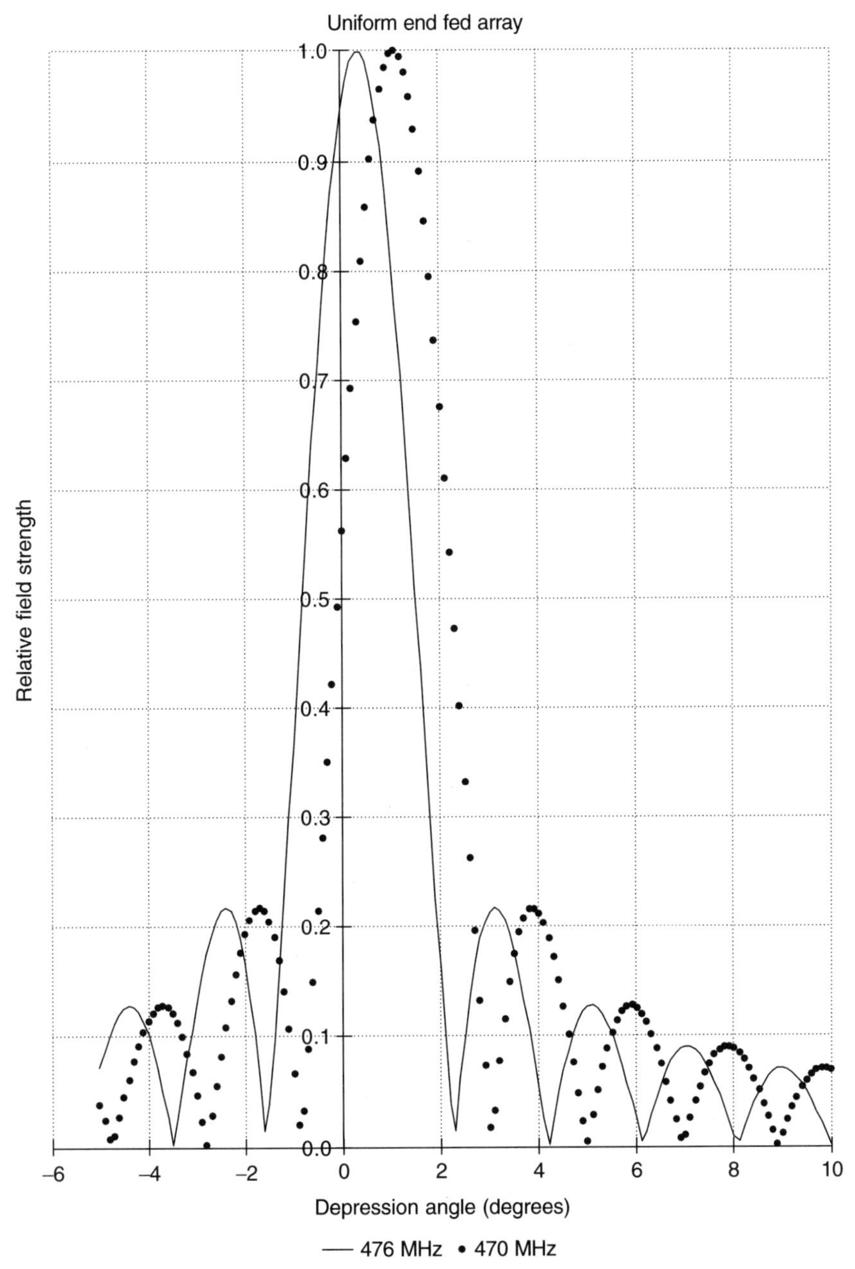

Figure 7-5. Pattern versus frequency.

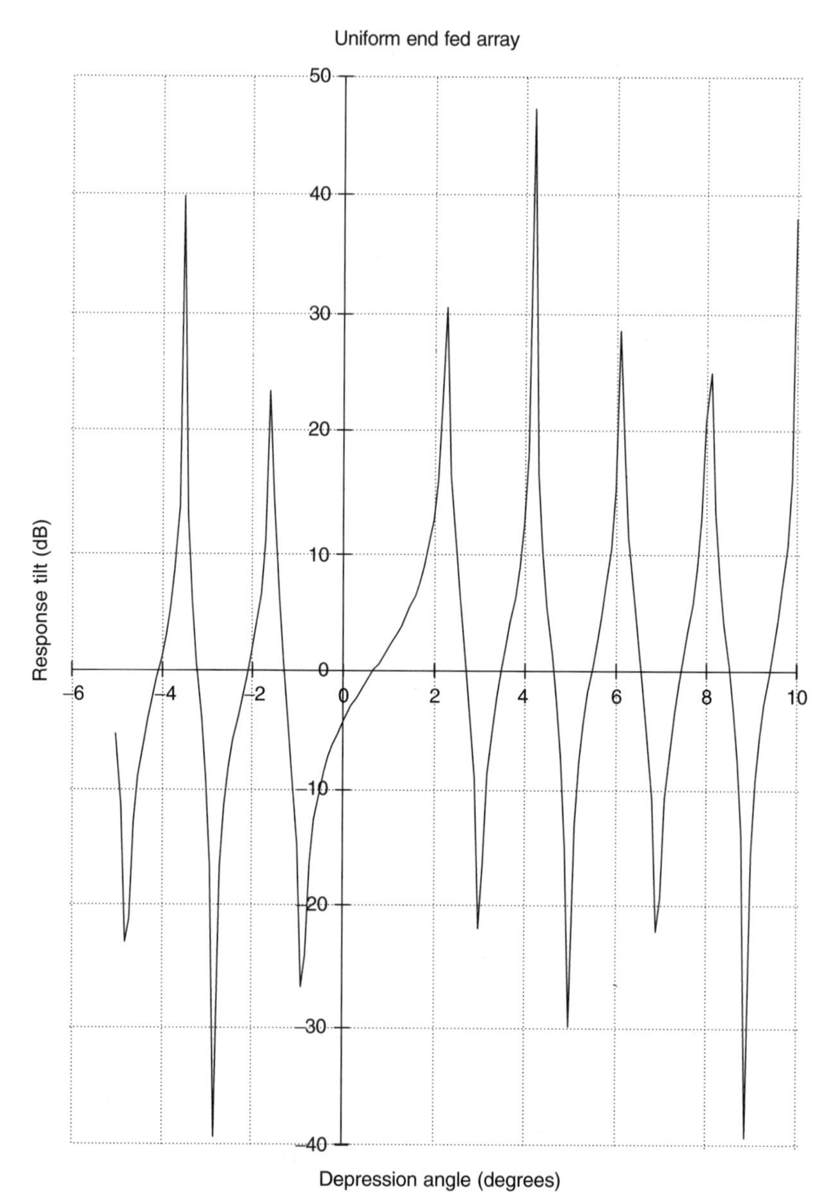

Figure 7-6. Frequency response tilt.

The results obtained by compensating the beam tilt by 0.4° are illustrated in the pattern plots of Figure 7-7 and the plot of response tilt in Figure 7-8. A substantial improvement in response tilt is evident. When null fill is used, the effects of beam shift on response tilt in the null and sidelobe regions are reduced even further. The benefits and means of implementing null fill are discussed later.

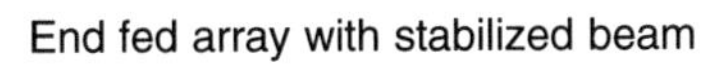

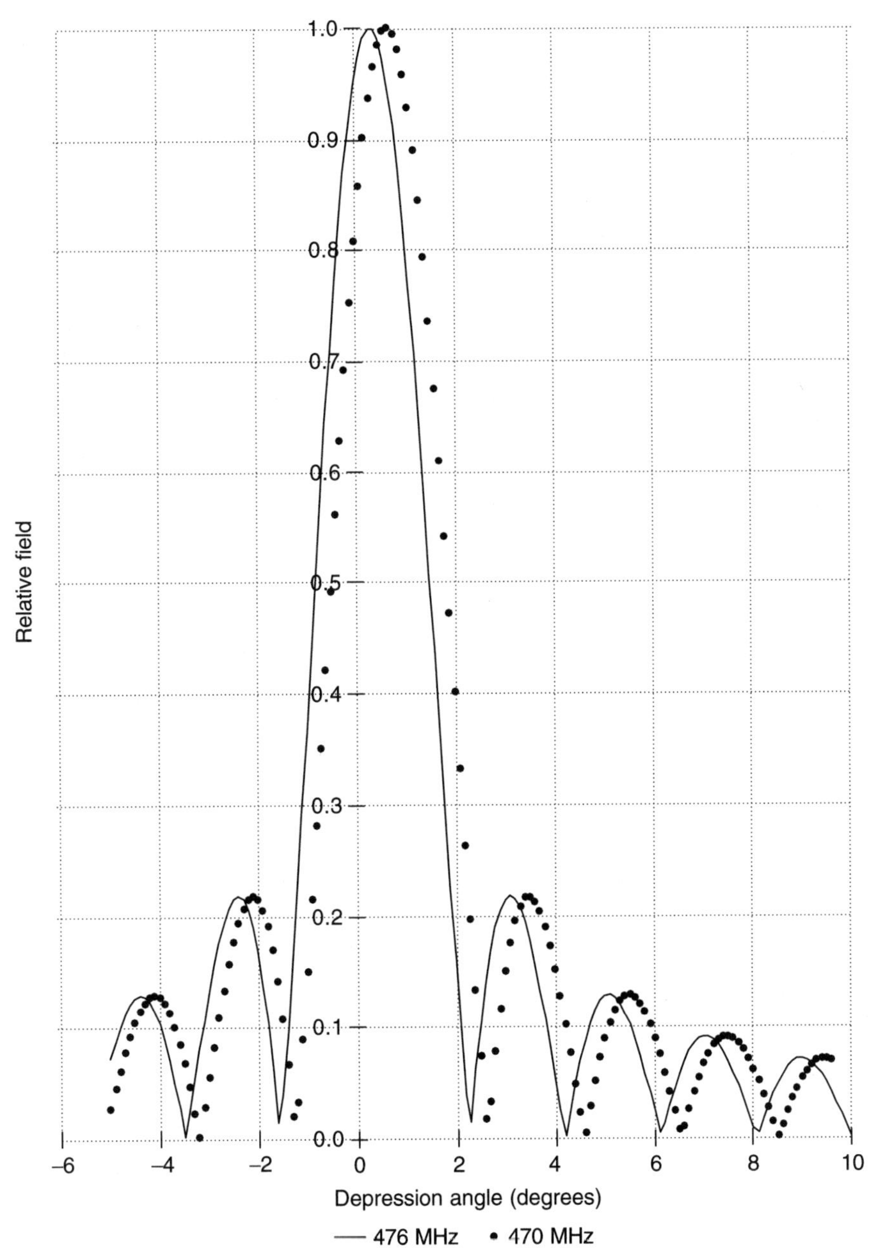

Figure 7-7. Pattern versus frequency.

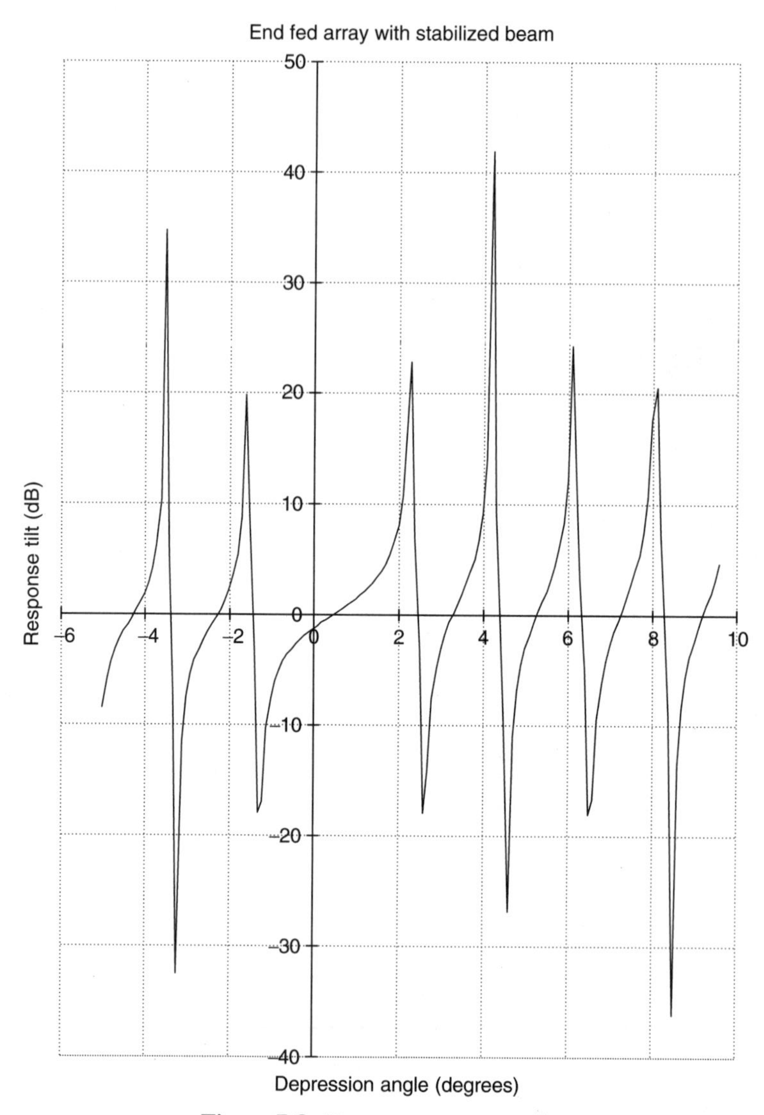

Figure 7-8. Frequency-response tilt.

Another approach to electrical beam stabilization is the use of a center-fed array as depicted in Figure 7-9. If all elements are excited with equal-amplitude currents, the array factor may be written as

$$\mathrm{AF} = \frac{1}{2M}[e^{j\psi/2} + e^{j3\psi/2} + e^{j5\psi/2} + \cdots + e^{j2M-1\psi/2} \\ + e^{-j\psi/2} + e^{-j3\psi/2} + e^{-j5\psi/2} + \cdots + e^{-j(N-1)\psi/2}]$$

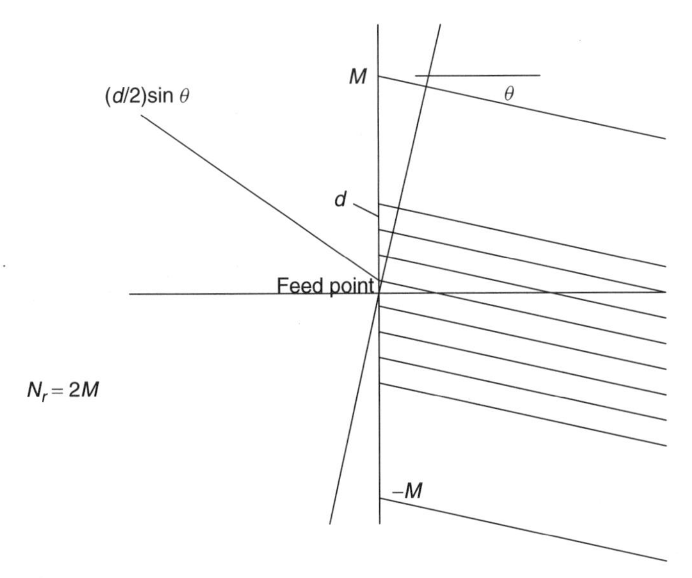

Figure 7-9. Geometry of center-fed array.

where $M = 2N_r$. This expression may be simplified to

$$\mathrm{AF} = \frac{1}{2M}\sum_{n=1}^{M}\left[e^{j(2n-1)\psi/2} + e^{-j(2n-1)\psi/2}\right]$$

which, by Euler's equation, may be rewritten as

$$\mathrm{AF} = \frac{1}{M}\sum_{n=1}^{M}\cos\frac{(2n-1)\psi}{2}$$

The array factor for a 30-element center-fed array ($M = 15$) is plotted in Figure 7-10 for the upper and lower frequencies of U.S. channel 14. The beam shift is approximately one-half that of the uncompensated end-fed array. As a consequence, the response tilt is less than half, as shown in Figure 7-11. As with the end-fed array, the response tilt is quite acceptable near the peak of the beam, but increases to unacceptable levels at angles below the main beam, in the null regions, and much of the sidelobe regions. Thus the center-fed array, while providing improved performance over the end-fed array, does not eliminate the effects of beam shift vs. frequency entirely.

As with the end-fed array, the reactive properties of slot elements and other techniques may be used to compensate for the change in spacing and associated element-to-element phase shift. Since there is less beam tilt to compensate, this compensation can be more effective than for the end-fed array. The computed results obtained by compensating the phase shift of a center-fed array are

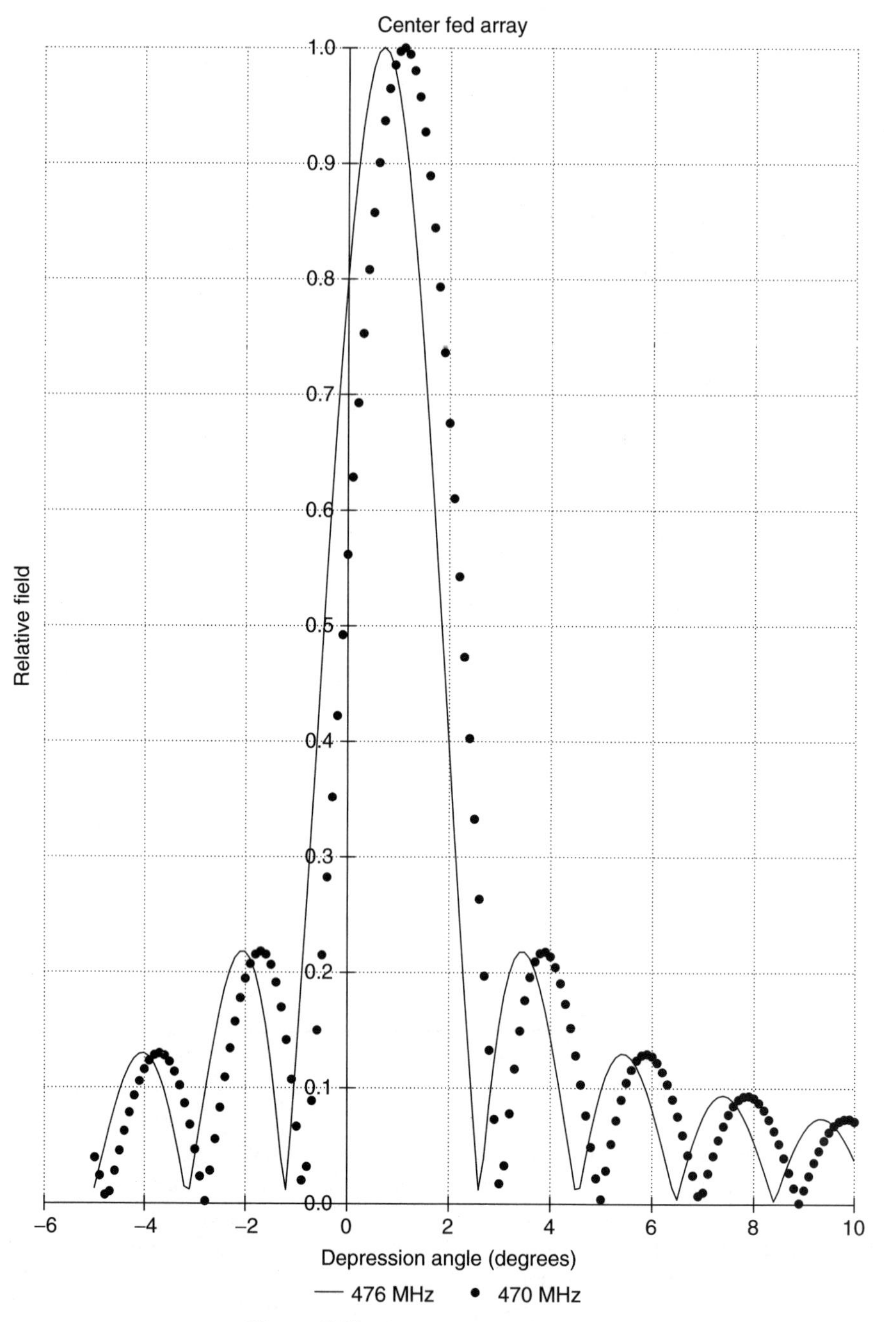

Figure 7-10. Pattern versus frequency.

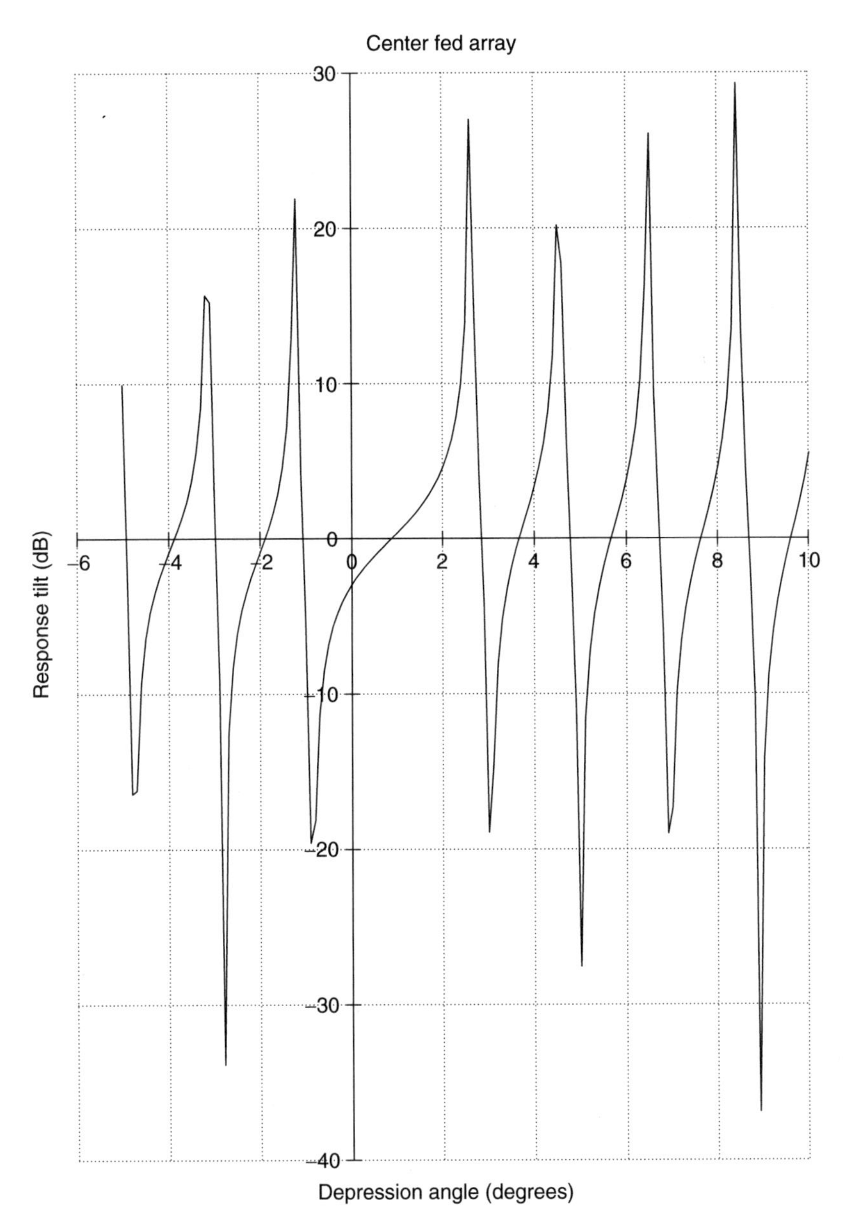

Figure 7-11. Frequency response tilt.

illustrated in Figures 7-12 and 7-13. The pattern differences are due primarily to the change in beamwidth due to the change in wavelength. The response tilt is quite acceptable except in the null regions. Center feeding plus the use of null fill reduces array response tilt to an acceptable level at all elevation angles.

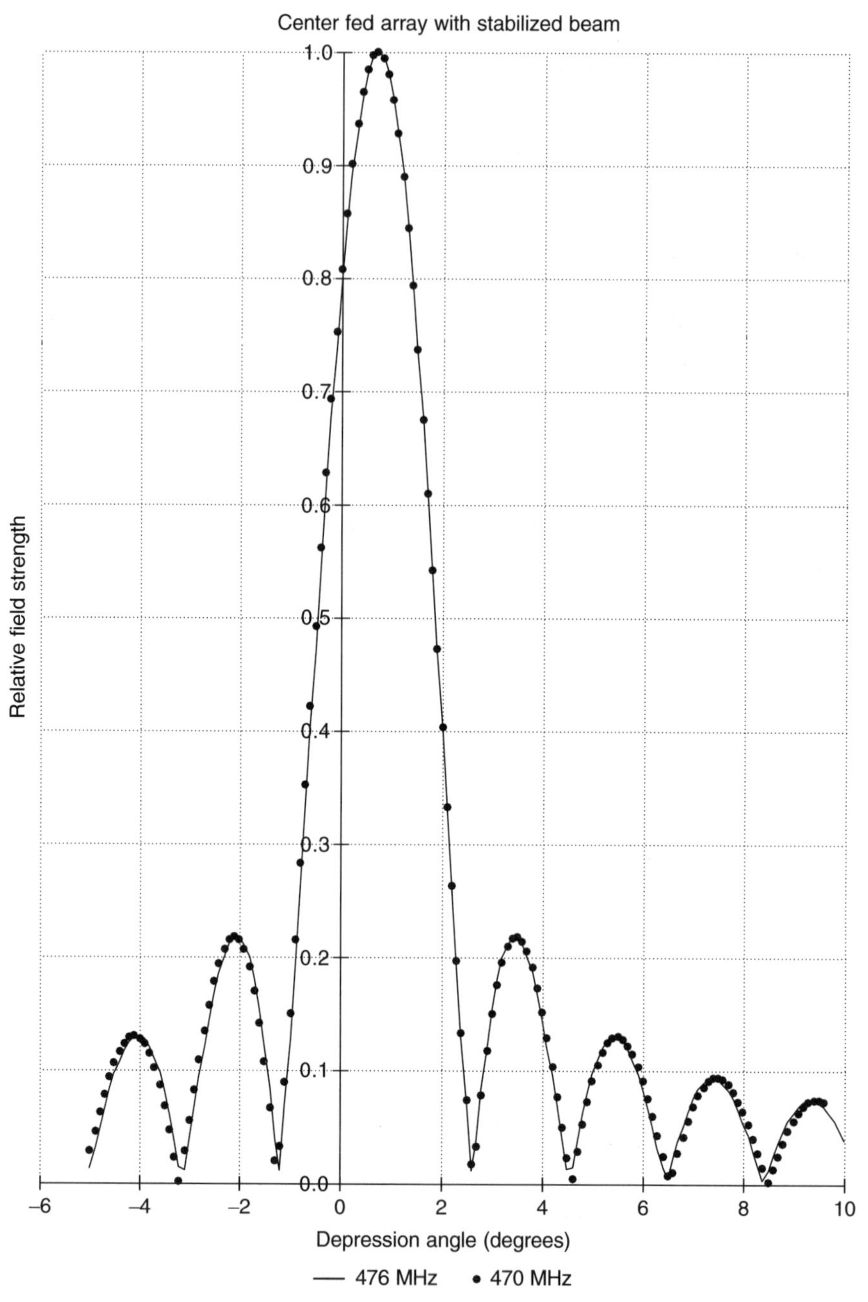

Figure 7-12. Pattern versus frequency.

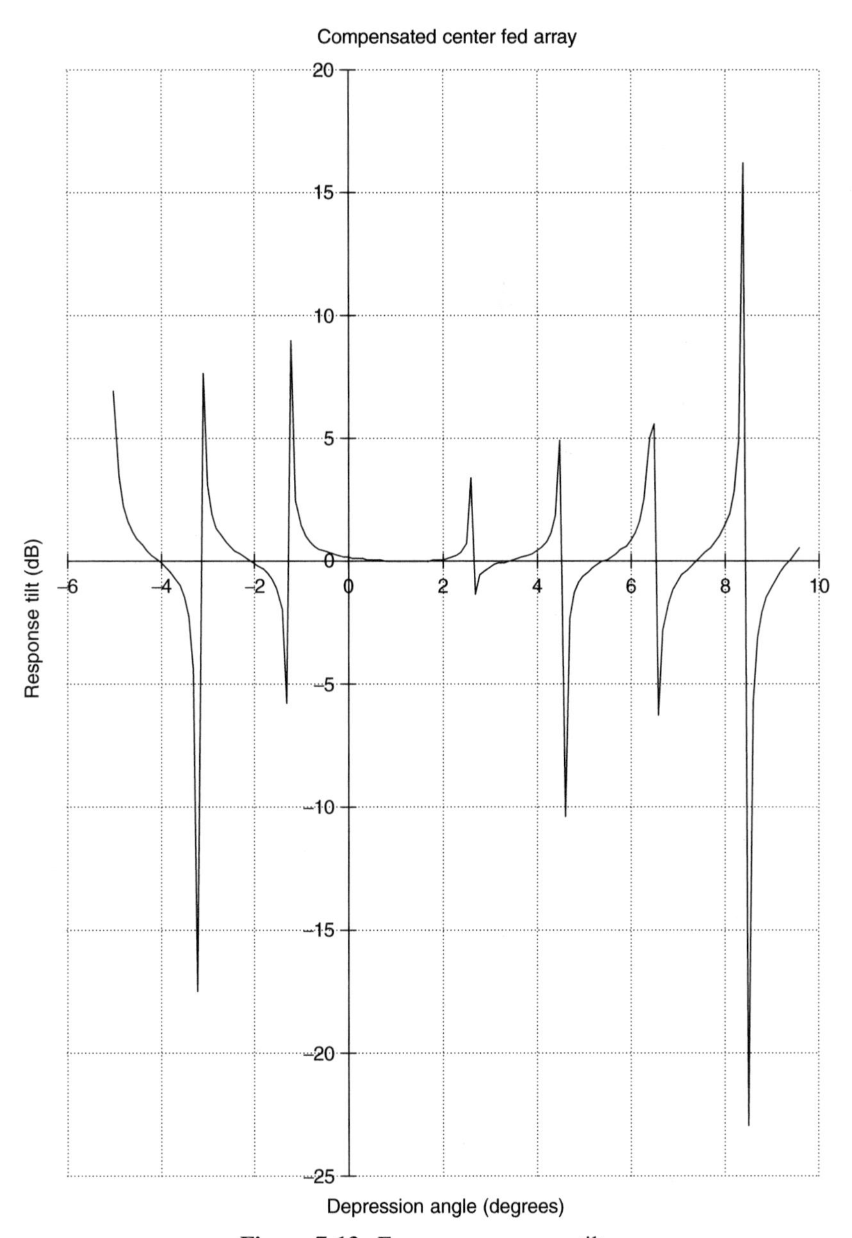

Figure 7-13. Frequency-response tilt.

Although it is not evident in the patterns as plotted, 180° phase changes occur in the null regions of both the end- and center-fed arrays, so that each lobe is out of phase with each of its adjacent neighbors. When these phase changes are considered along with pattern amplitude and beamwidth changes with frequency, the result is a nonlinear phase change versus frequency or group delay. Therefore, means must be found to reduce both linear distortions, amplitude and phase, to acceptable levels.

MECHANICAL STABILITY

The beam direction must also be stable with respect to time, so that both the signal strength and the frequency response are relatively constant. This implies a certain degree of mechanical stiffness in the structural design so that the antenna is stable under wind load. If the deformation of the structure is known, this can be translated to an equivalent nonuniform phase distribution and subsequent changes in beam direction.

Deformation of the antenna structure under wind load is difficult to generalize, since structural designs tend to vary widely depending on electrical and mechanical requirements. To reduce weight and wind load, load-bearing members are usually larger near the base and smaller near the top of the structure. For this type of design, the actual structure must be evaluated to determine the amount of deflection at any specific point on the structure due to wind.

NULL FILL

Null fill is used to assure solid near-in coverage and to mitigate the effects of variations in beam direction for broadcast arrays. As has been shown in the computed patterns (Figures 7-5, 7-7, 7-10, and 7-12), for a uniform amplitude and phase current distribution, the radiated signal will precisely cancel at certain angles, periodically producing nulls or zeros in the pattern. If the antenna is located a substantial distance from populated areas and close-in coverage is not important, this may be acceptable, even for narrow beam antennas. It also may be acceptable in the case of VHF antennas, for which the beam is very broad. However, for moderate- to high-gain antennas located close to receiving locations, near-in coverage is important and null fill is usually necessary. Null fill is evident in the elevation pattern shown in Figure 7-3. The first null is filled to a level of 22%; the second, to a level of 9%. Common amounts of first null fill range from 5 to 35%. Unlike beam tilt, inclusion of null fill in the elevation pattern reduces the antenna directivity and gain in proportion to the null fill. This is illustrated in Figure 7-14, which shows the gain of a typical six-element antenna as a function of null fill. Directivity and gain are discussed in greater detail later.

Implementation of null fill can be accomplished by making adjustments in the antenna current amplitude or phase distribution or both. The results achieved depends on the distribution used. One way is to feed the elements of the array with a non-constant-amplitude distribution. This results in incomplete field cancellation in the null regions of the pattern. There are many variations on this theme. These variations include excitation of the upper and lower halves of

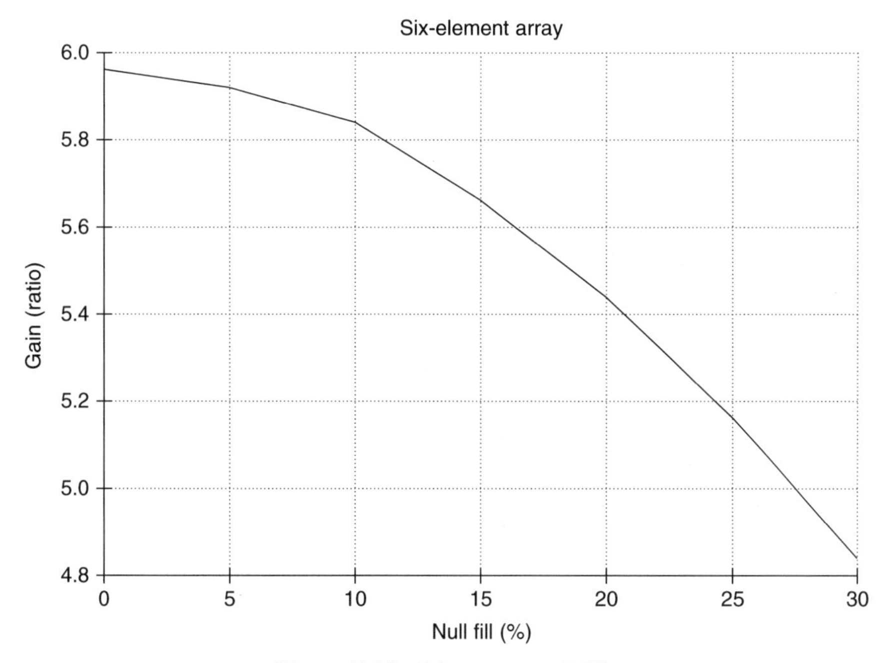

Figure 7-14. Gain versus null fill.

the array with different but constant current amplitudes or use of an exponential distributions. A second method uses a parabolic phase distribution over the length of the antenna. These phase and amplitude distributions may also be combined. A more general method makes use of a technique called pattern synthesis. This technique begins with the desired far-field pattern to compute the required phase and amplitude distributions.

To illustrate the use of non-constant-amplitude distribution to obtain null fill, consider the N-element center-fed array with an exponential amplitude distribution. The array geometry is the same as for the center-fed array shown in Figure 7-9. The only difference is that the amplitude of each element above the center is reduced from the amplitude of its next-lower adjacent neighbor by a fixed percentage; the amplitudes of the elements in the lower half of the array are similarly tapered. The resulting array factor is[4]

$$\mathrm{AF} = \frac{1}{M}\sum_{n=1}^{M} A_n \cos\frac{(2n-1)\psi}{2}$$

The patterns computed at the lower and upper frequency limits of a 30-element array for U.S. channel 14 are shown in Figure 7-15. In this example, the current amplitude

[4] Balanis, op. cit., p. 242.

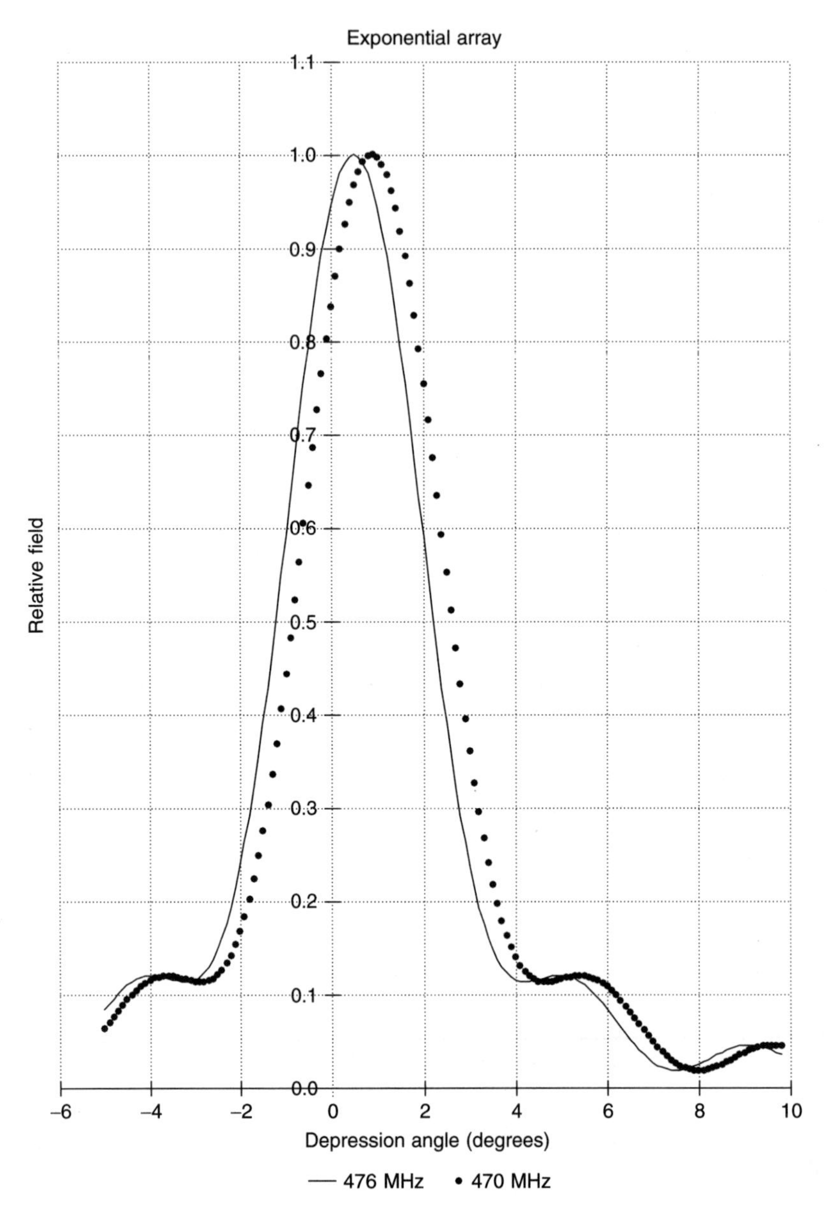

Figure 7-15. Pattern versus frequency.

of the element $n + 1$ is reduced from that of the element n by 1.3 dB. Thus the amplitude taper for the entire array is approximately 20 dB with the maximum level in the center. The result is a very smooth pattern in the null and sidelobe regions. Despite evident beam tilt variation, the response tilt is moderate to low as shown in Figure 7-16. If the beam is stabilized by careful radiating element design, the

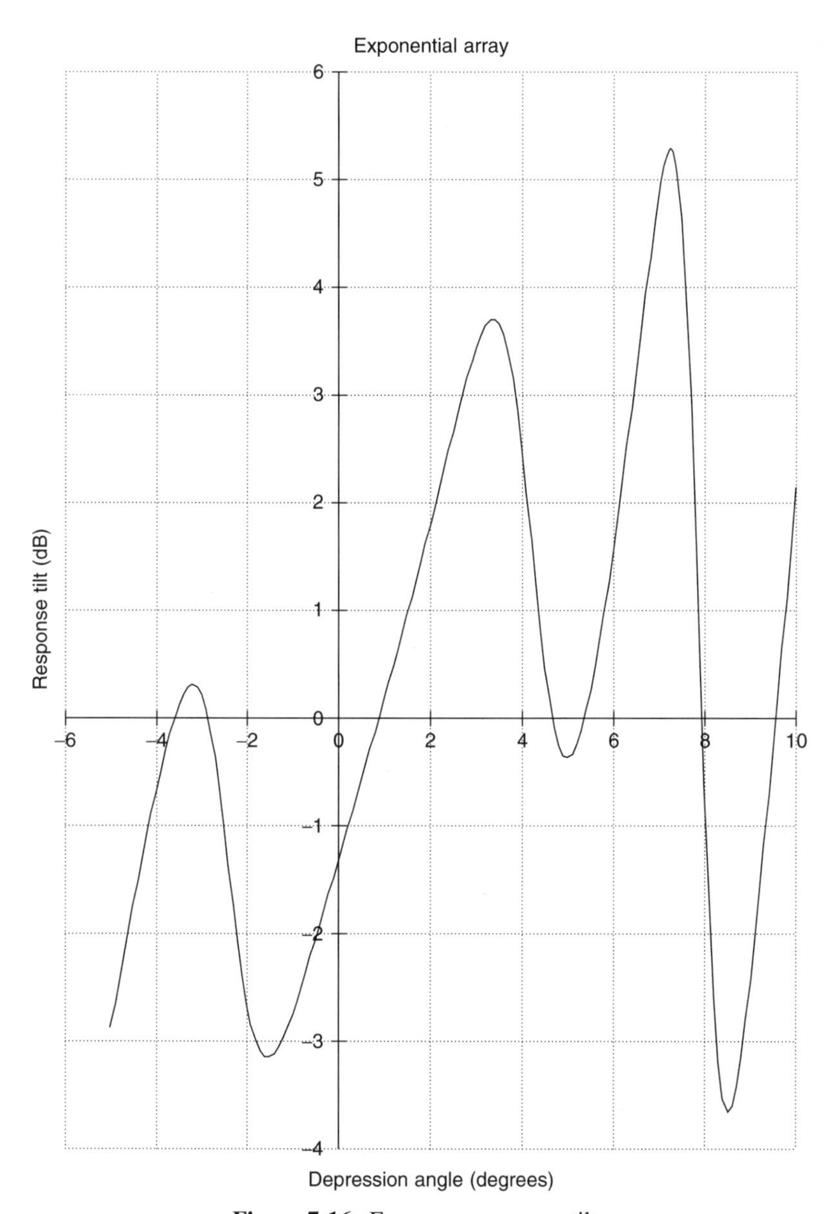

Figure 7-16. Frequency-response tilt.

patterns at opposite ends of the channel may be made almost identical, as seen in Figure 7-17. The pattern differences are due primarily to the change in beamwidth resulting from changes in wavelength. The resulting frequency response tilt is very low for depression angles of interest, as shown in Figure 7-18.

Another benefit of the exponential distribution is the possibility of the absence of phase changes between adjacent lobes in the far-field pattern. As a result there

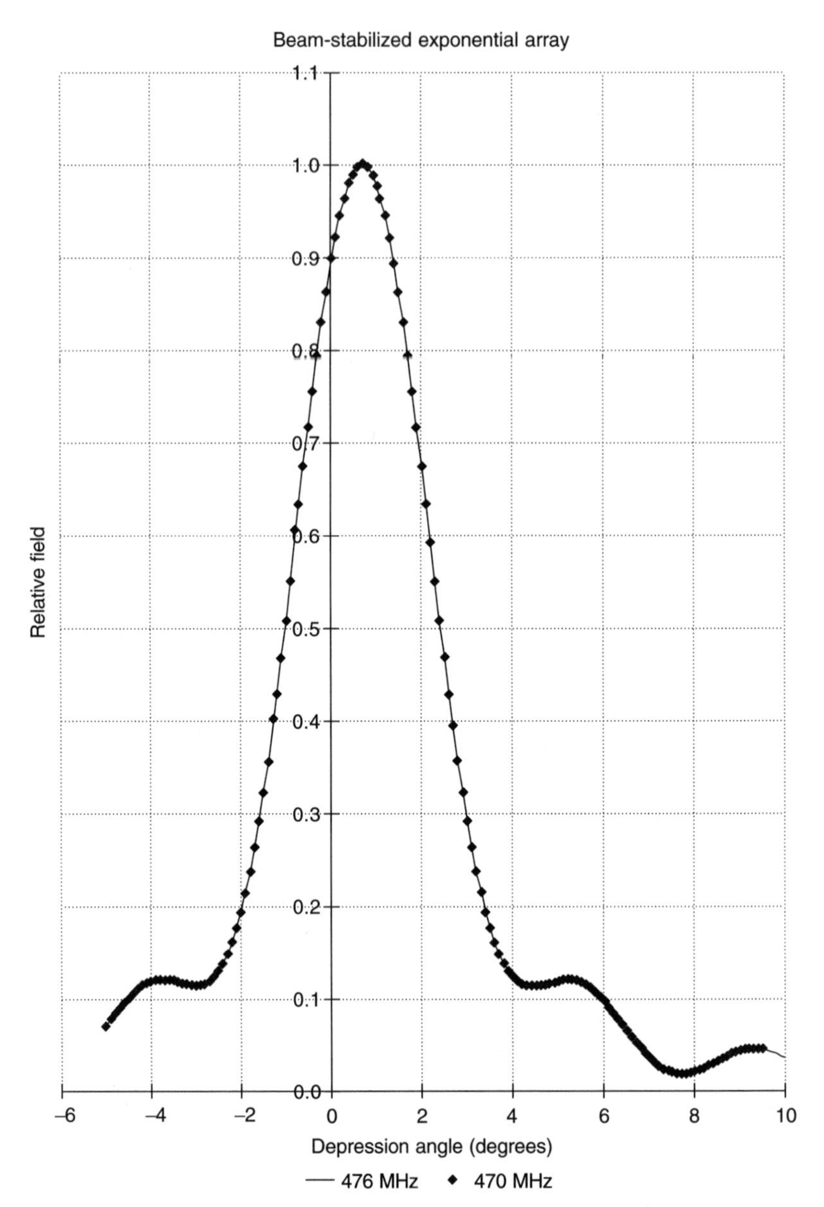

Figure 7-17. Pattern versus frequency.

is very little nonlinear phase versus frequency and group delay, even in a practical embodiment of this design. The absence of phase changes is dependent on the method of obtaining null fill and the amount of taper in the aperture distribution. Even the exponential distribution does not produce a pattern free of phase changes unless the aperture amplitude taper is greater than a minimum

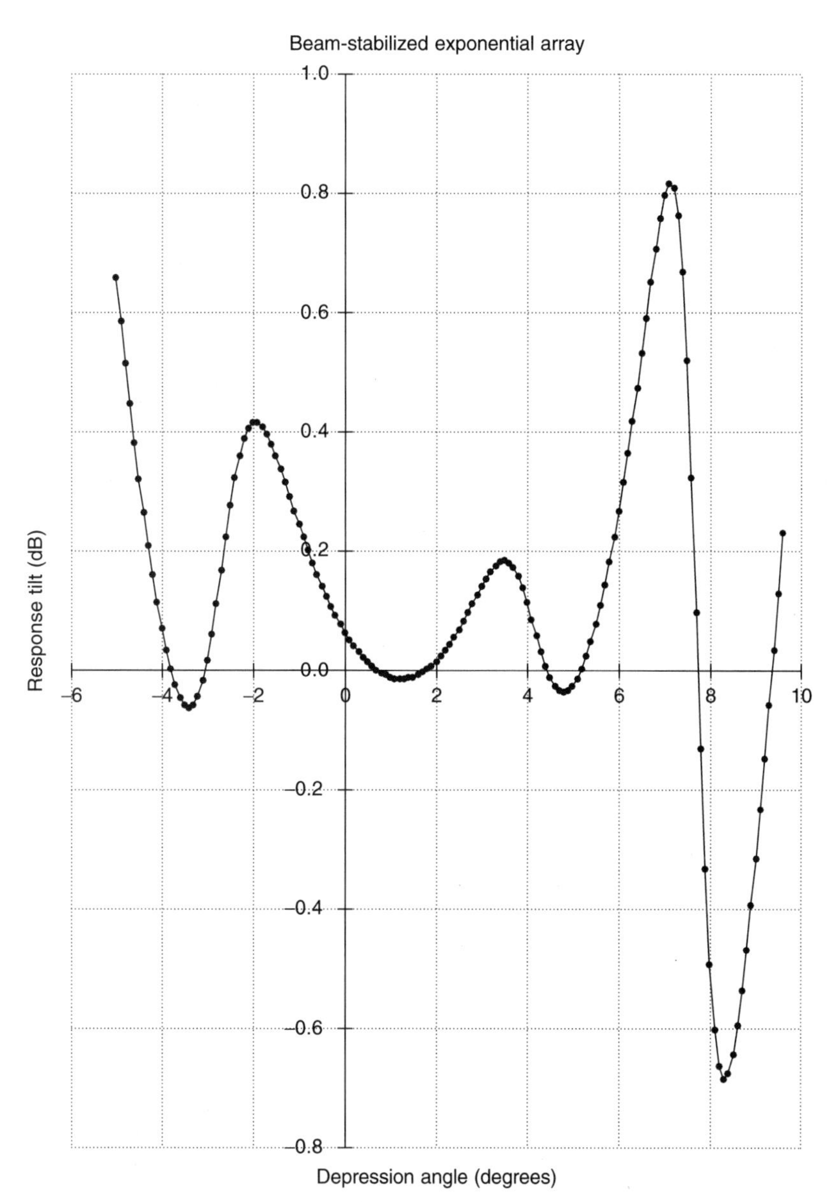

Figure 7-18. Frequency-response tilt.

value. For the example cited, if the relative excitation of adjacent elements were 0.7 dB instead of 1.3 dB, a phase change would be present between adjacent lobes.

The quest for beam stability, good frequency response, and solid near-in coverage has led from a discussion of end-fed arrays to the phase compensation of radiating elements to center-fed arrays and finally to the use of null fill.

Although these techniques are effective and are in widespread use, the benefits do not come without costs. Phase compensation involves some small increase in antenna complexity. Use of center-fed arrays for top-mounted slot antennas requires the use of a triaxial transmission line in the bottom half of the antenna to accommodate the feed system and the radiating elements.[5] Use of nonuniform aperture distributions to obtain null fill also has the effect of increasing the beamwidth, which in turn reduces directivity and gain. Comparing the patterns of the end-fed array with those of the exponential array, it is evident that the 3-dB beamwidth has increased from 1.7° to about 2.5°, despite the fact that both antennas are of the same length. This represents a substantial decrease in directivity.

Antenna gain is an important specification in achieving the desired AERP. If it is desired to achieve the directivity of the end-fed array while providing the near-in coverage and frequency response of the exponential array, the antenna length must be increased. The increase in length is proportional to the ratio of the beamwidths. For the example under discussion, the antenna length must be increased to about 44 wavelengths to maintain a beamwidth of 1.7°. Both the acquisition cost and antenna wind load can be expected to increased in proportion to the antenna length. Depending on tower structural capacity, the additional loads could have an impact on tower design and cost. Thus all relevant parameters must be evaluated carefully when specifying an antenna design.

AZIMUTH PATTERN

The shape of the horizontal or azimuth pattern is just as important as the vertical or elevation pattern. If the coverage area is concentrated in one or more distinct directions, a cardioid, peanut, or trilobe directional pattern might be used. Each of these provide meaningful directive gain and can help to reduce the TPO required for the desired coverage. On the other hand, if population is more or less evenly distributed around the tower, an omnidirectional pattern is usually best.

The shape of the azimuth pattern is dependent on many factors. These factors include the number, location, and type of radiating element used as well as the amplitude and phases of the excitation currents. The most common broadcast antennas are comprised of one, two, three, or four radiating elements around the axis of the antenna. For antennas of only one element, the array pattern is reduced to that of the radiating element. For all other values of N, both the array factor and the element factor must be considered to determine the complete azimuth pattern. Unlike the elevation pattern, for which only the pattern at small angles is important, the azimuth pattern is of interest for all angles. Because the element pattern is quite broad in the azimuth plane, it is necessary to know both components of the pattern and perform pattern multiplication to determine

[5] Ernest H. Mayberry, "Slotted Cylinder Antenna Design Considerations for DTV," *NAB Broadcast Engineering Proceedings*, 1998, pp. 33–39.

the complete pattern. The array factor and the element pattern will be examined separately in order to understand the role of each.

For most broadcast antennas, the array factor for the azimuth pattern is that of a circular array of radius a, with N isotropic, equally spaced radiators given by[6]

$$\mathrm{AF} = \sum_{n=1}^{N_r} I_n e^{j[ka \sin\theta' \cos(\phi - \phi_n) + \alpha_n]}$$

where I_n is the amplitude of the current exciting the nth element, α_n is the phase of this current relative to the center of the array, and ϕ_n is the angular position of the nth element, equal to $2\pi n/N_r$.

The azimuth pattern is of interest primarily near the horizon so that $\theta' = 90 \pm 5^\circ$ and $\sin\theta' \sim 1$. For omnidirectional antennas, the current amplitudes and phase are equal. With these simplifications and normalization of the pattern, the array factor may be written as

$$\mathrm{AF} = \frac{1}{N_r} \sum_{n=1}^{N_r} e^{j[ka \cos(\phi - \phi_n)]}$$

In this expression, the array factor is dependent only on the size of the array and the number of array elements. In the limit when a approaches zero, this expression approaches a constant, $1/N_r$; the azimuth pattern is independent of the angle. Obviously, an antenna of zero radius is not physically realizable. However, this is the condition for achieving a perfect omnidirectional pattern. This is one reason that the azimuth patterns of all practical "omni" antennas deviate somewhat from the ideal.

The array factor for two-, three-, and four-element circular arrays, each with a diameter of 0.5 wavelength, are shown in Figures 7-19, 7-20, and 7-21. Since each array has a diameter greater than zero, the azimuth patterns are not perfectly circular. The deviation from a perfect circle, or the circularity, is less as the number of elements increases. This is shown clearly in Figure 7-22, which includes a plot of the peak-to-RMS, value for each array as a function of array size. Up to a critical radius of about 0.3 wavelength, the peak-to-RMS ratio increases directly with array radius. In every case the peak-to-RMS value is less for larger N. The ratio of the RMS level to the null level shows a similar increase. Taken together, these two ratios define the pattern circularity. The critical radius at which the circularity reaches a maximum indicates an abrupt change in the pattern shape. At this radius, an additional lobe appears in the pattern. This is illustrated in Figure 7-23, a plot for an array with diameter equal to $\frac{3}{4}$ wavelength. Thus it is seen that even though an antenna pattern is said to be omnidirectional, it has directional properties.

[6] Balanis, op. cit., p. 275.

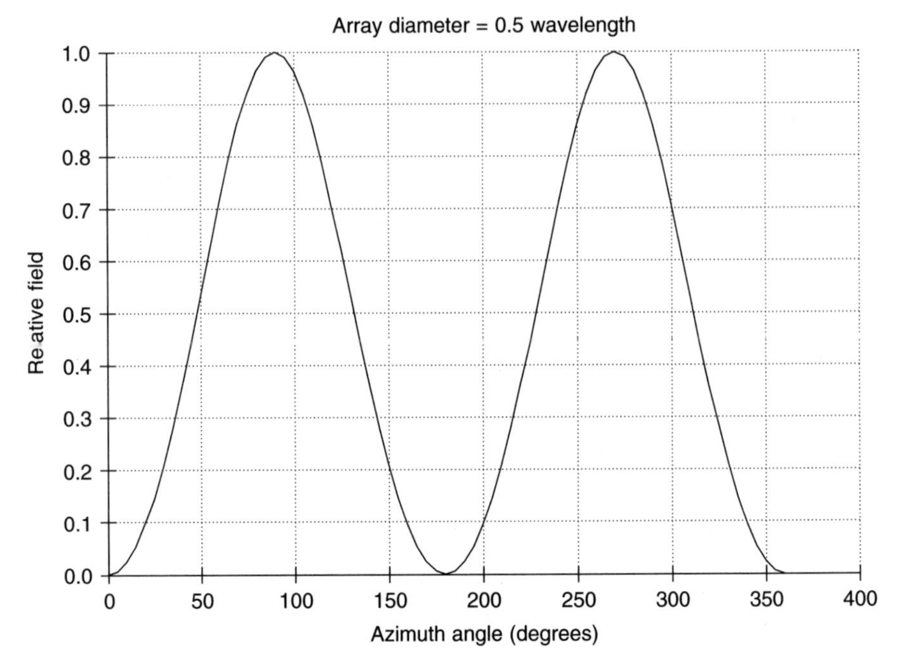

Figure 7-19. Two-around array factor.

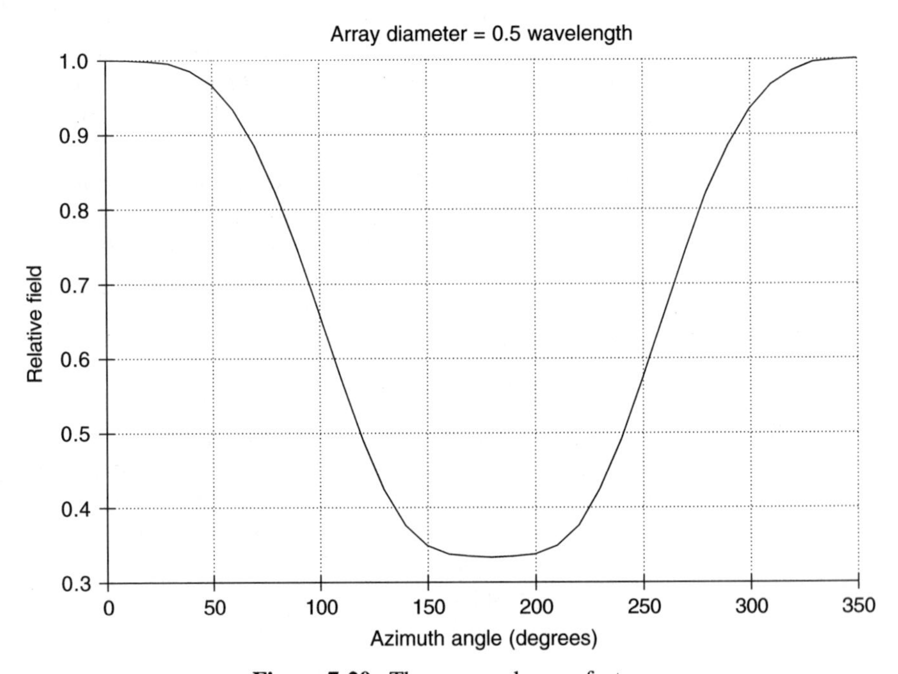

Figure 7-20. Three-around array factor.

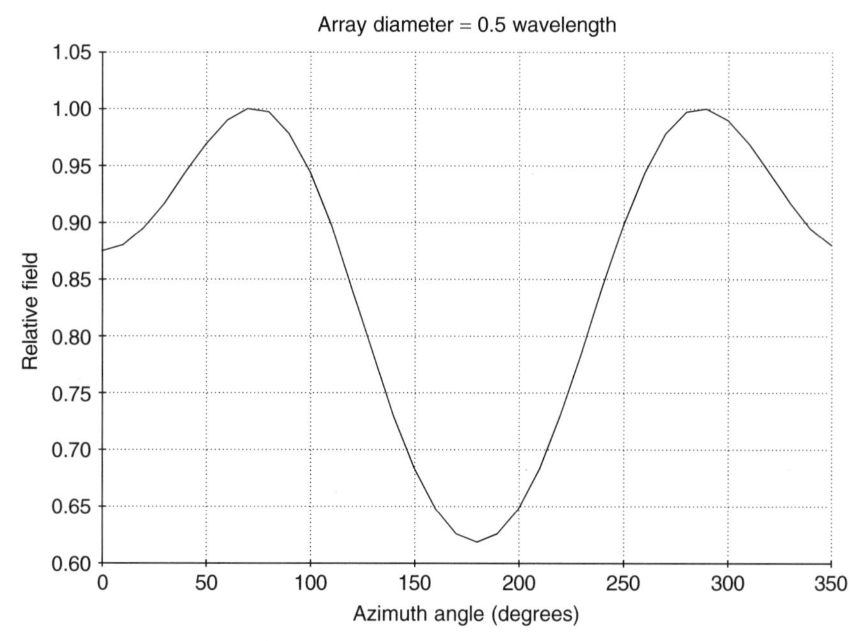

Figure 7-21. Four-around array factor.

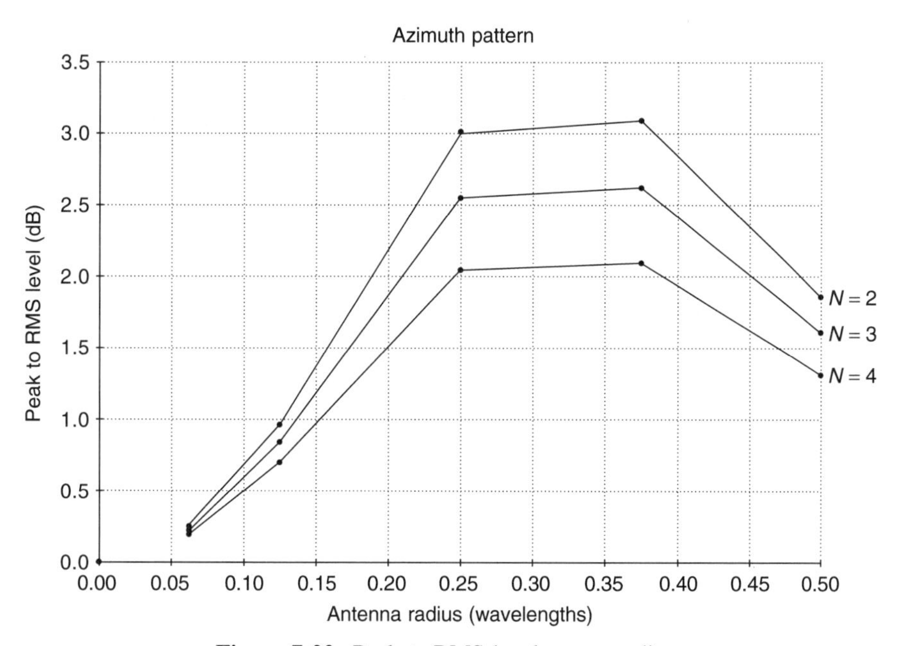

Figure 7-22. Peak-to-RMS level versus radius.

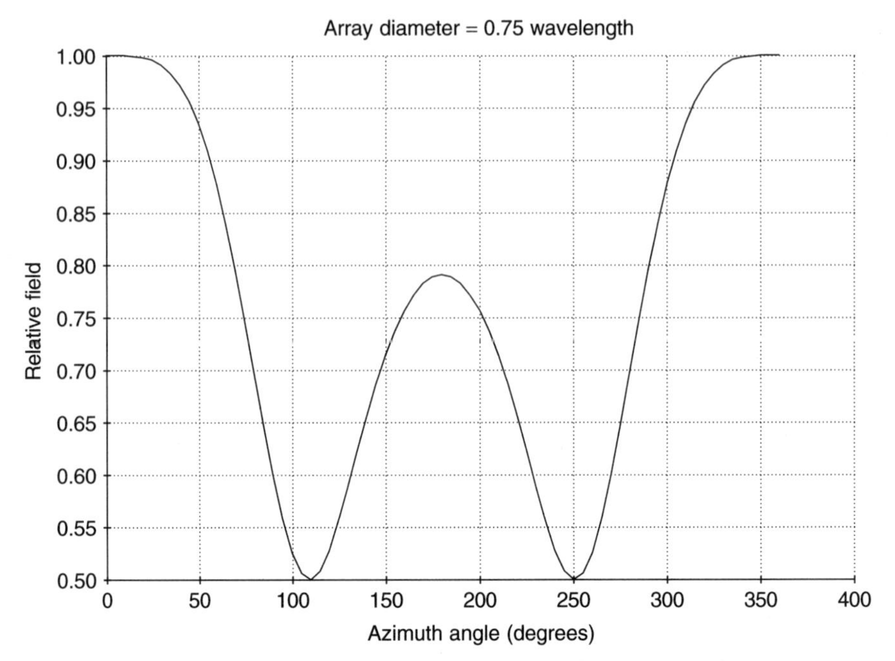

Figure 7-23. Four-around array factor.

Depending on station requirements, the directional characteristic of a nominally omnidirectional antenna may be undesirable, or it may be used to enhanced coverage. For NTSC transmission in the United States, there is no mandated specification for pattern circularity or orientation of omni antennas. If there is a preferred direction, the peaks of the omni pattern may be oriented toward this azimuth and a stronger signal provided. For example, if the circularity of the horizontal pattern was ± 2 dB, the signal strength in one or more preferred directions might be increased by up to 2 dB. The disadvantage was that there might be some directions in which the actual field strength was reduced. Whether or not the FCC will allow this practice to continue for DTV is not clear at the time of this writing.

Although the array factor has a major impact on the shape of the azimuth pattern, the element pattern must also be considered to determine the complete directional characteristic. Radiating elements for broadcast applications are most often electrically small, center-fed antennas. These include dipoles of various designs and resonant slots. Dipole elements are mounted over ground planes or in cavities. Slot radiators are usually cut in the surface of a cylindrical pipe.

To illustrate the effect of the radiating element on the azimuth pattern of a circular array, consider a center-fed dipole. This antenna may be thought of as an open-circuited parallel-wire transmission line. Thus a sinusoidal current distribution with zero current at the ends may be assumed when computing the radiation pattern and impedance. This assumption is exactly true for infinitely thin wires, but it is also a reasonable assumption for the plates, rods, and wires of finite size used to build practical antennas. A center-fed dipole in free space produces maximum radiation in the plane normal to its axis. This is a consequence of the symmetrical sinusoidal current distribution, that is,

$$I = \begin{cases} I_m \sin\left[\beta\left(\dfrac{l}{2} - z\right)\right] e^{j\omega t} & \text{for } z > 0 \\ I_m \sin\left[\beta\left(\dfrac{l}{2} + z\right)\right] e^{j\omega t} & \text{for } z < 0 \end{cases}$$

where I_m is the value of the maximum current and l is the overall dipole length.

Generally, there is interest only in the distant field patterns. The far electric field pattern, E_θ, for a thin vertically polarized dipole at a distant point is[7]

$$E_\theta = \frac{\cos[(\beta l/2)\cos\theta - \cos\beta l/2]}{\sin\theta}$$

This is related to the magnetic intensity, H_ϕ, of the radiated field by

$$E_\theta = \zeta_0 H_\phi$$

where ζ_0 is the characteristic impedance of free space. Thus the magnetic field pattern is of the same shape as the electric field. Note that E_θ and H_ϕ are also in phase. Computed patterns in the plane of the electric field for dipoles of two lengths of interest are shown in Figure 7-24. The 3-dB beamwidth varies from 90° to 47.8° for lengths varying from very short to a full wavelength. Dipoles shorter than a halfwave in length exhibit a beamwidth not much different than that of a half-wave dipole. Each lobe is of opposite sign, indicating that they are out of phase. Although the pattern of a dipole in free space is of some interest for broadcast applications, the more usual interest is in radiating elements with a single lobe, such as a dipole over a ground plane or a dipole exciting a cavity. To illustrate, consider the dipole over a ground plane.

For analytical purposes, a dipole at a distance, h, above and parallel to an infinite ground plane may be approximated by a pair of dipoles fed out of phase and separated by a distance, $2h$. In this case, the pattern is determined by multiplying the pattern of the dipole in free space by[8]

$$\cos\left(\frac{2\pi h}{\lambda}\sin\theta'\sin\phi - \frac{\pi}{2}\right)$$

so that the complete pattern is

$$E_\theta = \frac{\cos[(\beta l/2)\cos\theta - \cos\beta l/2]}{\sin\theta}\cos\left(\frac{2\pi h}{\lambda}\sin\theta'\sin\phi - \frac{\pi}{2}\right)$$

[7] Balanis, op. cit., p. 120.

[8] Samuel Silver, *Microwave Antenna Theory and Design*, Boston Technical Publishers, Lexington, Mass., 1963, p. 102.

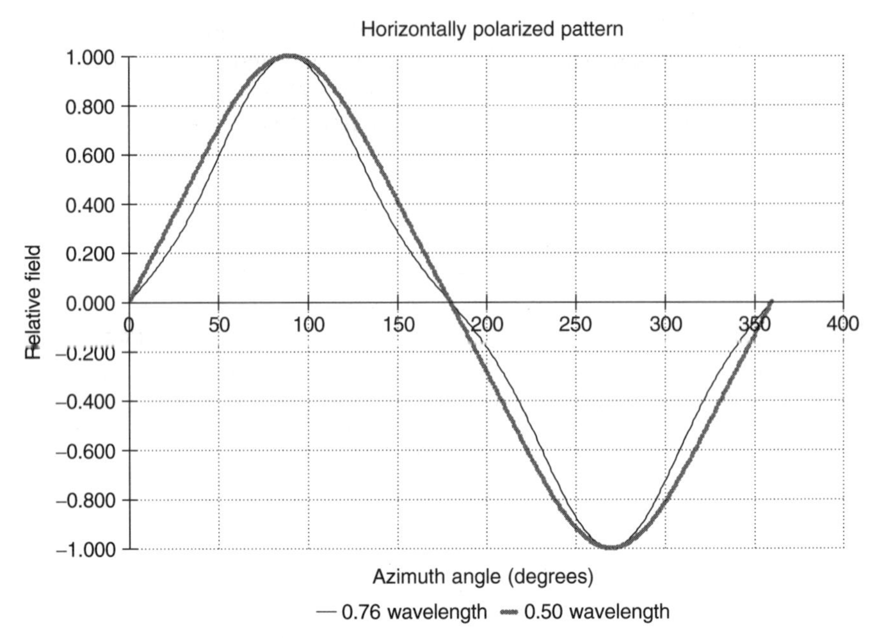

Figure 7-24. Dipole in free space.

Although this equation was derived for a vertically polarized dipole above ground, it may be used to compute the azimuth pattern for a horizontally polarized antenna in the principal plane. To do so it is only necessary to set $\phi = 90^\circ$ and consider θ' to be the azimuth angle. Patterns of dipoles a quarter-wavelength over a ground plane are shown in Figure 7-25. The beamwidth variation is somewhat less than that of the dipole in free space, ranging from 81° for a very short dipole to 50° for a full-wave dipole. As with dipoles in free space, dipoles over ground of length equal to a half wave and shorter exhibit a beamwidth not much different than for that of a half-wave dipole.

The computed patterns indicate the presence of two lobes, each normal to the ground plane. Both lobes are present in the mathematical formula, but only one exists physically since the ground plane shields the dipole from the rear half-plane. Unlike the dipole in free space, the horizontally polarized lobes of the dipole over ground are of the same sign, indicating that they are in phase. This is a consequence of multiplying the dipole pattern by an array factor, both factors being cosine functions. When computing the complete patterns of a circular array, this in-phase condition is quite useful. The computed pattern is equivalent to a pair of back-to-back dipoles over ground, so it may be used "as is" to compute the pattern of a two-around array. A pair of these patterns may be pointed at right angles to each other and added to produce the element pattern for a four-around array.

Patterns of a vertically polarized dipole over ground may be computed from the same equation. In this case, θ' is set to 90° and ϕ is the usual azimuth angle. Since the azimuth pattern of a vertically polarized dipole in free space

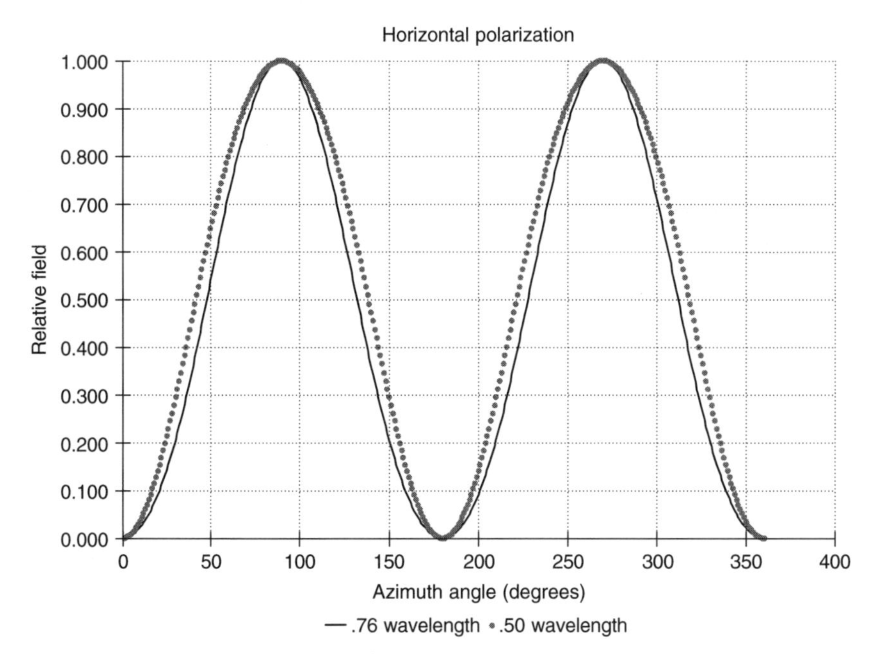

Figure 7-25. Dipole over ground.

is omnidirectional, the resulting pattern is just the array factor of the dual-dipole array. The result for a dipole one quarter-wavelength above ground is plotted in Figure 7-26. It is evident that there is significant difference from the horizontally polarized pattern. The pattern shape is independent of dipole length and is more "flat topped." The half-power beamwidth is approximately 120°. This result indicates the difficulty of simultaneously optimizing the vertical and horizontally pattern shapes of circularly polarized antennas. In addition, the lobes for vertical polarization are of different sign, indicating an out-of-phase condition.

For a two-around array, the magnitude of the patterns shown in Figures 7-25 and 7-26 are the element patterns required to compute the complete antenna pattern. This is done simply by multiplying the circular array factor by the element pattern. By inspection, it is apparent that a major effect of the element pattern is to narrow the lobes of the array pattern. Thus the shape and circularity of the azimuth pattern is dependent on both the array factor and the element pattern.

On very large structures such as triangular towers, the pattern circularity becomes unacceptable for most omnidirectional applications. This situation is most often encountered with panel antennas, such as dipole over ground or cavity elements that are side mounted on a tower that is large with respect to wavelength. Some improvement in circularity may be obtained by using a tangential fire array.[9]

[9] J. Perini, "A Method of Obtaining a Smooth Pattern on Circular Arrays of Large Diameter," *IEEE Trans. Broadcast.*, Vol. 14, No. 3, September 1968, pp. 126–136.

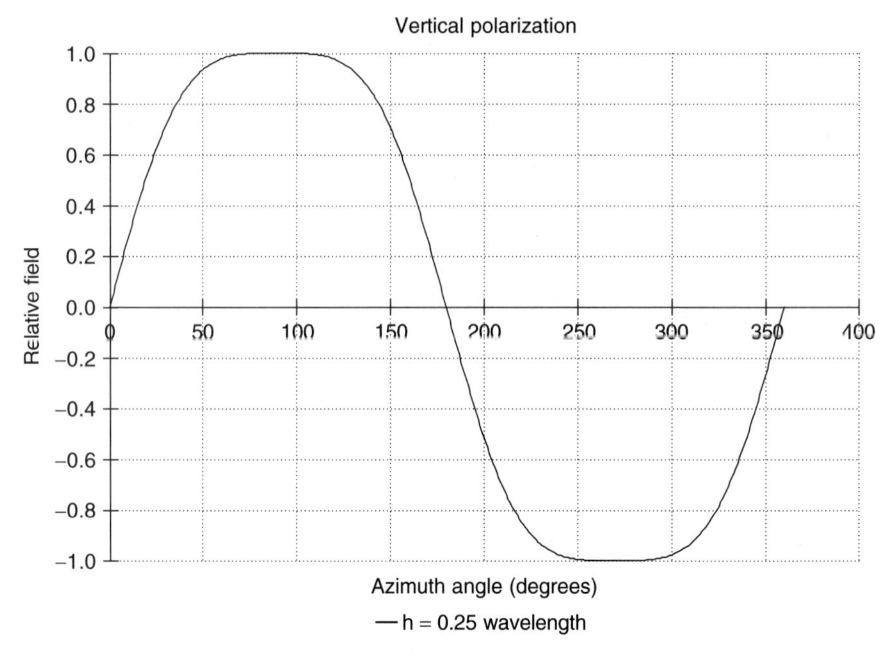

Figure 7-26. Dipole over ground.

SLOTTED CYLINDER ANTENNAS

Slotted cylinder antennas are used primarily for top-mounted applications, although side-mounted applications are sometimes made when top-mounted tower space is not available. The mathematical treatment of their patterns[10,11] is somewhat more complicated than the dipole over ground and is not treated indepth here. For a uniformly distributed horizontally polarized field in the slot, the far-field pattern of the slot on a cylinder of radius a is approximated by a Fourier series

$$E_\phi \sim \frac{b_0}{2} + \sum_{n=1}^{\infty} b_n \cos n\phi$$

where $b_n = j^n / H_n^{(2)'}(ka)$ and $H_n^{(2)'}(ka)$ is the first derivative of the Hankel function. Both the amplitude and phase versus azimuth angle are strong functions of cylinder diameter. Curves illustrating the amplitude of the radiated field versus azimuth angle for slots of small diameter relative to the wavelength are reproduced in Figure 7-27. As might be expected, the pattern becomes more narrow as the cylinder diameter increases. Similar plots for cylinders of

[10] George Sinclair, "The Patterns of Slotted-Cylinder Antennas," *Proc. IRE*, December 1948, pp. 1487–1492.

[11] J.R. Wait, "Radiation Characteristics of Axial Slots on a Conducting Cylinder," *Wireless Eng.*, December 1955, pp. 316–323.

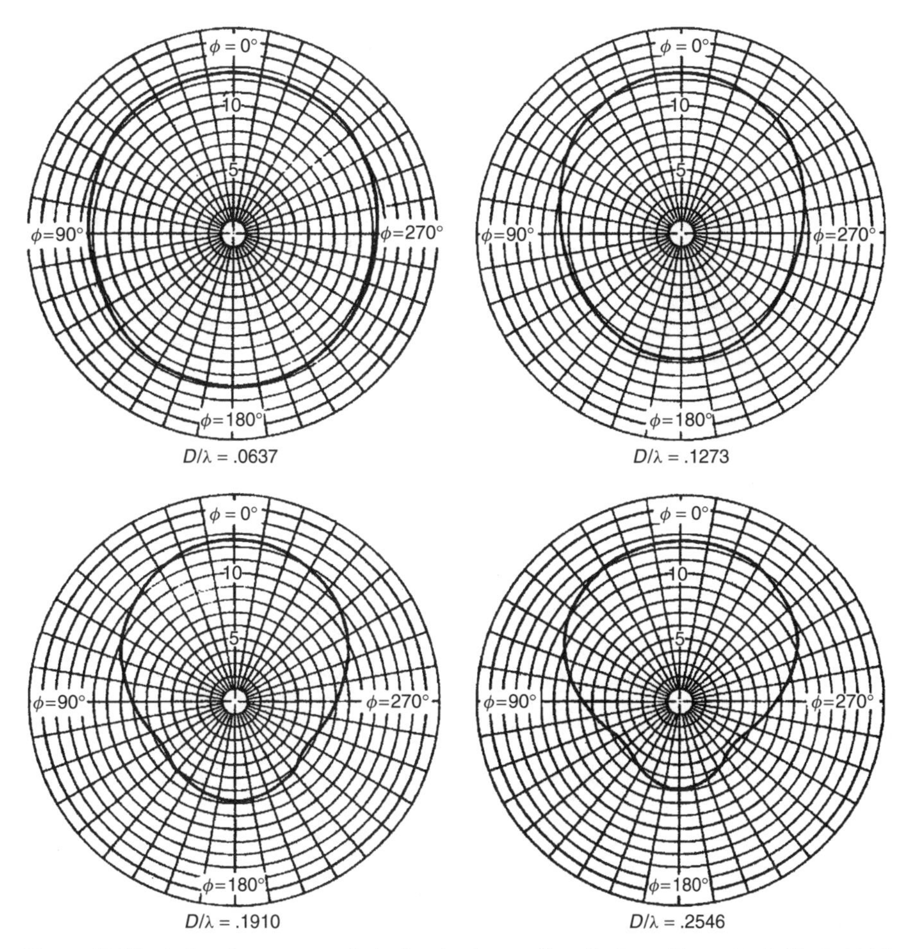

Figure 7-27. Azimuth patterns of single slot in small cylinders. [From Ref. 11 © 1994 IRE (now IEEE); used with permission.]

even greater diameter are shown in Figure 7-28. In a similar fashion, the phase variation as a function of azimuth becomes greater as the cylinder size increases. This is illustrated for small-diameter cylinders in Figure 7-29.

Multiple slots may be combined to produce a variety of directional and omnidirectional patterns. Examples using two slots on various-sized cylinders are shown in Figure 7-30. The result is a two-lobed or "peanut"-shaped pattern. The depth of the nulls is dependent on cylinder diameter. As with any circular array, the circularity of slot arrays is a strong function of the number of slots and the cylinder diameter. This is illustrated by the plot shown in Figure 7-31. It is evident that the pattern become less circular as the cylinder becomes larger, but the circularity improves with an increasing number of slots around the cylinder.

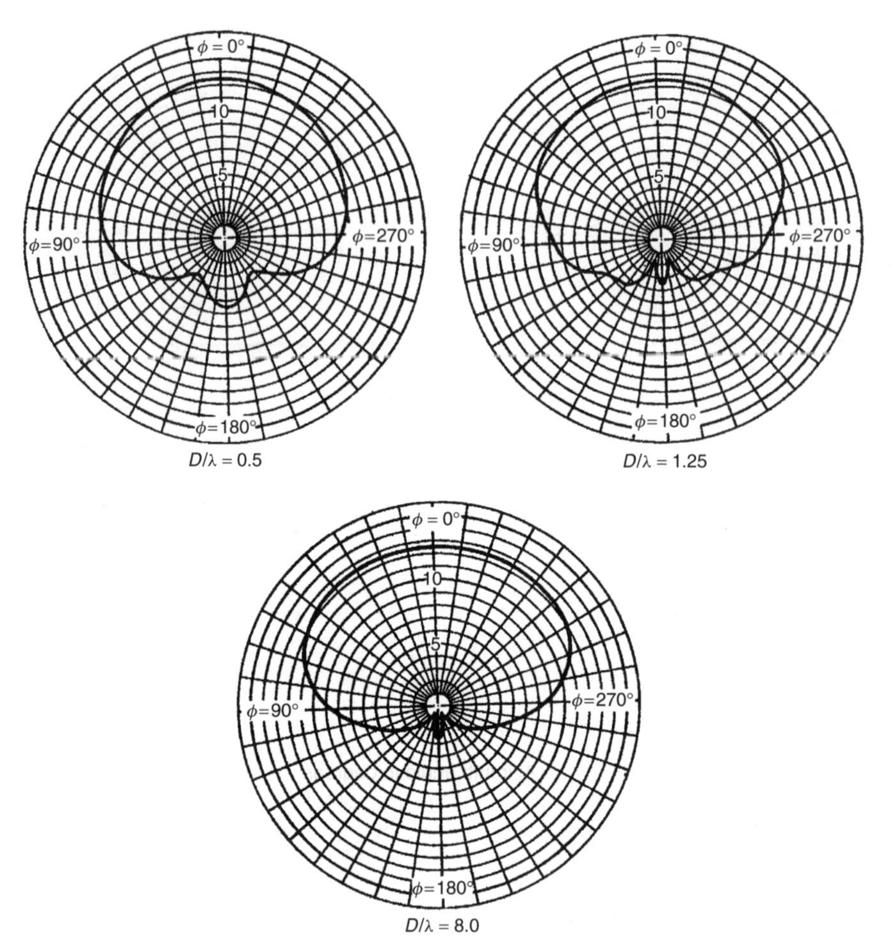

Figure 7-28. Azimuth patterns of single slot in large cylinder. [From Ref. 11 © 1994 IRE (now IEEE); used with permission.]

GAIN AND DIRECTIVITY

In the classical sense, the gain of an antenna in a given direction is defined as 4π times the ratio of the radiation intensity in that direction to the total power input.[12] The maximum gain may be defined as the ratio of the power radiated by an isotropic antenna to the power radiated by the actual antenna when both are producing the same maximum radiation or field intensity. That is, gain is referenced to a completely nondirectional or omnidirectional radiator, one that radiates uniformly in all directions. As a matter of fact, there is no such thing

[12] Edward C. Jordan, *Electromagnetic Waves and Radiating Systems*, 2nd ed., Prentice Hall, Upper Saddle River, NJ, 1968, p. 413.

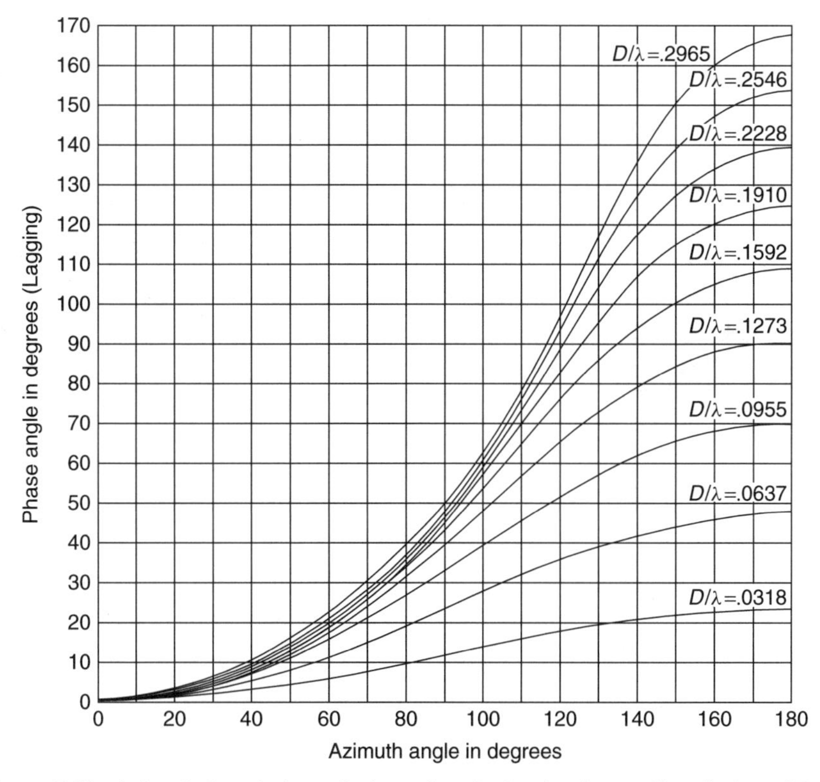

Figure 7-29. Azimuthal variation of phase for single slot in small cylinders. [From Ref. 11 © 1994 IRE (now IEEE); used with permission.]

as a practical isotropic radiator. However, this is a useful theoretical concept for gain-comparison purposes. The unit of gain is classically stated in decibels above isotropic (dBi). When gain is expressed in decibels, it is denoted by G, where

$$G = 10 \log g_a$$

In television broadcasting the definitions of gain and directivity are modified somewhat from that of the classical sense. First, it is common practice to reference the gain of broadcast antennas to a half-wave dipole. Since the directivity of a half-wave dipole is 1.64 or 2.15 dBi, the stated gain for television antennas is 2.15 dB less than it would be if the gain were referenced to an isotropic radiator. For ease of presentation, the concepts associated with gain are often described in the classical sense. However, when numerical specifications are presented by broadcast antenna manufacturers, the gain is referenced to a dipole.

It is also the practice in broadcasting to state gain in numeric ratio as well as decibels. For example, the gain of a half-wave dipole may be stated as a numeric ratio of unity rather than 0 dBd (decibels above a half-wave dipole). The gain

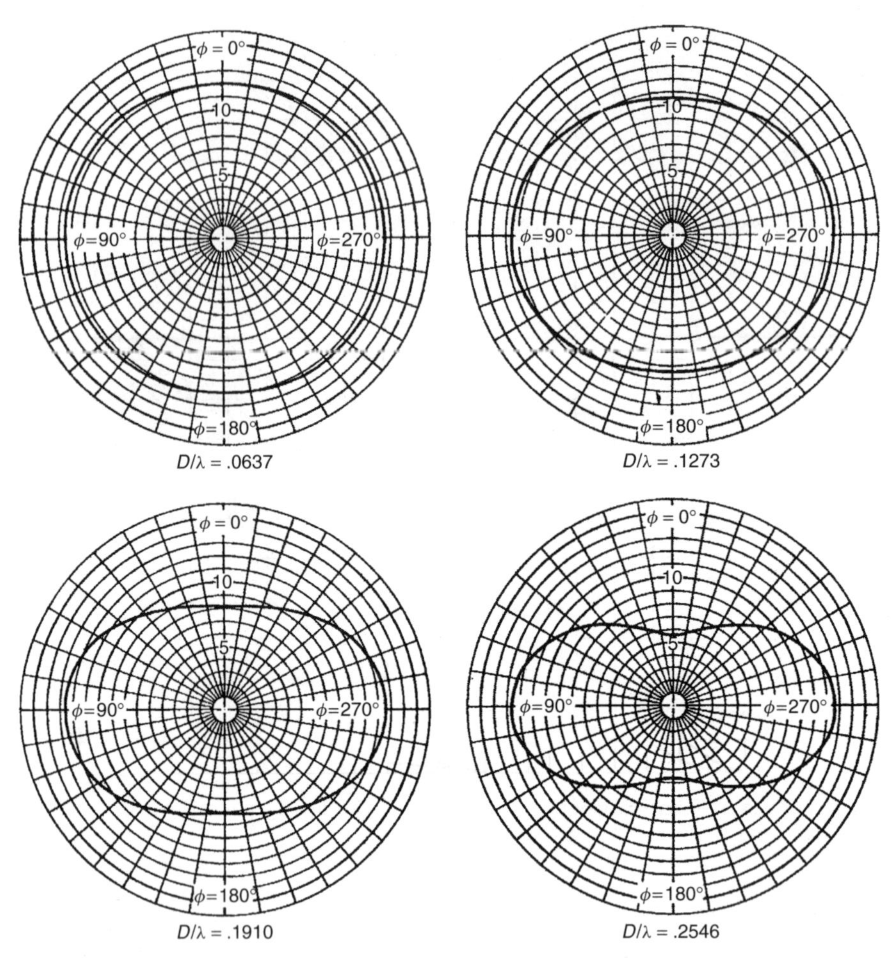

Figure 7-30. Azimuth patterns for a pair of opposed slots in small cylinders. [From Ref. 11 © 1994 IRE (now IEEE); used with permission.]

of broadcast antennas is always referenced to the horizontally polarized signal. If the antenna is elliptically polarized, the gain is reduced by a factor equal to 1 plus the vertical-to-horizontal power ratio. For circular polarization, which is just a special case of elliptical polarization with a vertical-to-horizontal power ratio of unity, the gain is reduced by a factor of 2.

The antenna gain is strongly dependent on the directivity as well as any losses in the antenna. Consider first the directivity, which is closely related to the directional characteristic or shape of the antenna pattern. Directivity is a measure of the degree to which an antenna concentrates the radiated energy in the direction of the peak. The directivity of an omnidirectional antenna in the classical sense is, by definition, unity. Thus any value of directivity

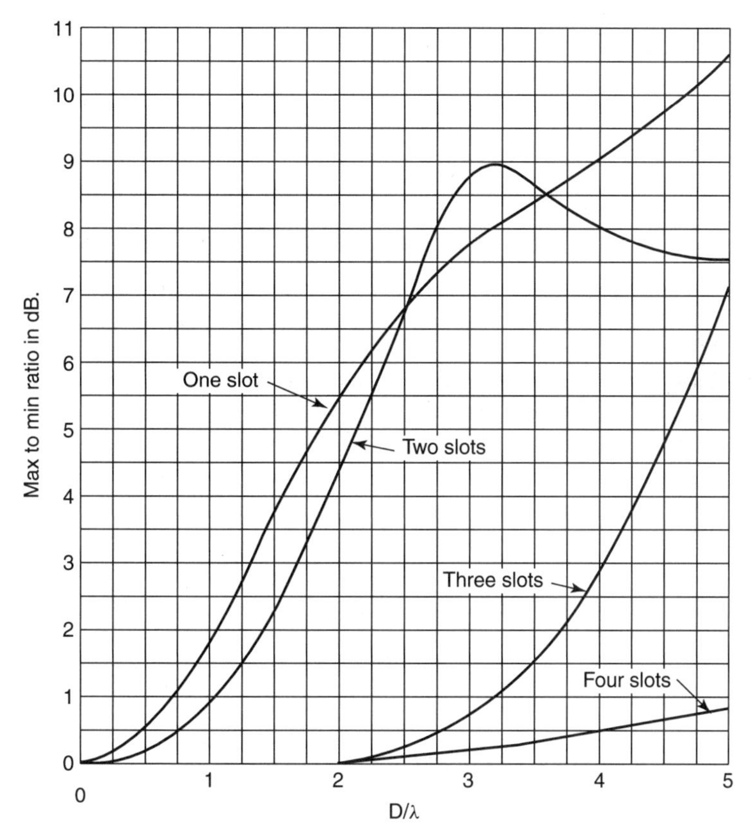

Figure 7-31. Circularity of slotted cylinders with multiple slots. [From Ref. 11 © 1994 IRE (now IEEE); used with permission.]

greater than unity implies a directional pattern. Consequently, it is important to consider carefully the requirement for antenna directivity and the impact on the directional characteristics. If the vertical pattern is too narrow, there is some risk of sacrificing near-in coverage. As discussed earlier, this can often be remedied by providing sufficient beam tilt and null fill, although at some sacrifice of directivity. If near-in coverage is not important and the vertical pattern is unnecessarily broad, the directivity will be lower than it could be, thereby decreasing gain and increasing the required TPO to achieve the desired AERP.

The most common system designs require an omnidirectional, horizontally polarized antenna. In this case, the RMS horizontal (or azimuthal) directivity is unity with respect to a horizontally polarized dipole, so that only the directivity of the vertical (or elevation) pattern must be considered. Common values of gain in the United States due to the vertical pattern are from 2 to 6 for low-band VHF, 4 to 12 for high-band VHF, and up to 25 or 30 for UHF.

With a directional horizontal pattern, the total gain is the product of the horizontal and vertical pattern directivities, d_h and d_v, and the antenna efficiency, η,

$$g_a = \eta d_h d_v$$

where the efficiency accounts for any ohmic losses within the feed lines and on the radiating elements. Efficiency may be expressed in numeric ratio or percent. The efficiency can be a significant factor, depending on feed system design and associated losses. For example, a typical 12-bay turnstile antenna for the high-VHF band may have an efficiency of only 0.83 or 83%. The RMS horizontal directivity may be as low as unity for an omnidirectional antenna to up to nearly 6 for some highly directive UHF antenna types. More common values are in the range of 2.

In addition to its relationship to the antenna directional pattern, choice of the antenna gain involves several other trade-offs. As discussed earlier, the antenna beamwidth is inversely proportional to aperture length. Since decreasing beamwidth implies increasing directivity, it follows that gain is proportional to length. The relationship between gain and length leads to many of these trade-offs.

One consequence of greater vertical directivity and length is higher wind load. Figure 7-32 shows the wind shear for typical omnidirectional VHF and UHF transmitting antennas as a function of gain. The linear relationship is evident. Overturning moment is another important antenna structural parameter, which rises even faster with increasing gain. This is a consequence of the overturning

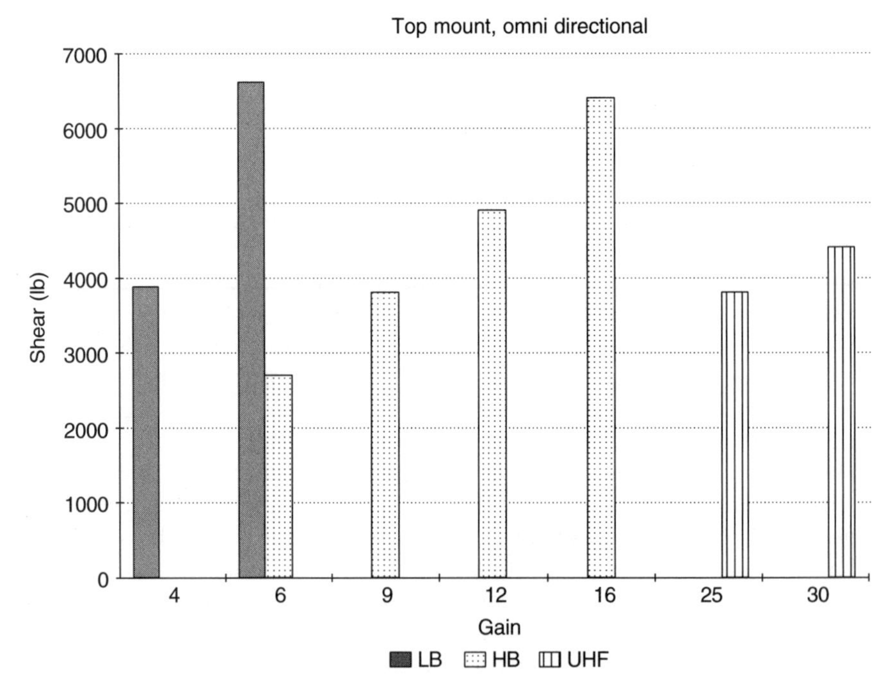

Figure 7-32. Antenna wind load versus gain.

moment being the product of the mechanical center of pressure and shear. Since the shear of a uniform cylinder is proportional to length, the overturning moment is approximately proportional to the square of length.

In some cases, the antenna gain may be limited by the available aperture space. Since, at a specified channel, gain is proportional to antenna length, the available aperture space may become a key consideration. Also, cost of the antenna is often proportional to length. Actual values of these interrelated specifications vary greatly by antenna type and manufacturer. Specific values for various types of antenna should be obtained from the manufacturers.

POWER HANDLING

Transmitting antennas for digital television broadcast must handle the transmitter power remaining at the output of the transmission line. For the high power levels required for many installations, reasonably high currents and voltages are present; the antenna design must be implemented to handle these currents and voltages properly. In general, large current densities require large conductors made of high-conductivity materials to minimize losses. High voltages require widely spaced conductors and insulators with high insulation strength to avoid voltage breakdown. The presence of high voltages also requires smooth, rounded corners on metal parts. The antenna impedance must also be well matched to the transmitter and transmission line for minimum standing wave ratio and maximum power transfer.

These considerations are obviously most important at the input to the antenna, where the power level and associated currents and voltages are highest. However, sound design criteria must be applied at other points in the antenna to assure reliable distribution of energy to the radiating elements. Many antennas use coaxial power dividers, rigid coax lines, and semiflexible coaxial cables to distribute power. Thus, the principles related to efficiency and power handling capability described in Chapter 6 apply to these antenna components.

ANTENNA IMPEDANCE

In the design of a digital television transmission system, the antenna is but one link in a complex chain that leads from the original baseband signal to an estimate of the signal at the output of the receiver. From this point of view, the antenna may be considered to be just another circuit element that must be properly matched to the rest of the system for efficient power transfer. The input or terminal impedance of the antenna is of primary concern. In general, antenna input impedance is a complicated function of frequency that cannot be described in any simple analytical form. However, over a narrow frequency band such as encountered for digital TV transmission, the impedance may often be accurately modeled by a resistance in series with a reactance.

The impedance of an ideal broadband lossless antenna would be defined by its radiation resistance, R_{rad}, which is related to the radiated power, P_{rad}, and the effective current, I_{eff}, that is,

$$P_{rad} = |I_{eff}|^2 R_{rad}$$

The effective current is not necessarily the input current or even the peak current on the antenna structure. If the antenna is lossless, the radiated power is the same as the transmission line output power.

Even though it finds no practical application as a broadcast antenna, it is instructive to consider the short dipole antenna in order to understand the parameters that affect radiation resistance. For the short dipole of length, l, the radiation resistance is given by

$$R_{rad} = 20\pi^2 \left(\frac{l}{\lambda}\right)^2 \quad \Omega$$

This formula is derived using the Poynting vector method and is strictly true only for very short antennas, but it is approximately correct for dipoles of length up to a quarter-wavelength. The important point is that radiation resistance is proportional to the square of the length of the antenna in wavelengths. For an antenna $\lambda/4$ in length, the radiation resistance is 12.8 Ω.

To calculate the radiation resistance of longer antennas it is necessary to know the current distribution on the antenna. In general, this is a difficult theoretical problem. In the absence of knowledge of the actual current distribution, a sinusoidal current distribution is assumed. The accuracy of the resulting calculations depends on how well the assumed current distribution matches reality. When this computation is made for the half-wave dipole, the radiation resistance is found to be 73 Ω. Thus, it should be expected that for efficient transfer of power from the transmission line to the antenna, antenna elements in the neighborhood of a half wavelength or slightly less are required. For an ideal, lossless, half-wave dipole, the input resistance is equal to the radiation resistance.

Consideration of the radiation resistance of linear antennas has been under the assumption of infinitely thin conductors. This assumption yields useful results because the radiation resistance depends only on the distant fields. To determine the reactance of the antenna, the shape and thickness of the radiators must be considered. The reactive power, and hence the reactance, depends on the fields close to the antenna. The strength of these fields depends on the specific geometry of the antenna.

The complete impedance of a dipole antenna may be computed using the induced emf method. The expressions for the radiation resistance and reactance resulting from this method are forbidding equations involving sine and cosine integrals. Because of their complexity, these formulas will not be shown here. The interested reader is referred to either Jordan or Balanis. The results of such calculations are shown in Figure 7-33, which shows graphs of resistance and

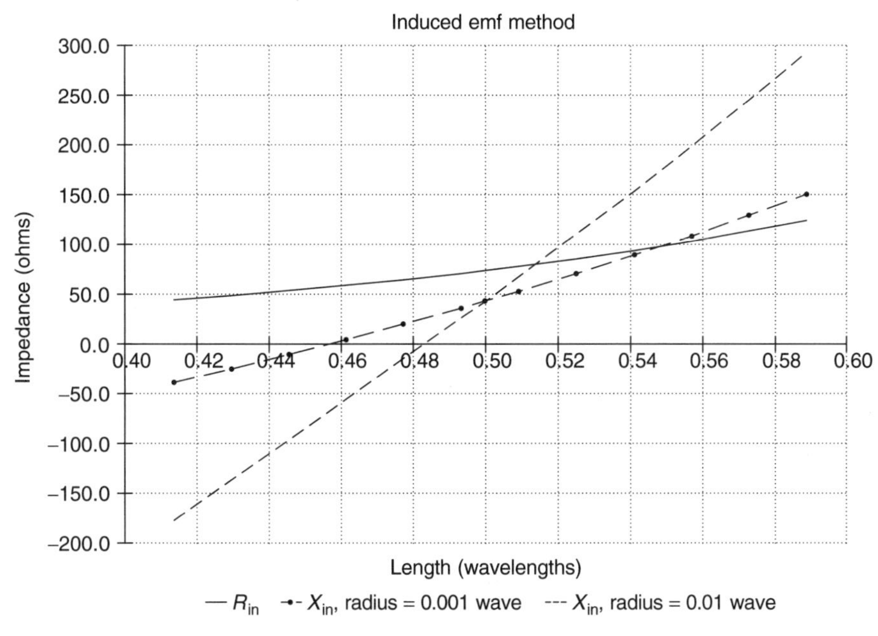

Figure 7-33. Dipole impedance.

reactance of short dipole antennas with radii of 0.01 and 0.001 wavelength. Note that the half-wave dipole has a reactance of $+j42.3\ \Omega$, independent of diameter. The reactance of dipoles of other lengths is very much dependent on diameter, with the lowest reactance for the largest diameters. This demonstrates the desirability of large-diameter radiating elements for wider band applications, such as digital television systems operating in the VHF bands. To maintain a good impedance match over the full channel bandwidth, a low ratio of reactance to resistance or low Q is required. Fortunately, this is consistent with the need to use large structural members for their current carrying capacity and structural integrity.

It is apparent that the antenna reactance goes through zero for a dipole length somewhat less than a half wavelength. This length is called the resonant length. In free space, the resonant length is always less than a half wavelength, being shorter with increasing diameter. This effect is often called the end effect in that the antenna appears to be longer than its physical length. For larger conductor sizes, the resonant length is even shorter and the input resistance is closer to 50 Ω. This is desirable from the standpoint of obtaining a good match to a 50 Ω transmission line.

For a resonant antenna, the input impedance, Z, may be approximated over the bandwidth of a single digital television channel by a series combination of resistance, R_r, a capacitive reactance, and an inductive reactance, that is, a series RLC circuit. Below the resonant frequency the reactance is capacitive; above resonance it is inductive. The general expression for the impedance of this circuit is

$$Z = R_r + j\left(\omega L - \frac{1}{\omega C}\right)$$

At the resonant angular frequency

$$\omega_0 L = \frac{1}{\omega_0 C} \quad \text{and} \quad Z = R_r$$

The unloaded Q is given by

$$Q = \frac{\omega_0 L}{R_r} = \frac{1}{\omega_0 C R_r} = \frac{\omega_0}{\omega_h - \omega_l}$$

where ω_h and ω_l are the upper and lower angular frequencies, at which the current has dropped to one-half of its maximum value.

The values of the equivalent circuit elements RLC and Q are the quantities of interest for dipoles made of wires of different sizes. When the transmitter is loaded by a properly matched antenna, the total loaded Q is one-half the unloaded Q:

$$Q_l = \tfrac{1}{2} Q_u$$

When measurements are made on an actual antenna, the impedance is somewhat different from that which would be computed using the simple RLC model. In reality, R, L, and C are functions of frequency, not constants as is assumed for the simple model.

As was noted in the discussion of array element patterns, a dipole over a ground plane represents a more practical antenna for television broadcast applications. As might be expected, the presence of the ground plane affects the dipole impedance, including the radiation resistance. It can be shown that the resistance is multiplied by[12]

$$1 - \frac{\sin 2kh}{2kh} - \frac{\cos 2kh}{(2kh)^2} + \frac{\sin 2kh}{(2kh)^3}$$

This factor is plotted in Figure 7-34 for heights above ground up to one wavelength. For dipoles very close to ground the resistance is reduced, becoming a short circuit at zero height. At a height above ground near a quarter-wavelength, the resistance is 1.15 times that of free-space value. This illustrates the necessity of using both dipole length and height above ground to control the antenna impedance as well as the pattern shape. For a 50-Ω input resistance, one should expect that a dipole $\frac{1}{4}$ wavelength over ground to be somewhat shorter than the resonant length of the dipole in free space.

[12] Balanis, op. cit., pp. 145.

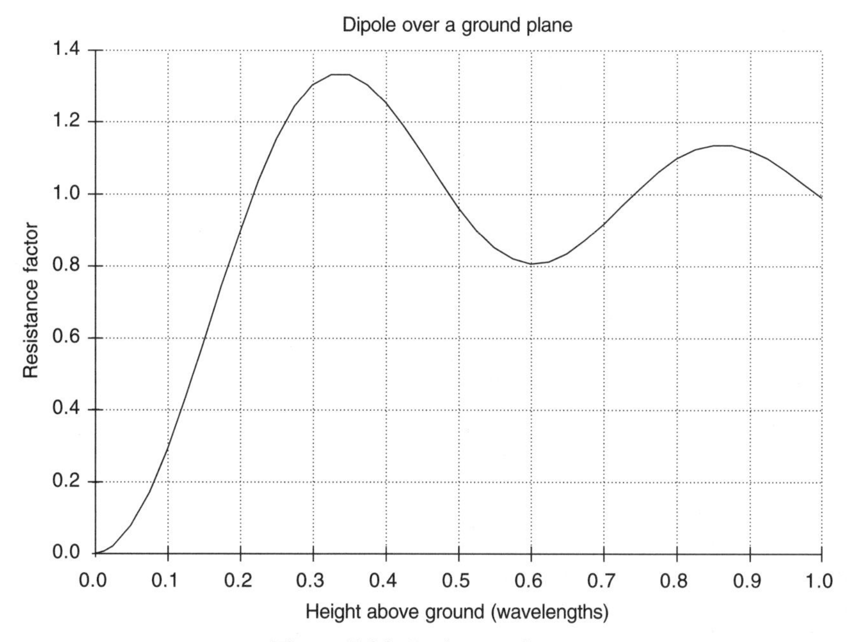

Figure 7-34. Resistance factor.

Modification of the radiation resistance of a dipole in the presence of a ground plane is equivalent to the effect of the mutual impedance of a pair of parallel dipoles driven out of phase by equal currents. In general, the driving-point impedance of an antenna in the presence of another antenna is

$$Z_1 = Z_{11} + Z_{12}\frac{I_2}{I_1}$$

where Z_{11} is the self-impedance of the antenna, Z_{12} is the mutual impedance between the pair of antennas, and I_2/I_1 is the ratio of the driving currents. Similarly, the input impedance of the second dipole is

$$Z_2 = Z_{22} + Z_{21}\frac{I_1}{I_2}$$

Since antennas are linear, bilateral devices,

$$Z_{12} = Z_{21}$$

In the case of the dipole over ground,

$$\frac{I_2}{I_1} = \frac{I_1}{I_2} = -1$$

The mutual impedance of dipoles has been derived for many common configurations. Like the self-reactance formulas, these expressions involve the sine and cosine integrals and will not be repeated here. The interested reader is again referred to Jordan or Balanis. For the sake of illustration, the results of calculation of the mutual impedance of parallel half-wavelength dipoles as a function of separation are plotted in Figure 7-35. Note that the peak values of mutual resistance and reactance tend to be diminished as the separation increases. For a separation of one-half wavelength, the mutual impedance is $-11.1 - j\ 29.9$. Subtracting the mutual impedance from the self-impedance of $73 + j\ 42.3$ results in a driving-point impedance for a half-wave dipole a quarter-wave above ground of $84.2 + j\ 72.2\ \Omega$. Thus the effect of placing the dipole over a ground plane is to enhance the end effect. To assure a resistive input impedance the dipole must be shortened further to compensate for the end effect.

This example of mutual impedance is for parallel arrays of dipoles and represents a worst-case configuration. This high level of coupling is often encountered in horizontally polarized antennas using a vertical stack of dipole elements. For vertical stacks of slot array elements, the mutual impedance is much lower. Similarly, vertically polarized dipole elements exhibit much less mutual impedance when arranged in vertical stacks. However, horizontal arrays of horizontally polarized slots and vertical dipoles may couple quite strongly, depending on specific design details. In circular polarized antennas, both weak and strong mutual effects may exist simultaneously. Complete understanding of

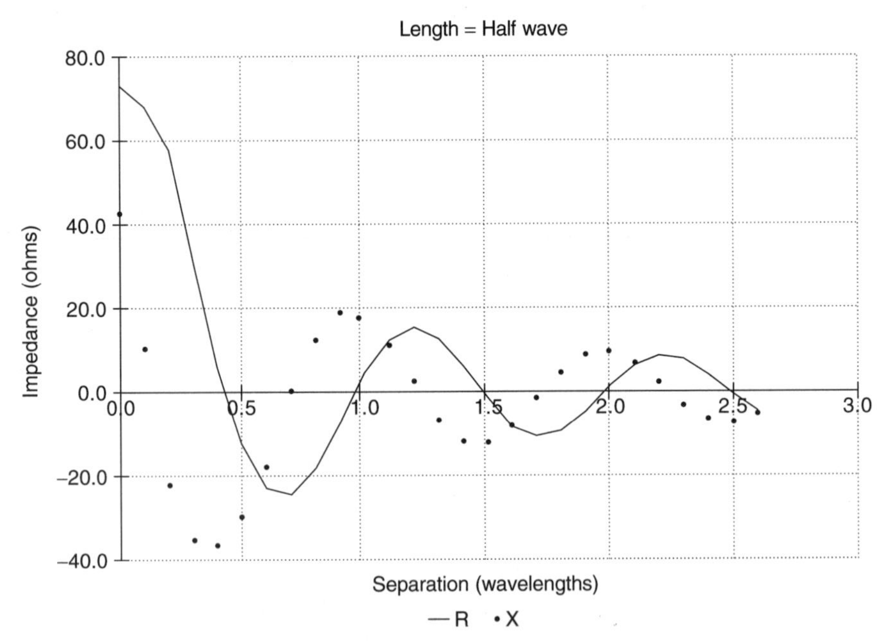

Figure 7-35. Mutual Z of parallel dipoles.

the mutual impedance characteristics of the array element is the key to successful impedance control in a broadcast array.

In a high-gain array, the issue of mutual impedance becomes very complex because the driving-point impedance of any one element is affected by the currents flowing in and the mutual impedance of every other element in the array. Furthermore, those elements in the center of the array are affected differently than those on the ends, since they are in close proximity to more elements. Fortunately, the influence of mutual impedance is reduced with increasing separation, so that only those elements nearby have a large affect. Even so, to account properly for mutual impedance requires careful design, analysis, and measurement. Antennas for broadcast applications are available in a wide variety of elevation patterns, horizontal patterns, and gains. Adding elements, deleting elements, or changing the current distribution introduces changes to the driving-point impedance of each element. For this reason, antenna manufacturers attempt to reduce to a manageable level the effect of mutual impedance in their products.

One approach is to offer a limited number of antenna configurations and to manufacture only those standard models. For example, standard values of gain, null fill and beam tilt, and standard azimuth patterns might be offered. With this approach, the design activity is completed at the close of a well-defined product development cycle, and the manufacturer can then focus on production. This approach tends to reduce cost since much less continuing engineering effort is required. If the manufacturer has defined the standard product properly, this approach should be acceptable for many, even most, digital television stations. However, when a custom radiation pattern or gain is required, this approach is not very accommodating.

Accommodation of custom radiation patterns and gain can be achieved to some degree by implementation of mini arrays of standard elementary antennas. In this approach, a small number of array elements, say a group four dipole or slot elements, are combined to form a standard mini array. The directional, polarization, and impedance properties of the mini array are optimized to provide desirable performance. The result is a super array element with lower mutual impedance and greater element-to-element separation than that of the constituent array elements. Larger arrays may then be built using this super element. Since the mutual impedances are low, custom arrays may be built without undue engineering and manufacturing labor.

BANDWIDTH AND FREQUENCY RESPONSE

Antenna bandwidth is a somewhat elusive concept unless it is defined very carefully. A complete definition adequate for antennas to be used for transmission of digital television must account for both the directional and impedance properties over at least one channel. The earlier discussion of frequency response as a function of elevation angle illustrates one aspect of bandwidth as it relates to the directional properties. It is important that the amplitude and phase of the radiated signal be maintained within desired limits so that the adaptive equalizer capacity in the receiver is reserved for removal of linear distortions occurring due to propagation from transmitter to receiver.

The previous discussion focused on linear distortions due to elevation pattern variations. Similar control must be exercised over the azimuth pattern. As was demonstrated, variation in the azimuth pattern at any particular frequency is minimum for small-diameter arrays. Since the key parameter in determining array size is the radius-to-wavelength ratio, it follows that minimum-size arrays tend to exhibit minimum variations with respect to frequency. This would naturally tend to favor the use of top-mounted arrays for best frequency response. In many cases, use of side-mounted and wraparound arrays is unavoidable. This will be especially true during the transition period or when community towers are used. Wraparound antennas will provide satisfactory frequency response provided that the tower size is not excessive. This suggests that stacked antenna arrangements will perform best if the highest channels are top mounted, with lower channels located at progressively lower levels. This general configuration allows for a large tower face where it is needed for structural reasons, while maintaining reasonable electrical size for pattern shape and bandwidth purposes. For circularly polarized antennas, the concept of pattern bandwidth must be applied to both the horizontally and vertically polarized components of the field.

Impedance bandwidth is just as important as pattern bandwidth for digital television. Unlike analog transmission, in which the luminance content is concentrated around the visual carrier, the digital TV signal fills the full channel bandwidth. Thus minimization of linear distortions due to impedance mismatch is important throughout the channel. As discussed in Chapter 6, the antenna mismatch introduces both amplitude and nonlinear phase distortions, the effect of both being essentially independent of line length. The nonlinear phase introduces group delay, which is dependent on line length. Therefore, the antenna mismatch should be minimized across the channel bandwidth. In the ideal case, the antenna impedance would also be constant or at least known and predictable with some precision. In this event, knowledge of the line length and attenuation could possibly permit preequalization of the antenna mismatch and associated linear distortions at the transmitter.

MULTIPLE-CHANNEL OPERATION

Standard antennas are usually manufactured for single-channel applications. However, some designs are adaptable to dual- or even multichannel operation. For VHF, these designs include certain panel types and batwing antennas. For UHF, slotted antennas and broadband panels are available. The possibility of operating a digital television transmitting antenna on more then one channel depends on the pattern and impedance bandwidth. If adjacent channels are involved, a minimum of 12 to 16 MHz continuous bandwidth is required. For nonadjacent assignments, two or more bands, each with at least 6 to 8 MHz bandwidth is required. An acceptable level of pattern variation and good impedance match must be achieved in each band of operation.

The requirement for dual- or multichannel operation will generally exclude the use of end-fed antennas. As discussed earlier, end-fed antennas must be carefully designed to assure acceptable pattern variations over even a single channel.

Center-fed antennas were shown to exhibit much less pattern variation over a single channel. By the same reasoning it can be shown that acceptable pattern variations may be achieved over the bandwidth of two adjacent channels. Thus center-fed antennas designed specifically for dual-channel operation can provide acceptable pattern bandwidth. Similarly, branch-fed designs with acceptable pattern bandwidth should be feasible. These antennas use power dividers and flexible coaxial cables to distribute power to a large number of elementary radiators. By distributing the power in a symmetrical manner to the upper and lower halves of the antenna, performance equivalent to that of a center-fed array may be achieved.

Good impedance match over both channels is important for the same reasons as those given for single-channel antennas. Elementary radiator impedance, the effects of ground planes and cavities, and mutual impedances each affect the overall antenna performance. The necessity for added bandwidth with essentially no change in performance only serves to make the designer's task more difficult.

The antenna peak and average power ratings must be adequate to handle the power radiated by both channels. The minimum average power rating is simply the sum of the average powers for each channel. The peak power rating must account for maximum possible input voltage from the combination of signals. Thus an assumption must be made with respect to the maximum peak-to-average power ratio. If the peak-to-average ratio is 5:1 (7 dB) and the average power is the same for each channel, the peak power could be 20 times the single-channel average power.

Turnstile or batwing antennas offer a unique opportunity to provide dual-channel operation and a means to combine the output of two transmitters. The impedance and pattern bandwidth of these antennas are well known to be adequate from extensive analog applications for European channel 2 (47 to 54 MHz), U.S. channels 2 and 3, U.S. channels 4 and 5, and for pairs of high-band VHF channels. Diplexing of the analog visual and aural signals has also demonstrated a means of combining signals of different frequencies. This technology can readily be applied to adjacent assignments of analog and digital stations or pairs of digital TV stations, provided that the channel separation is not too great.

To understand this technique, consider the turnstile antenna system shown in Figure 7-36. The elementary radiators are equivalent to broadband dipoles, producing a double-lobe azimuth pattern approximated by a cosine-squared function. When orthogonal pairs of these elements are fed with equal currents in phase quadrature, an omnidirectional azimuth pattern results. A convenient method for equal division of power in phase quadrature is the use of a quadrature hybrid. Two ports are used for inputs, the other two for outputs. As discussed in Chapter 5, a signal applied to one input results in equal voltages at the outputs

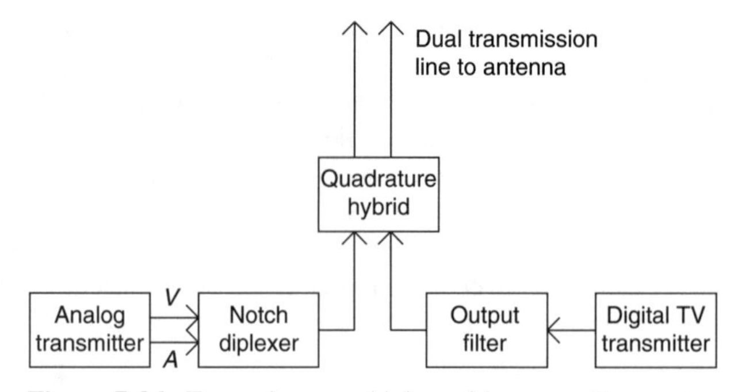

Figure 7-36. Transmitter combining with a turnstile antenna.

with a 0°, −90° phase relationship. Another signal applied at the second input provides equal voltages at the outputs with a −90°, 0° phase relationship. When these signals are applied to the inputs of a turnstile antenna, an omnidirectional azimuth pattern results for both signals. One signal may be the output of a digital television transmitter and the other the combined visual and aural output of an analog transmitter. Alternatively, the signals may be the outputs of digital transmitters operating on adjacent or closely spaced channels. Standard top-mounted antennas suitable for operation on adjacent channels at low band or high band are available. At high band, standard top-mounted designs are available that allow combining of any channel pair. Similar percentage bandwidths could be covered with antennas designed to operate for UHF. In the case of high-gain UHF antennas, the power distribution might complicate the mechanical configuration and result in low efficiency.

TYPES OF DIGITAL TELEVISION BROADCAST ANTENNAS

Several general types of antenna have been presented in a conceptual context. The purpose was to illustrate the factors that influence the directional and impedance properties of broadcast antennas. Batwing or turnstile antennas have been discussed to suggest possible means of operating a single antenna on multiple channels. There is a variety of other practical implementations that are offered by various manufacturers.

Other horizontally polarized antenna for VHF and UHF include slotted cylinder arrays and wraparound panels. Slotted cylinder antennas are available for single-channel applications. For VHF, slotted antennas are available with omnidirectional and a limited number of directional patterns; for UHF a wide variety of direction azimuth patterns are available. An exponential aperture distribution results in a smooth elevation pattern. Slotted antennas provide low wind load and are best used in top-mounted applications.

Wraparound panel antennas are available for triangular and square towers. For three-sided towers, the individual panels have a 6 dB beamwidth of 120°; for four-sided towers, the beamwidth is 90°. With these panels, omnidirectional azimuth patterns may be obtained from either tower cross section. In addition, a wide variety of directional azimuth patterns can be obtained by using various unequal power and/or phase distributions. Panel antennas may be either top or side mounted. Multichannel capability may be provided if the panel bandwidth is adequate. Panel antennas are often a good approach for digital television if an existing tower with open aperture space can be upgraded to support the additional loads. Good pattern circularity for omni patterns is possible provided that the tower face size is approximately equal to the panel size. Optimum tower face sizes range from 12 ft for channel 2 to 4 ft for high-band VHF (band III) to 2 ft for UHF (bands IV and V).

Elliptical (EP) and circularly polarized (CP) antennas are also available in slotted cylinder and panel designs, with other properties similar to those of horizontally polarized designs. VHF panel designs include cavity radiators excited by crossed dipoles. In addition, helical or spiral designs for CP and EP are available for single-channel omnidirectional stations. These are top-mounted designs with low wind load. Another top-mounted single-channel design for VHF uses a three-around array of crossed vee dipoles. A variety of directional azimuth patterns as well as omni patterns are available.

ANTENNA MOUNTING

The position of choice for any transmitting antenna is obviously the top of the broadcast tower. From this position, optimum control of the antenna characteristics may be exercised with no obstructions to the radiation pattern or affect on impedance. Top-mounted antennas usually provide the lowest wind load. Unfortunately, the top position on an existing lower is probably already occupied by an analog television antenna. In order to use this position during the transition period, it therefore may be necessary to replace the existing antenna with a broadband design with which the digital TV signal may be combined.

In a some cases it may be possible to stack a pair of top-mounted antennas. For example, a low-band analog station with a UHF digital assignment might stack a UHF slotted cylinder array atop a batwing antenna. This probably would require replacement of the existing analog antenna as well as purchase of the new digital unit. Some sacrifice of performance can be expected. For example, it would probably be necessary to pass the feed line for the upper antenna through the aperture of the lower. This will usually result in some pattern distortion to the lower antenna. The mechanical stability of both antennas may also suffer. The limited mechanical strength of the lower will allow more flexing at the base of the upper. The added loads of the upper will result in more flexing of the lower. The capability of the tower to support the necessary overturning moment may limit the feasibility of this approach. Even if feasible, it may be necessary to reduce the gain of the analog antenna to reduce overall length and maintain satisfactory mechanical stability.

If top mounting of both antennas is not desirable or feasible, a combination top-mounted and wraparound arrangement might be considered. Although this is almost never an ideal configuration, it may be the only realistic choice from structural considerations. For example, a VHF wraparound panel antenna might be used for the analog TV, supporting a slotted cylinder array for UHF. This arrangement has the advantage of providing space on the inside of the wraparound to hang the transmission line for the upper antenna. Azimuth and elevation patterns of the wraparound antenna can be expected to suffer somewhat due to the presence of tower members, guy wires, transmission lines, and other tower accessories. For an omni azimuth pattern, the degradation will produce less than ideal circularity. Directional patterns may also be expected to be distorted but to a lesser extent. Null fill and beam tilt of elevation patterns may be affected. Pattern distortions will be frequency dependent, so the radiated signal may suffer additional linear distortions. These degradations may be difficult to quantify due to the physical size of wraparound configuration and the attendant difficulty in making pattern measurements. The methods of determining the extent of these effects range from analytical methods to scale models and full-scale measurements, depending on the capabilities of the antenna manufacturer.

Side mounting of the digital television antenna is yet another option. If the tower can be strengthen to provided the needed support, a side-mounted slot antenna might be located below the existing analog antenna. This approach has the advantage of not requiring replacement of the analog antenna. The major disadvantage for omnidirectional and some directional azimuth patterns is the severe distortion due to the presence of the tower.

In summary, mounting of the digital television antenna while maintaining analog service may be one of the most difficult issues to resolve. Virtually every situation is different. However, in nearly every case the available options for a conventional tower are included in the categories discussed (i.e., top mounting, stacking, side mounting, wraparound, and multiplexing). In a few cases it may be possible to interleave two antennas within the same aperture.

The remaining option is to use a tee bar or candelabra atop the tower. This arrangement allows two or three antennas to be mounted at the top location. Stacking, side mount, wraparound, and multiplexing may be combined with this approach. It should be noted that all antennas on a tee bar or candelabra will suffer pattern distortions due to the presence of the other antennas in their apertures. This will have an especially severe impact on the vertical component of circularly polarized antennas. Extensive design studies are recommended before finalizing the design of a candelabra or tee-bar system.

8

RADIO-WAVE PROPAGATION[1]

The energy radiated from the transmitting antenna may reach the receiving antenna over several possible propagation paths, as illustrated in Figure 8-1. For VHF and UHF signals, the space wave, which is composed of the direct wave, reflected waves, and tropospheric waves, is the most important. As the name implies, the direct wave travels the most direct path from the transmitter to receiver. Reflected waves arrives at the receiver after being reflected from the surface of the earth and other reflecting objects. Tropospheric waves are reflected and refracted at abrupt changes in the dielectric constant of the lower atmosphere ($<$10 km) and may produce propagation beyond the horizon. Energy may also be received beyond the horizon as a result of diffraction around the spherical surface of the earth or other obstacles. In addition to the various signal paths and spatial variations, the propagation model must account for time variations in the signal level or fading. In this chapter we consider the factors that affect these parameters. Examples from digital television field testing are discussed to clarify and illustrate these concepts.

In most respects propagation of digital TV signals is identical to that of their analog counterparts. However, an important difference is the signal bandwidth; the broad continuous spectrum of digital television brings special concern for the effect of multipath on frequency response. Beyond this concern, consideration of digital TV propagation gives an opportunity to review the factors that affect the terrestrial channel, compare theoretical concepts with measured data, and assess the effectiveness of various prediction methods.

[1] The material presented in this chapter relating to free-space propagation, multipath, and the effect of the earth's curvature is adapted from an article by the author which appeared in *Microwave J.*, Vol. 41, No. 7, July 1998, pp. 78–86.

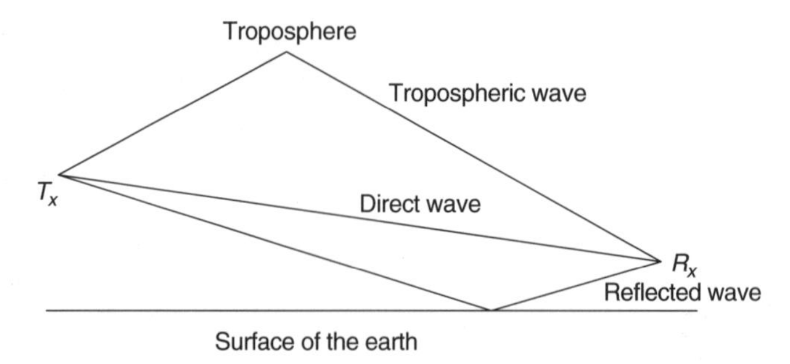

Figure 8-1. Propagation paths affecting digital television coverage.

Application of the methods discussed will not yield an exact prediction of the signal strength or frequency response at any particular location or along any particular radial; no method is this precise. The objective is to understand the factors that affect the performance of the channel well enough to make useful estimates.

FREE-SPACE PROPAGATION

In the absence of multipath and blockage by obstacles in the propagation path, the power available at a receive location depends only on the average effective radiated power and the free-space line-of-sight path attenuation. The maximum AERP is assigned by the applicable regulatory agency and is usually expressed in dBK or dB above 1 kW. Note that 0 dBK is equivalent to 60 dBm.

Because of the law of conservation of energy, the total power flowing through a sphere surrounding an antenna in a lossless medium in which no power is added or removed is constant, independent of the diameter of the sphere. Thus the power density, $\mathcal{P}$, is reduced with increasing distance from the antenna. Consider an antenna that radiates power, P_0, uniformly in all directions through a pair of concentric spheres of radius R_1 and R_2 as shown in Figure 8-2. Since the surface area of a sphere of radius R is $4\pi R^2$, the power density, $\mathcal{P}_1$, at R_1 is

$$\mathcal{P}_1 = \frac{P_0}{4\pi R_1^2}$$

At R_2, the power density is

$$\mathcal{P}_2 = \frac{P_0}{4\pi R_2^2}$$

The ratio of the power densities is

$$\frac{\mathcal{P}_1}{\mathcal{P}_2} = \frac{P_0/4\pi R_1^2}{P_0/4\pi R_2^2} = \left(\frac{R_2}{R_1}\right)^2$$

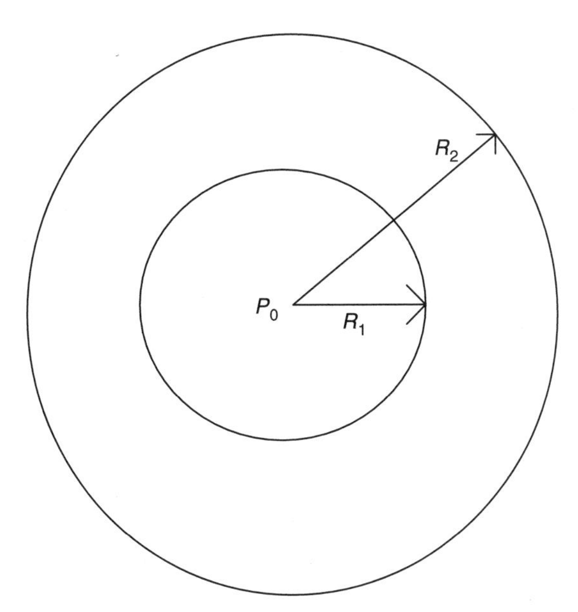

Figure 8-2. Concentric spheres surrounding an antenna.

Thus, the power density is inversely proportional to the square of the distance from the source. For example, if the distance is doubled, the power density is reduced by $\frac{1}{4}$, or 6 dB.

The power available at the output of a receiving antenna is dependent on the effective area of the antenna. This is defined as the ratio of power available at the terminals of the antenna to the power density. Thus the received power is the power density times the effective area, A_a:

$$P_r = \mathcal{P} A_a$$

Effective area may also be defined in terms of the gain of the antenna and wavelength at the channel of interest:

$$A_a = \frac{g_a \lambda^2}{4\pi}$$

Since losses are not included and the antenna impedance and polarization are assumed to be properly matched, all available power is delivered to the antenna terminals. When $g_a = 1$, the antenna is isotropic and the effective area, A_i, is

$$A_i = \frac{\lambda^2}{4\pi}$$

Thus the available power at any distance R from the transmitter is

$$P_r = \frac{P_0}{4\pi R^2}\frac{\lambda^2}{4\pi} = P_0\left(\frac{\lambda}{4\pi R}\right)^2$$

The free-space path attenuation, L_s, in decibels is defined to be $20\log(4\pi R/\lambda)$, R and λ being expressed in common units. The power received by an antenna in a linearly polarized field of intensity E is[2]

$$P_r = \frac{g_a E^2 \lambda^2}{480\pi^2} \qquad \text{watts}$$

so that the formula for field strength is

$$E = \frac{21.9\pi (P_r)^{1/2}}{\lambda} \qquad \text{volts/meter}$$

or

$$E = 5.475\frac{P_0^{1/2}}{R}$$

The free-space field strength is thus independent of frequency and inversely proportional to distance.

If the free-space path loss were the only factor, the received signal power for an ERP of 1 kW would be given by

$$P_r = 60 - 20\ \log\frac{4\pi R}{\lambda} \qquad \text{dBm}$$

However, the signal is partially blocked or attenuated by urban clutter, trees, and other obstacles. Multipath propagation occurs because of reflections from the ground as well as other reflecting, refracting, and diffracting objects. In addition, the earth is curved, preventing line-of-sight propagation to great distances even in the absence of other obstacles.

DISTANCE TO THE RADIO HORIZON

To determine the distance to the horizon as viewed from the transmitting antenna, consider the geometry representing the broadcast station over spherical earth shown in Figure 7-2. By the Pythagorean theorem,

$$R^2 + (R_e + h_r)^2 = (R_e + h_t)^2$$

[2] Edward C. Jordan, *Electromagnetic Waves and Radiating Systems*, 2nd ed., Prentice Hall, Upper Saddle River, N.Y. 1968, p. 417.

Expanding both sides gives

$$R^2 + R_e^2 + 2R_e h_r + h_r^2 = R_e^2 + 2R_e h_t + h_t^2$$

Canceling R_e^2 from both sides yields

$$R^2 + 2R_e h_r + h_r^2 = 2R_e h_t + h_t^2$$

For the terrestrial broadcast case, $2R_e h_t \gg h_t^2 \gg h_r^2$, so that

$$R \sim \sqrt{2}\, R_e h_t$$

The line-of-sight distance is determined by the sum of the antenna and tower heights and the radius of the earth. Assuming the earth's radius to be 6250 km, the distance to the horizon from a 300-m antenna and tower is about 61 km or 38 miles.

REFRACTION

In the troposphere, variations in the temperature, pressure, and water vapor content cause variations in the dielectric constant, index of refraction, and velocity of propagation. Waves passing from one medium to another having a different velocity of propagation are bent or refracted. The index of refraction, n, is defined as the ratio of the velocity of the wave in a vacuum, c, to the velocity in the medium, v; that is,

$$n = \frac{c}{v}$$

or

$$n = \varepsilon_r^{1/2}$$

An expression for the dielectric constant of air was given in the Chapter 5. Applying the binomial expansion to obtain the square root, the index of refraction is

$$n = 1 + 103.4 \times 10^{-6} \left(\frac{P_a}{T}\right) + 84.6 \times 10^{-6} \left(1 + \frac{5880}{T}\right) \frac{P_w}{T}$$

This is often written as the modified index or refractivity, N_r, where

$$N_r = (n - 1) \times 10^6$$

so that

$$N_r = 103.4\frac{P_a}{T} + 84.6\left(1 + \frac{5880}{T}\right)\frac{P_w}{T}$$

$$= \frac{103.4\ [P_a + 0.818P_w(1 + 5880/T)]}{T}$$

At normal air temperatures, $5880/T \gg 1$, so that

$$N_r \sim \frac{103.4\ (P_a + 4811P_w/T)}{T}$$

When the partial pressures are expressed in millibars instead of mmHg, the expression for N_r is

$$N_r = 77.6\frac{P_a}{T} + 3.73 \times 10^5\frac{P_w}{T^2}$$

Pressure, humidity, and temperature vary with time and location. However, for a standard atmosphere, each of these quantities is assumed to decrease exponentially with increasing altitude, h, so that N_r also decreases exponentially. For the CCIR standard radio atmosphere,[3]

$$N_r = N_r(h) = N_s e^{-Bh}$$

where h is measured in kilometers, $N_s = 315$, and $B = 0.136\ \text{km}^{-1}$. The index of refraction as a function of height is therefore

$$n = n(h) = 1 + 3.15 \times 10^{-4} e^{-0.136h}$$

For low-elevation angles as normally encountered in broadcasting, the radius of curvature of the propagation path, ρ, is equal to the inverse of the slope of $n(h)$, or

$$\frac{1}{\rho} \sim -\frac{dn}{dh}$$

Normally, dn/dh is negative, in which case waves are refracted downward toward the surface of the earth. This causes an apparent increase in the radius of curvature. The equivalent earth radius factor, K, is

$$K = \frac{R_{\text{eff}}}{R_e}$$

[3] CCIR Study Group 5 Document, Vol. V, *Radiometeorological Data*, XVth Plenary Assembly, 1982.

where R_{eff} is the effective earth radius. K is related to $1/\rho$ by[4]

$$K = \frac{1}{1 - R_e/\rho}$$

For the standard atmosphere, $n = 1.000315$ at the surface; at an altitude of 1 km, $n = 1.000275$, so that the average value of dn/dh in the lower portion of the atmosphere is $-4 \times 10^{-5}\ \text{km}^{-1}$. Thus, $K = \frac{4}{3}$; the earth's radius normally appears to be about 33% greater than actual. This is referred to as a $\frac{4}{3}$ earth model.

For an effective earth radius of 4(6250)/3 or 8330 m, the distance to the radio horizon from a 300-m tower and antenna is 71 km or 44 miles. Thus, the effect of standard refraction and all atmospheric conditions for which $1 < K < \frac{4}{3}$ is to extend the distance for line of sight propagation. In practice, $N_r(h)$ is a function of location and time; both daily and seasonal variations are observed. Tables for N_s and B for different months and locations are available in the referenced CCIR document.

MULTIPATH

In the case of a single direct wave propagating in free space, the voltage at the output of a properly terminated receive antenna is proportional to

$$\frac{e^{-jkR}}{R}$$

where k is the propagation constant given by $2\pi/\lambda$. When a wave encounters a discontinuity, part of the energy may be reflected. Reflected signals may reinforce or interfere with the direct signal. Since the velocity of the wave does not change, the angle of reflection, θ_r, is equal to the angle of incidence, θ_i. For a wave reflected from any object, be it the ground or other structure, the voltage at the receive antenna terminals is proportional to

$$\frac{\Gamma e^{-jk(R+\delta R)}}{R + \delta R}$$

where Γ is the complex reflection or scattering coefficient and δR is the incremental distance traveled by the reflected wave. Any phase shift at the reflecting boundary is accounted for in the complex reflection coefficient. The reflection coefficient may also, for convenience of calculations, include the effects of variations in the response of the antennas with respect to angle or polarization

[4] Jordan, op. cit., pp. 647–650.

and the divergence factor of the reflecting object. In the event of multiple reflections, additional terms may be considered with $\rho = \rho_n$ and $\delta R = \delta R_n$.

Waves arriving at the receive antenna by other means, including troposcatter and transmission through partially opaque objects, may be considered in a similar manner. For example, troposcatter may occur in the regions of the troposphere where local indices of refraction are different from that of the surrounding atmosphere. Since the scattering occurs in all directions, the signal strength is usually low. This effect occurs primarily for frequencies above 100 MHz. Like reflected signals, the signal strength available at the receive antenna for troposcatter and partially transmitted waves is proportional to

$$\frac{B_n e^{-jk(R+\delta R_n)}}{R + \delta R_n}$$

where B_n is the net amplitude of the nth wave and δR_n is the effective incremental distance traveled, including any phase shifts due to diffraction around or transmission through the region.

The total voltage at the terminals of the receive antenna is given by the vector sum of all signals,

$$\sum_{n=0}^{N} \frac{A_n e^{-jk(R+\delta R_n)}}{R + \delta R_n}$$

A_n represents the amplitude of the nth wave, whether direct, reflected, refracted, or diffracted. The index $n = 0$ represents the direct wave for which $A_0 = 1$ and $\delta R_0 = 0$. All other values of n represent waves arriving by other paths. N is the total number of waves arriving by other paths. It is assumed that the transmitter, receiver, and all scatterers are in fixed locations.

This expression may be simplified somewhat by including any effect of δR in the denominator in the value for A_n and by rewriting as follows:

$$\frac{e^{jkR}}{R} + \sum_{n=1}^{N} \frac{A_n e^{-jkR - jk\delta R_n}}{R}$$

Since e^{jkR}/R is common to both terms, the normalized voltage at the receive antenna terminals may be expressed as

$$\frac{V}{V_0} = 1 + \sum_{n=1}^{N} A_n e^{-jk\delta R_n} \qquad (*)$$

This expression describes the sum of a unit vector and a set of vectors of magnitudes A_n with phases $k\delta R_n$. It may be used to compute the complex value of the signal as a function of frequency at any location. For the special case of reflection from a perfectly reflecting earth, this equation may be simplified by assuming that $A_1 = \Gamma = -1$, all other $A_n = 0$, and $N = 1$.

For specific reflecting objects such as tall buildings, an expression for δR_n may be written in terms of the distance from the transmitter to the echo, R_r, and the relative bearing of the echo and the receiver, $\Delta\phi$. Using the law of cosines, it can be shown that

$$\delta R_n = R_r - R + (R_r^2 + R^2 - 2R_r R \cos \Delta\phi)^{1/2}$$

GROUND REFLECTIONS

Although signals reflected from the ground are fundamentally no different than any other reflected signal, they are considered separately simply because they contribute to the received signal strength at virtually all terrestrial receiving locations. In the presence of a single reflection, the total signal strength at a receiving location is simply the vector sum of the direct-wave field intensity, E_d, and the reflected-wave field intensity, E_r:

$$E = E_d + E_r$$

Following Jordan,[5] consider the case of a flat earth with transmitter antenna height, h_t, and receive antenna height, h_r, separated by a large distance as shown in Figure 8-3. This is the geometry of the classic ground-level antenna range. The sum of a direct wave of unit value and the reflected waves is

$$|E| = 1 + \Gamma e^{-jk\delta R}$$

where

$$\delta R = R_2 - R_1$$

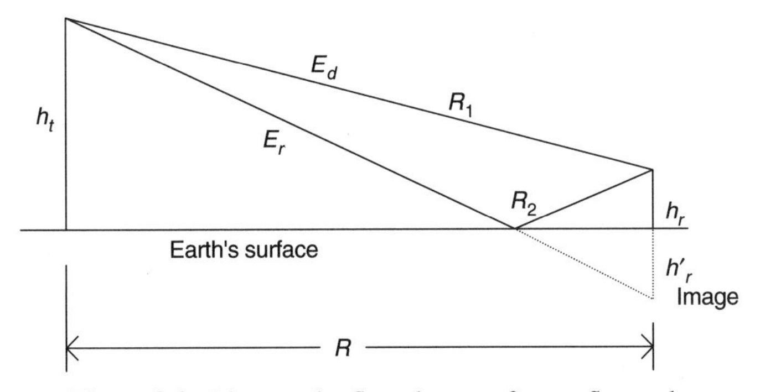

Figure 8-3. Direct and reflected waves from a flat earth.

[5] Jordan, op. cit., pp. 610–635.

The equations for the earth's reflection coefficient for vertical and horizontal polarization are complex expressions; the interested reader is referred to Jordan's text. For both polarizations, the magnitude of the reflection coefficient varies between 0 and 1. At the Brewster angle the magnitude of the vertically polarized reflected wave is at a minimum. It is readily shown that for grazing angles much less than the Brewster angle, the reflection coefficient is approximately equal to -1. For a broadcast station, the grazing angle is typically less than 2.5°. Thus, the assumption of $\Gamma = -1$ is very good, especially for horizontal polarization over smooth terrain.

For the assumed conditions, the sum of the direct and reflected signals may be written

$$|E| = 1 - e^{-jk\delta R}$$

By the Pythagorean theorem

$$R_1^2 = R^2 + (h_t - h_r)^2$$

Therefore,

$$R_1 = R\left[1 + \left(\frac{h_t - h_r}{R}\right)^2\right]^{1/2}$$

Similarly, using the Pythagorean theorem with the image of the receive antenna yields

$$R_2 = R\left[1 + \left(\frac{h_t + h_r}{R}\right)^2\right]^{1/2}$$

By the binomial expansion, when $x \ll 1$

$$(1 + x)^{1/2} \sim 1 + x/2$$

Then

$$R_2 - R_1 \sim R\left[1 + \frac{1}{2}\left(\frac{h_t + h_r}{R}\right)^2\right] - R\left[1 + \frac{1}{2}\left(\frac{h_t - h_r}{R}\right)^2\right]$$

$$\sim \frac{2h_t h_r}{R}$$

Therefore, $k\delta R = 4\pi h_t h_r / \lambda R$.

Now the relative field strength may be rewritten as

$$|E| = 1 - \cos k\delta R + j \sin k\delta R$$

When $k\delta R$ is small, $\cos k\delta R \sim 1$ and $\sin k\delta R \sim k\delta R$; this approximation is very good for $4\pi h_t h_r/\lambda R < \pi/8$ and is useful for $4\pi h_t h_r/\lambda R < 1$. With this approximation,

$$|E| \sim \sin k\delta R = \sin \frac{4\pi h_t h_r}{\lambda R} \sim \frac{4\pi h_t h_r}{\lambda R}$$

This expression may be considered as a ground reflection attenuation factor, α_{gr}, and may be used to estimate the relative signal level as a function of the height of the receiving antenna for fixed transmitter antenna height, wavelength, and distance. Alternatively, it may be used to calculated signal strength versus wavelength or frequency for fixed receiving antenna height and distance. By way of illustration, assume a transmitter antenna and tower height of 600 m and a receiver located 60 km distant. The signal strength is plotted as a function of receiving antenna height in Figure 8-4. At ground level, the signal level is zero, increasing linearly as the antenna height is increased. At low- and high-band VHF, the signal never reaches a maximum value for receiving antenna heights up to 9 m. At UHF, the signal strength reaches a peak for a receiving antenna height of about 6 m and begins to decrease at greater heights. Peak signal strength may be as great as twice the free-space value.

Similar plots could be made for other transmitter antenna heights, distances, and specific channels. The shape of the curves would remain essentially unchanged, although specific values at each height would change. The key point

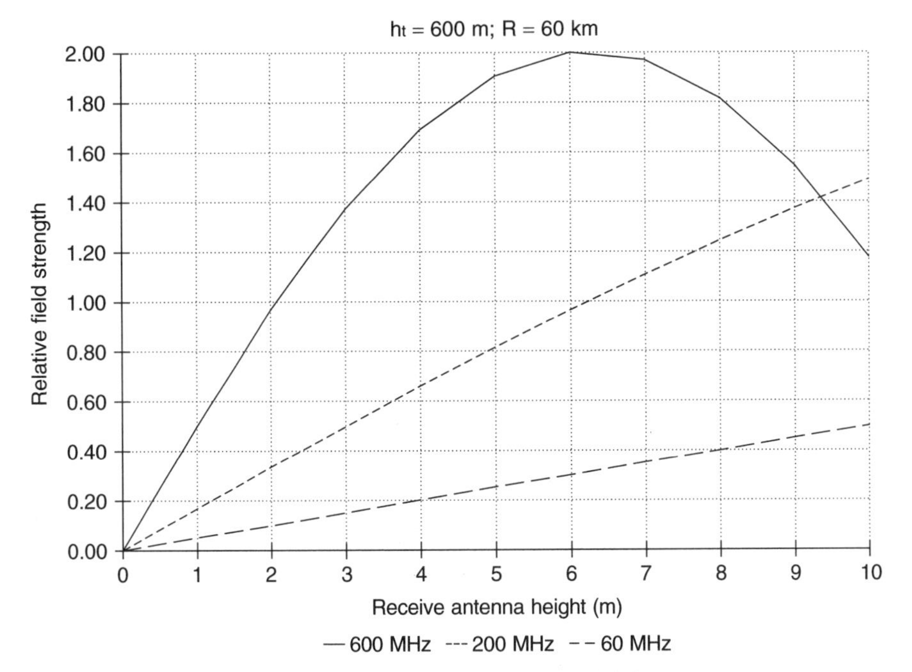

Figure 8-4. Field strength versus height.

is that ground reflections have a substantial effect on signal strength. In general, signal strength reductions below free-space values should be anticipated. This is especially true near ground level, such as for indoor antennas on the first floor of a building. For the conditions assumed for Figure 8-4, the signal strength at 2 m above ground is 12 to 13 dB lower than that at 9 m for the VHF bands. At UHF, the signal strength at 2 m above ground is 3 dB below the maximum value. The actual loss will obviously vary by site. The Implementation Guidelines to DVB-T provides for a height loss of 10 dB for bands I and III and 12 dB for bands IV and V. An additional 7 to 8 dB is allowed for building penetration loss.

By multiplying the expression for free-space attenuation by the ground reflection attenuation factor, simultaneous account may be taken of these effects. In the presence of a single ground reflection, the relative power received by an isotropic antenna near a flat earth is

$$\frac{P_r}{P_t} = \frac{\lambda}{4\pi R}\,\frac{4\pi h_t h_r}{\lambda R} = \left(\frac{h_t h_r}{R^2}\right)^2$$

This expression is valid when the grazing angle and the path-length difference between direct and reflected waves are small. Under the assumed conditions, the received power is dependent only on the transmitted power, antenna heights, and separation distance; there is no frequency dependence.[6] The signal is reduced by 12 dB whenever the distance is doubled,[7] rather than 6 dB as would be expected for free space propagation. These considerations may be extended to show that the field strength in the presence of ground reflections is inversely proportional to wavelength. For larger values of $k\delta R$,

$$\begin{aligned}\alpha_{gr}^2 &= (1 + \Gamma \cos k\delta R)^2 + \Gamma^2 \sin^2 k\delta R \\ &= 1 + 2\Gamma \cos k\delta R + \Gamma^2(\cos^2 k\delta R + \sin^2 k\delta R) \\ \alpha_{gr}^2 &= 1 + \Gamma^2 + 2\Gamma \cos k\delta R\end{aligned}$$

SURFACE ROUGHNESS

The full effect of ground reflections on signal strength is obtained only if the surface is "smooth." The is called specular reflection, for which the foregoing expressions are useful for estimating signal strength. For digital television broadcast frequencies, the surface may appear somewhat rough to the eye, yet still be considered smooth for the purpose of estimating reflected signal level.

[6] C. R. Burrows, A. Decino, and L.E. Hunt, "Ultra-Short-Wave Propagation over Land," *Proc. IRE*, Vol. 23, No. 12, December 1935, p. 1509.

[7] B. Trevor and P.S. Carter, "Propagation of Waves," *Proc. IRE*, Vol. 21, No. 3, March 1933, p. 1509.

A surface is usually consider smooth if the height difference between peaks and valleys, Δh, meets Lord Rayleigh's criteria,[8] developed originally for optics:

$$\sin \psi < \frac{\lambda}{M_s}$$

where M_s may range from 8 to 32. This criterion applies within a region analogous to the first Fresnel zone, where the reflection occurs. For broadcast systems this region is defined by an ellipse whose major axis, a, and minor axis, b, are given approximately by

$$a = \frac{12(1 \pm 0.943)h_r^2}{\lambda}$$

and

$$b = 2\sqrt{2}h_t$$

For surfaces whose height variation exceeds this criterion, the reflection becomes diffused rather than specular. "Rough surfaces" may have a reflection coefficient in the range 0.2 to 0.4. On any general propagation path, both specular and diffused reflection may occur simultaneously.

The FCC field strength charts are published for an assumed surface roughness of 50 m. This value does not meet the Rayleigh criterion anywhere in the TV broadcast bands, even for $M_s = 8$. To be considered smooth under this, the least stringent Rayleigh criterion, the surface roughness, should be less than 10 m for low band, 4 m for high band, and 1.2 m for UHF. For a perfectly smooth earth ($\Delta h = 0$) the field strength values on the FCC charts must be raised by 1.9 dB for low band, 2.5 dB for high band, and 4.8 dB for UHF.

EFFECT OF EARTH'S CURVATURE

The curvature of the earth further affects the propagation of the space wave because the ground-reflected wave is reflected from a curved surface. The energy diverges more than it does from a flat surface, and the ground-reflected wave reaching the receiver is weaker than for a flat earth. The divergence factor that describes this effect is therefore less than unity. Again, consider only the case for elevated antennas within the line of sight (LOS).

It can be shown[9] that the divergence factor, D, is given by

$$D = \frac{1}{(1 + 2d_t d_r / R_e R \tan \psi)^{1/2}}$$

[8] Donald E. Kerr, *Propagation of Short Radio Waves*, Boston Technical Publishers, Lexington, Mass., 1964, pp. 411–416; Y.T. Lo and S.W. Lee, eds., *Antenna Handbook*, Van Nostrand Reinhold, New York, 1988, pp. 29–38, 32–22.

[9] Kerr, op. cit., pp. 406, 422–428.

where ψ is the grazing angle. All quantities are defined in Figure 7-2. This form of the divergence factor is useful only when expressions for d_t, d_r, and ψ are available. Since this is not usually the case, further geometrical considerations are necessary. From trigonometry,

$$\tan\psi = \frac{h_t' + h_r'}{R}$$

where h_t' and h_r' are the apparent tower heights, given by

$$h_t' = h_t - \frac{d_t^2}{2R_e} \qquad \text{and} \qquad h_r' = h_r - \frac{d_r^2}{2R_e}$$

Adding the expressions for h_t' and h_r' yields

$$h_t' + h_r' = h_t + h_r - \frac{d_t^2 + d_r^2}{2R_e}$$

Therefore,

$$\tan\psi = \frac{h_t + h_r}{R} - \frac{d_t^2 + d_r^2}{2R_e R}$$

By proportionality,

$$d_t = \frac{h_t R}{h_t + h_r}$$

and

$$d_r = \frac{h_r R}{h_t + h_r}$$

Using these additional relationships, the divergence factor may easily be computed using known quantities. For small values of R, D approaches unity; it approaches zero as R approaches the distance to the radio horizon. The divergence factor may be combined with the ground reflection coefficient so that the attenuation due to ground reflections becomes

$$\alpha_{\text{gr}} = [1 + (\Gamma D)^2 - 2\Gamma D \cos k\delta R]^{1/2}$$

The reflection coefficient has simply been multiplied by the divergence factor.

FRESNEL ZONES

To fully understand the effect of ground reflections, further consideration must be given to the concept of Fresnel zones. As described earlier, a pair of rays propagate from the transmitter to the receiver by direct and reflected paths. For small grazing angles, a 180° phase reversal occurs at the ground reflection point. As the receiving antenna is raised above the earth, alternating minima and maxima of signal strength are observed. When the receiving antenna is at ground level, the direct and reflected path are equal in length and the total signal is zero. As the receive antenna is raised, the difference in path length increases until the path of the reflected wave is one-half wavelength longer than the path of the direct wave. At this point, the reflected wave arrives in phase with the direct wave and there is constructive interference. The radius of the base of a cone defined by the ray paths within which the path lengths differ by one-half wavelength or less is called the first Fresnel zone radius, F_1. When the direct path clears the earth's surface by a distance equal to the radius of the first Fresnel zone, the received signal is twice as strong as the direct wave, or 6 dB greater than the free-space value. As the receive antenna is raised further, the path difference increases to one wavelength and the resultant signal is zero. The radius of the base of the cone defined by these ray paths is called the second Fresnel zone radius, F_2. Additional Fresnel zones are numbered outward from the center as shown in Figure 8-5.

The radius of the nth Fresnel zone, F_n, can be expressed in terms of the first by

$$F_n = \sqrt{n} F_1$$

where the radius of the first Fresnel zone is given by

$$F_1 = \left(\frac{\lambda d_t d_r}{R} \right)^{1/2}$$

All quantities are expressed in common units.

In general, if the reflected signal is due to reflection from within the first Fresnel zone, the total signal is maximum. If the reflected signal is due to reflection from an odd number of Fresnel zones, the signal level reaches secondary maxima, since a greater proportion of the reflected wave arrives in phase with the direct wave. In contrast, if the signal is reflected from an even number of Fresnel zones, the signal strength is reduced.

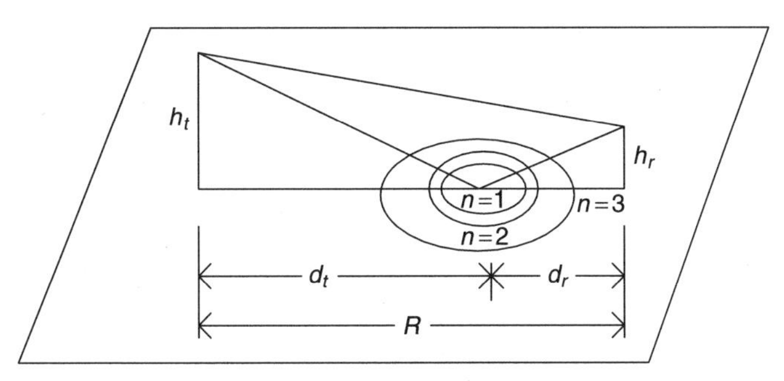

Figure 8-5. Fresnel zone geometry.

LINEAR DISTORTIONS

The expression for relative signal strength due to multipath may be used to estimate the propagation induced linear distortions across a digital TV channel. From this result the degradation of EVM, C/N or the tap values for an equalizing filter may be estimated. This effect is similar in nature to that of a mismatched transmission line.

To compute frequency response and group delay, the starred equation in the "Multipath" section is written in rectangular form:

$$\mathrm{Re}\left[\frac{V}{V_0}\right] = 1 + \sum_{n=1}^{N} A_n \cos \frac{\omega \delta R_n}{c}$$

$$\mathrm{Im}\left[\frac{V}{V_0}\right] = \sum_{n=1}^{N} A_n \sin \frac{\omega \delta R_n}{c}$$

The amplitude of the frequency response is simply the magnitude of the vector $\mathrm{Re}[V/V_0] + j\,\mathrm{Im}[V/V_0]$, or

$$\mathrm{Mag}\left[\frac{V}{V_0}\right] = \left[\mathrm{Re}\left[\frac{V}{V_0}\right]^2 + \mathrm{Im}\left[\frac{V}{V_0}\right]^2\right]^{1/2}$$

The phase is the angle of this vector

$$\mathrm{Ph}\left[\frac{V}{V_0}\right] = \tan^{-1} \frac{\mathrm{Im}[V/V_0]}{\mathrm{Re}[V/V_0]}$$

Both amplitude and phase are proportional to echo magnitude. If the direct signal is obstructed, the echo magnitudes may be greater than unity.

Recall from Chapter 4 that group delay, GD, is the negative first derivative of phase with respect to angular frequency. Also, recall from calculus that

$$\frac{d(\tan^{-1} u)}{dx} = \frac{du/dx}{1 + u_2}$$

For the present calculation, let $u = \mathrm{Im}[V/V_0]/\,\mathrm{Re}[V/V_0]$ and $x = \omega$. It is now straightforward to find the derivatives of the real and imaginary parts, from which the group delay may be computed:

$$\frac{d\,\mathrm{Re}[V/V_0]}{d\omega} = -\sum_{n=1}^{N} \frac{\delta R_n A_n [\sin(\omega \delta R_n/c)]}{c}$$

$$\frac{d\,\mathrm{Im}[V/V_0]}{d\omega} = \sum_{n=1}^{N} \frac{\delta R_n A_n [\cos(\omega \delta R_n/c)]}{c}$$

The group delay, after considerable manipulation, is

$$\mathrm{GD} = -\left(1+\sum_{n=1}^{N} A_n \cos\frac{\omega\delta R_n}{c}\right)\sum_{n=1}^{N}\frac{\delta R_n A_n[\cos(\omega\delta R_n/c)]}{c}$$

$$+\frac{\sum_{n=1}^{N} A_n \sin(\omega\delta R_n/c)\sum_{n=1}^{N}[\delta R_n A_n/c][\sin(\omega\delta R_n/c)]}{\left[\sum_{n=1}^{N} A_n \sin(\omega\delta R_n/c)\right]^2+\left[1+\sum_{n=1}^{N} A_n \cos(\omega\delta R_n/c)\right]^2}$$

This complex expression has the dimensions of seconds, as expected. The group delay is proportional to both echo magnitude and delay. The components of this expression are similar in form to a Fourier series, with coefficients equal to the amplitudes of the interfering waves. The periods are proportional to the frequency and the incremental distance traveled by the waves.

To visualize the effect of multipath signals on the received signal, consider the vector diagram shown for times t_1, t_2, t_3, and t_4 in Figure 8-6. The unit vector representing the direct wave is assumed fixed. The multipath signals, represented by the smaller, rotating vectors, add to the direct wave, just like the interaction of incident and reflected waves on a transmission line. The magnitude and phase of the sum of these vectors represents the total voltage at the receive antenna terminals at a specific frequency. The maximum signal level occurs when all of the vectors add along the axis of the unit vector; the minimum occurs when they subtract. The maximum phase shift occurs when all the reflected-wave vectors are at right angles to the direct vector. The rate of change of phase is independent of the direct signal but is proportional to the delay of the interfering signals.

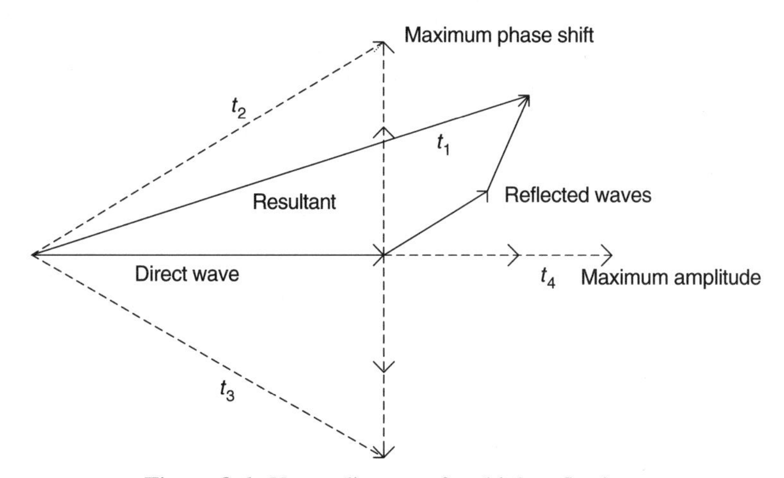

Figure 8-6. Vector diagram of multiple reflections.

Clearly, signal strength and linear distortions are dependent on the number of echoes and their strength and delay relative to the direct signal. The ground reflection is almost always present; usually, the incremental path length is short, and in many cases judicious selection of antenna location can maximize signal strength and minimize linear distortions. Unfortunately, the complete multipath environment is not under the direct control of the broadcast engineer. The general case includes multiple signals arriving at any given receive location. For example, even in rural areas it is likely that more than one echo will be present from low buildings, trees, overhead utilities, and the occasional tower. In suburban areas, the number of echoes may increase due to the higher density of homes, businesses, and industry and other man made structures. In dense urban areas, a total number of propagation paths on the order of 100 might be expected. The resulting frequency-dependent fading produces linear distortions that vary from channel to channel.

For a single echo, the group delay expression simplifies to

$$GD = \frac{-(A_1 \delta R_1/c)(\cos kR_1 + A_1)}{1 + A_1^2 + 2A_1 \cos kR_1}$$

As the strength of the multipath increases, the peak-to-peak signal variation and maximum phase change increase, independent of echo delay. As echo magnitude and delay increase, the group delay increases. The receiver equalizer compensates for these distortions by adjusting the tap weights. The overall effect is to decrease the effective signal level at the receiver. In general, echoes with time delays much less than a symbol period and magnitude of 10 to 15% of the direct signal degrade the threshold C/N value by less than 0.5 dB.[10] Unfortunately, echoes due to obstates such as buildings are often much stronger with longer time delay.

A theoretical study[11] of an urban area such as New York City concluded that as many as 90 echoes might be present, some within 3 or 4 dB of the direct signal and with delays ranging from 200 to more than 2000 ns. The large amount of phase shift and group delay across a pair of low-band channels for a single echo with an amplitude of −3 dB and a delay of 200 nS is shown in Figure 8-7. Peak-to-peak amplitude variations are approximately 15 dB. The random effect on the response at any specific channel is evident.

The study cited suggested that it may be possible to reduce the overall effect of multipath on C/N by using circularly polarized transmit and receive antennas. This is a consequence of the tendency for right-hand circularly polarized waves to be reflected as left-hand circularly polarized waves. This occurs for any surface for which the reflection coefficients of the parallel and perpendicular components of the wave are equal. For example, waves incident on many dielectric materials at low grazing angles are reflected at nearly full amplitude with 180° phase

[10] Carl G. Eilers and G. Sgrignoli, "Echo Analysis of Side-Mounted DTV Broadcast Antenna Azimuth Patterns," *IEEE Trans. Broadcast.*, Vol. BC-45, No. 1, March 1999.

[11] H. R. Anderson, "A Ray-Tracing Propagation Model for Digital Broadcast Systems in Urban Areas," *IEEE Trans. Broadcast.*, Vol. 39, No. 3, September 1993, p. 314.

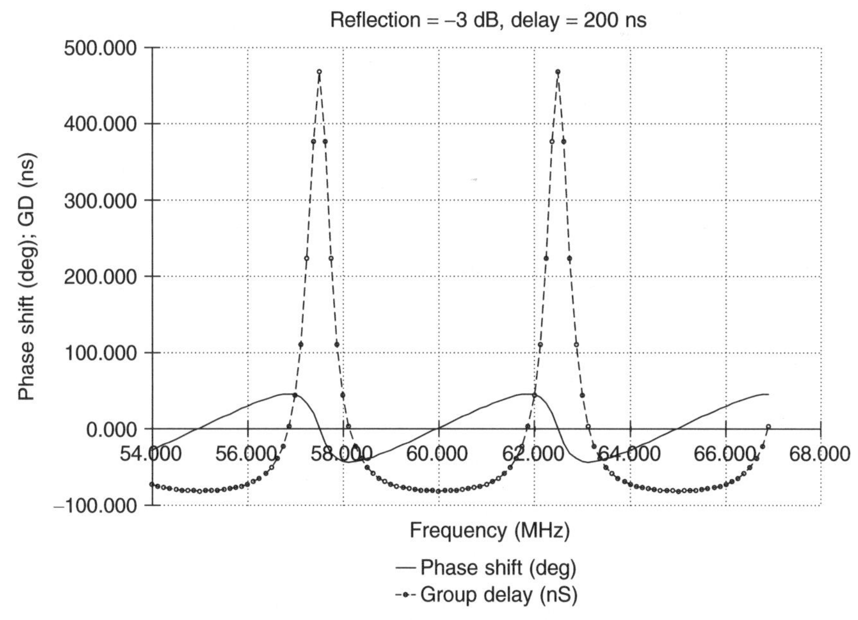

Figure 8-7. Phase and group delay.

shift for both components. This would include the earth's surface and many nonmetallic building materials. Similarly, good conducting materials exhibit reflection coefficients of -1. Since the circular polarized receiving antenna responds primarily to right-hand circular polarization, echoes from a single surface are rejected by the antenna. The result is a reduction in echo strength.

Four multipath models have been used to evaluate adaptive equalizers for digital television systems.[12] The echo levels and delays are summarized in Table 8-1. It is convenient to display this information in the form of a magnitude–delay profile. Model D is shown in Figure 8-8.

DIFFRACTION

Diffraction is a phenomenon that produces electromagnetic fields beyond a shadowing or absorbing obstacle. As the wave grazes the obstacle, a diffraction field is produced by a limited portion of the incident wavefront. According to Huygens' principle, every point on the incident wavefront may be considered a new point source of secondary radiation which propagates in all directions. By the principles of geometric optics, the vector sum of the rays from the secondary

[12] Y. Wu, B. Ledoux, and B. Caron, "Evaluation of Channel Coding, Modulation and Interference in Digital ATV Transmission Systems," *IEEE Trans. Broadcast.*, Vol. BC-40, No. 2, June 1994, pp. 76–78.

TABLE 8-1. Multipath Models

	Model			
n	A (Typical)	B (Typical)	C	D
1	−19 dB	−14 dB	−26 dB	−9 dB
	450 ns	200 ns	70 ns	100 ns
2	−24 dB	−18 dB	−26 dB	−17 dB
	2300 ns	1900 ns	100 ns	250 ns
3		−24 dB	−31 dB	−14 dB
		3900 ns	150 ns	600 ns
4		−22 dB	−28 dB	−11 dB
		8200 ns	250 ns	950 ns
5			−28 dB	−11 dB
			400 ns	1100 ns

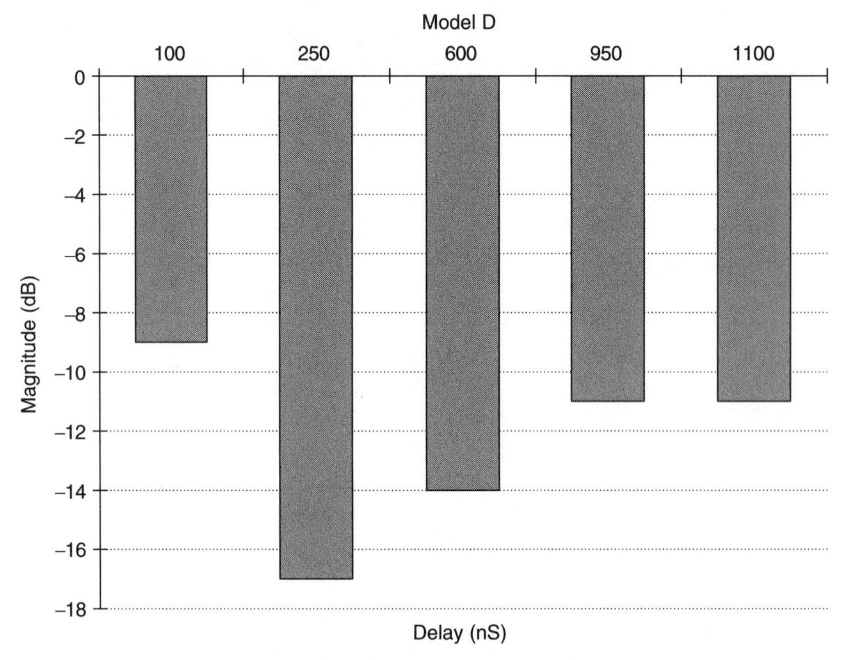

Figure 8-8. Magnitude delay profile.

sources create diffraction patterns with alternate peaks and nulls that propagate into the shadow region. This phenomenon is partially responsible for propagation of digital television signals beyond the radio horizon. The magnitude of the diffracted signal is dependent on the type of surface. For example, a smooth surface such as calm water on the curved surface of the earth produces minimum

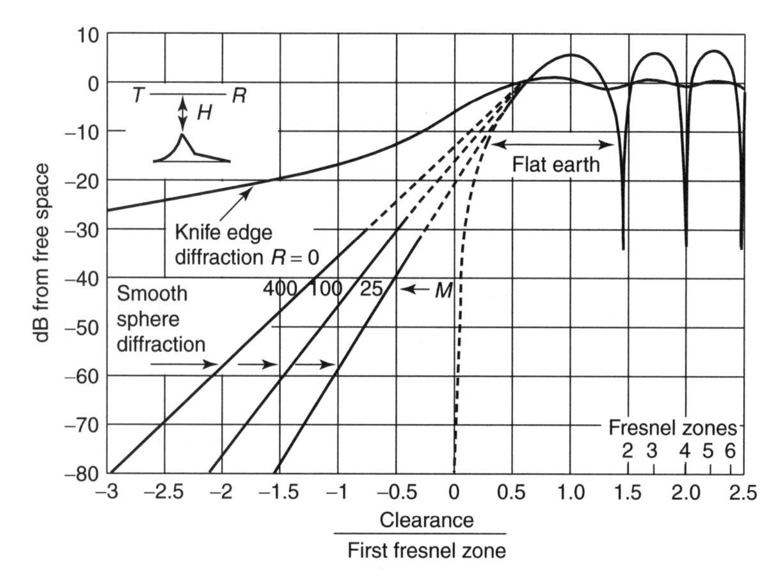

Figure 8-9. Diffraction loss for flat earth, smooth spherical earth, and knife edge. (From *Bell System Technical Journal*, May 1957, p. 608. Property of AT&T Archives. Reprinted by permission of AT&T.)

signal level beyond the horizon. A sharp projection such as a building, mountain peak, or tree may result in maximum diffracted signal. Most obstacles produce diffracted signals between these limits.

The signal strength available in the shadow of a diffracting object may be estimated from Figure 8-9. Graphs of the diffracted signal level relative to the free-space value are plotted for several types of idealized obstacles as a function of the ratio of clearance height, H, to first Fresnel zone radius. If the earth were flat, the signal strength would be zero for zero clearance. However, since the earth is actually curved, usable signal may be available at the radio horizon and beyond. The signal level for zero clearance may range from 6 to 19 dB below that of free space. Knife-edge diffraction is of particular interest in hilly and mountainous regions and the canyons of major cities. Smooth sphere diffraction is of interest in rural areas if the terrain can be considered smooth. The parameter, M, associated with smooth sphere diffraction is directly proportional to transmit antenna height and frequency to the $\frac{2}{3}$ power; that is,

$$M = \frac{h_t}{K^{1/3}} \left[\frac{1 + h_r/h_t)^{1/2}}{2} \right]^2 \left(\frac{f}{4000} \right)^{2/3}$$

The attenuation due to diffraction may be estimated by first calculating the Fresnel zone clearance at the location of interest, then reading the attenuation from the curve that best describes the obstacle. From the geometry of the curved earth displayed in Figure 7-2, it may be shown that the clearance height at any distance from the transmitter is given by

$$H = \frac{h_t'd_t + h_r'd_r}{R}$$

Use of these equations and graphs will be illustrated later in the analysis of digital television field tests.

The effect of an intervening hill is dependent on the extent to which it may be represented by a knife edge or a more rounded object. The hill may be represented by a cylinder of radius R_h on a pedestal with total height H_h as illustrated in Figure 8-10. The height is measured as the distance above the line connecting the transmitting and receiving antenna at the peak of the hill. The attenuation is a function of a height parameter, ν, which is the height measured relative to the first Fresnel zone radius in the absence of the hill.

$$\nu = \frac{\sqrt{2}H_h}{F_1}$$

The sharpness of the peak of the hill is represented by a contour parameter, p_h, which is proportional to the radius relative to the first Fresnel zone radius in the absence of the hill, given by

$$p_h = \frac{0.83R^{1/3}\lambda^{3/4}}{F_1}$$

For a sharp peak, $R_h = 0$, $p_h = 0$ and the knife edge condition applies. The knife edge diffraction loss, L_{ke}, is approximated by

$$L_{ke} = 6.4 + 20\log[(\nu^2 + 1)^{1/2} + \nu] \qquad \text{dB}$$

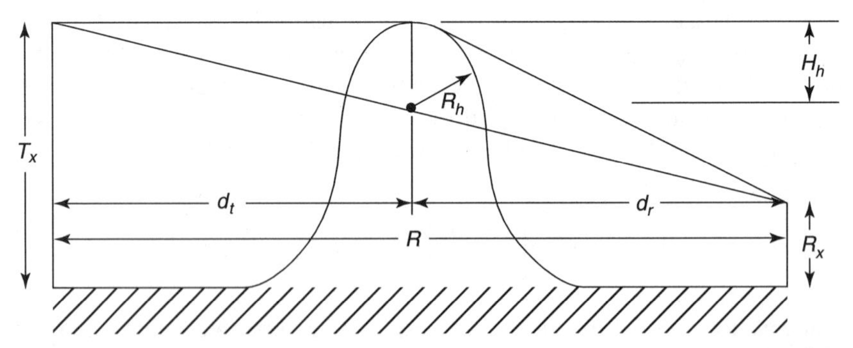

Figure 8-10. Idealized hill geometry. (From *NAB Engineering Handbook*, 9th edition; used with permission.)

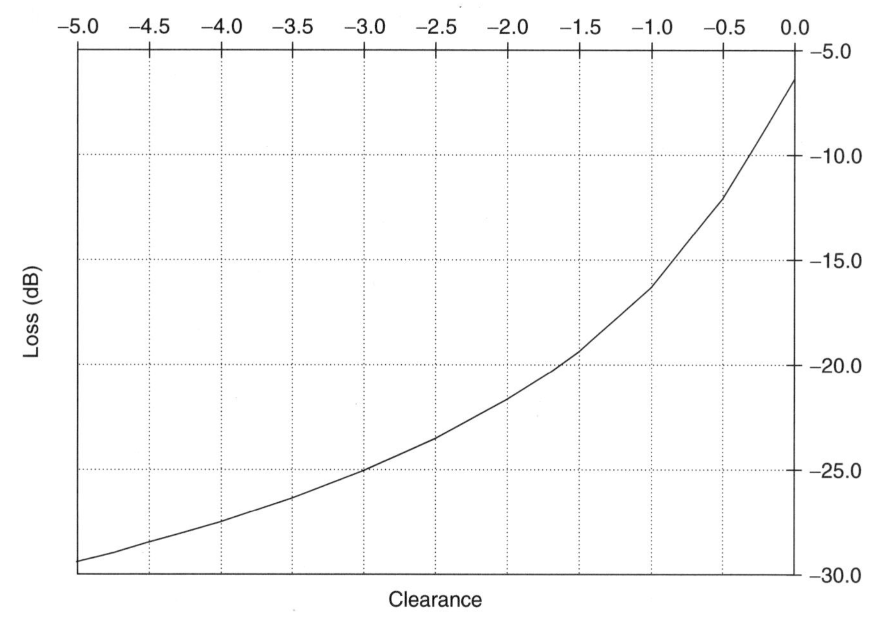

Figure 8-11. Knife-edge diffraction.

This equation is plotted as a function of H_h/F_1 in Figure 8-11. Not surprisingly, the loss increases as the shadowing increases. As the radius of the hill increases, p_h and the resulting attenuation increase at an even greater rate.

The effect of surface roughness on signal strength may partially be understood in terms of diffraction. As the surface roughness increases, the effective reflection coefficient of the surface is reduced[13] by a factor given by $e^{-2\delta}$, where $\delta = (4\pi\Delta h/\lambda)\sin\psi$. Some of the energy is scattered in the general direction of the source. If the obstacle is lossy, some of the energy may be absorbed. Some will propagated into the shadow region in accordance with Huygens' principle. If a reduction in effective reflection coefficient were the only phenomenon, the signal strength would be expected to drop at a rate closer to 6 dB per octave of distance in accordance with free-space propagation. Instead, signal strength is attenuated due to surface roughness. The FCC formula for the loss in signal strength relative to a perfectly smooth earth, ΔF, is[14]

$$\Delta F = -0.03\Delta h\left(1 + \frac{f}{300}\right) \qquad \text{dB}$$

This formula may be used in this form to compute loss for any specified height variation (in meters) and frequency. Alternatively, the elevation of shadowed

[13] Kerr, op. cit., p. 434; Anderson, op. cit., pp. 310–311.

[14] *FCC Rules*, Part 73, 73.684(i).

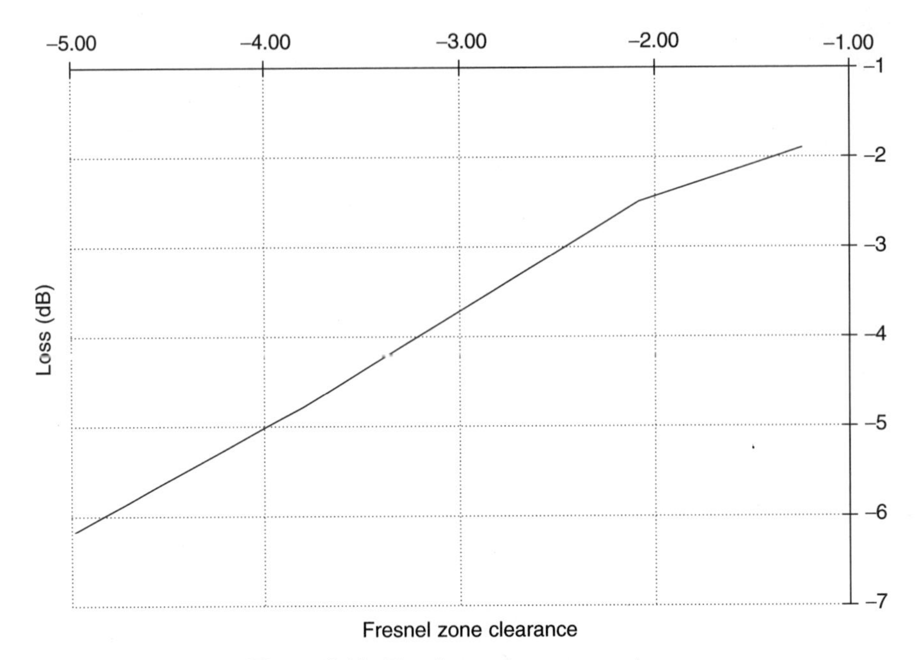

Figure 8-12. Terrain roughness correction.

regions may be "normalized"to the height of terrain peaks as measured in terms of the Fresnel zones radius at any specified location. The result is a relationship between attenuation due to surface roughness and the negative Fresnel zone clearance of the shadow region relative to the peak. Figure 8-12 is a representative plot of this relationship. The loss increases with increasing shadowing, in a manner that is qualitatively similar to diffraction. By normalizing the height to Fresnel zone radius, a single curve describes the attenuation for all frequencies.

FADING

In addition to frequency-dependent fades, the field strength may vary with respect to time due to changes in the propagation environment. These fades are caused by changes in factors that affect multipath and changes in the index of refraction of the atmosphere. Time-dependent fading due to refraction may be especially severe in hot, humid coastal, and tropical areas. Atmospheric temperature inversions can cause abnormal and time varying indices of refraction. In general, fading due to multipath may be expected to be more severe on longer propagation paths and at higher frequencies. The effect of fading is seen in the FCC curves. Curves are labeled FCC(50,10), FCC(50,50), and FCC(50,90), indicating signal strength at 50% of locations at 10%, 50%, and 90% of the time.

PUTTING IT ALL TOGETHER

The method used to predict signal strength is dependent on the purpose for which the prediction is needed. When filing regulatory license exhibits, the procedure specified in the rules of the regulatory agency must be followed. For FCC filings, the signal strength must exceed specified levels, as predicted using the terrain-dependent Longley–Rice[15] method. Digital systems are more sensitive to channel degradation due to multipath and fading than are analog systems. The transition from acceptable to unacceptable C/N is very abrupt; near threshold, a reduction in signal strength and/or increase in noise on the order of 1 dB can result in total loss of picture and sound. This phenomenon is referred to as the "cliff effect". To assure adequate signal within fringe areas, the FCC (50,90) curves are used for planning the extent of noise-limited coverage in the United States.

In general, use of the FCC and CCIR curves is preferred if a quick estimate of field strength is desired. Other methods that may be used to compute field strength include the Epstein–Peterson[16] and Bloomquist–Ladell[17] techniques. The accuracy and ease of use of these and other prediction models has been evaluated and compared.[18] In every case, accurate estimation of the loss due to surface roughness is the most difficult issue. None of these methods provide the accuracy required to guarantee a specific signal level at any particular point.

The method described in the following paragraphs applies the foregoing theoretical principles and provides an understanding of the factors affecting field strength and frequency response. Accurate treatment of the loss due to terrain roughness remains the most difficult issue. To account for the frequency dependence of the terrain loss, changes in elevation are normalized to the Fresnel zone radii. A spreadsheet with graphing capability expedites the calculation and graphical display of the data.

1. Using the transmitting antenna and tower height and effective earth radius, compute the distance to the radio horizon.
2. Using the carrier frequency, compute the free-space attenuation versus distance out to the radio horizon.
3. Compute the attenuation factor due to ground reflections. For locations for which the earth can be assumed to be flat, only the tower height at the transmitter and receiver and frequency need be known. To take the

[15] Rice, Longley, Norton, and Barsis, "Transmission Loss Predictions for Tropospheric Communications Circuits," *National Bureau of Standards Technical Note 101*. Also *OET Bulletin 69*.

[16] J. Epstein and D. W. Peterson, "An Experimental Study of Wave Propagation at 850Mc/s," *Proc. IRE*, Vol. 41, No. 3, May 1953, pp. 595–611.

[17] A. Bloomquist and L. Ladell, "Prediction and Calculation of Transmission Loss in Different Types of Terrain," *NATO AGARD Conference Proceedings*, 1974.

[18] F. Perez Fontan and J. M. Hernando-Rabanos, "Comparison of Irregular Terrain Propagation Models for Use in Digital Terrain Based Radiocommunications Systems Planning Tools," *IEEE Trans. Broadcast.*, Vol. 41, No. 2, June 1995, pp. 63–68.

effect of the curvature of the earth on reflection coefficient into account, the divergence factor should be computed.

4. Compute the diffraction loss, L_d, due to a spherical earth. This will require computating the diffraction parameter, M, and the Fresnel zone clearance.
5. Using the AERP, compute the available power at the receive location. The power, P_r, available at the output of an isotropic receive antenna (in dBW) is

$$P_r(\text{dBW}) = \text{ERP}(\text{dBK}) + 30 - L_s(\text{dB}) - L_{\text{gr}}(\text{dB}) - L_d(\text{dB})$$

where

$$L_{\text{gr}} = 20 \log \frac{1}{\alpha_{\text{gr}}}$$

6. Convert the receive power in dBW to watts.
7. Convert the receive power in watts to field strength in volts per meter. The formula for field strength is

$$E = \frac{21.9\pi (P_r)^{1/2}}{\lambda}$$

8. Convert the field strength in volts per meter to dBu using the formula

$$E(\text{dBu}) = 20 \log E + 120$$

9. If multiple points are of interest, such as a complete radial, plot field strength versus distance.
10. Compute any losses due to diffraction such as surface roughness, shadowing by buildings, hills and mountains, or shadowing due to the earth's curvature. This step requires an accurate topographical plot of the radial under consideration. Good judgment is required to characterize the topography and estimate the associated loss. Subtract the diffraction losses from the plot of field strength versus distance.
11. For specific reflecting objects such as tall buildings, estimate the magnitude and phase of the echo and the effect on the received signal strength.
12. As a reality check, compare the computed data to regulatory agency exhibits and/or field measurements.

To compute the carrier power at the receiver input, it is necessary to include the effect of receive antenna gain and down lead loss. These parameters vary by location; the FCC planning factors are listed in Table 2-1. The carrier power at the receiver input is

$$C(\text{dBm}) = P_r(\text{dBm}) + G_r(\text{dB}) - L(\text{dB})$$

UNDESIRED SIGNAL

In addition to the desired DTV signals, noise and interference will be present at the receiving site. These signals will corrupt the desired signal; their level will place a lower bound on the acceptable level for the desired signal. The level of the desired signal and the total of noise and interference combine to establish the carrier-to-noise plus interference ratio. Details of factors affecting these parameters and specific levels are discussed in Chapter 2.

FIELD TESTS

Analysis of field tests of the ATSC system at Charlotte and Raleigh, North Carolina and Chicago, Illinois serve to illustrate the application the principles of propagation as they apply to digital television signals. The foregoing process is applied to each of these experimental stations to illustrate the factors affecting the signal strength and linear distortions at a variety of receiving sites.

CHARLOTTE, NORTH CAROLINA

At Charlotte, tests were performed for both U.S. channels 6 and 53. The approximate antenna height above average terrain (HAAT) for both channels was 415 m. For channel 6, the AERP was 630 W (−2 dBK); at channel 53 the AERP was 31.6 kW (15 dBK). Tests were made with a receiving antenna height of 9 m.

In the analysis that follows, a $\frac{4}{3}$ earth's radius is assumed; the resulting distance to the radio horizon is 83 km. Considering the height of the receiving antenna over smooth earth, the radio horizon is extended another 12 km. The free-space loss and loss due to ground reflections were calculated, assuming a ground reflection coefficient of -1. The divergence factor was also calculated. In addition, the diffraction loss due to a spherical earth was computed. The diffraction parameter, M, is 30 for channel 6 and 124 for channel 53. The resulting field strength for each respective channel is plotted in Figures 8-13 and 8-14. For comparison purposes, the channel 6 field strength for the flat-earth model is also shown. To obtain these curves, the available power at the receive site was computed using the AERP and relevant attenuation factors; the received power was then converted to field strength. Also shown is the measured field strength data for selected radials.

For channel 6, there is little difference between the curved- and flat-earth models except at long range, where the curved-earth model shows the effect of the divergence factor approaching zero near the radio horizon; in this respect, the curved-earth model fits the measured data slightly better. On average, the measured field strength matches the predicted field strength rather well, especially at near range. At longer range, the calculated curve represents a

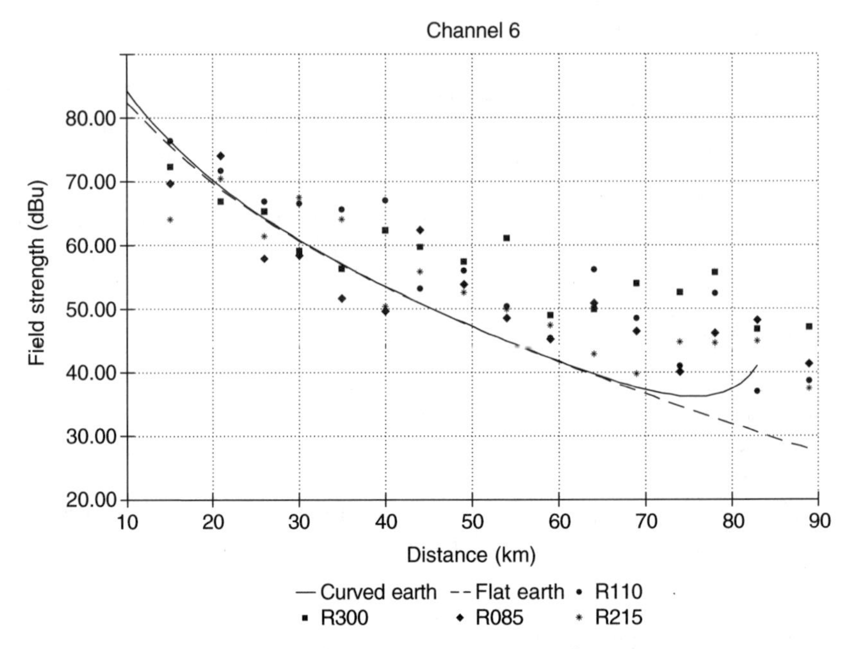

Figure 8-13. Field strength versus distance.

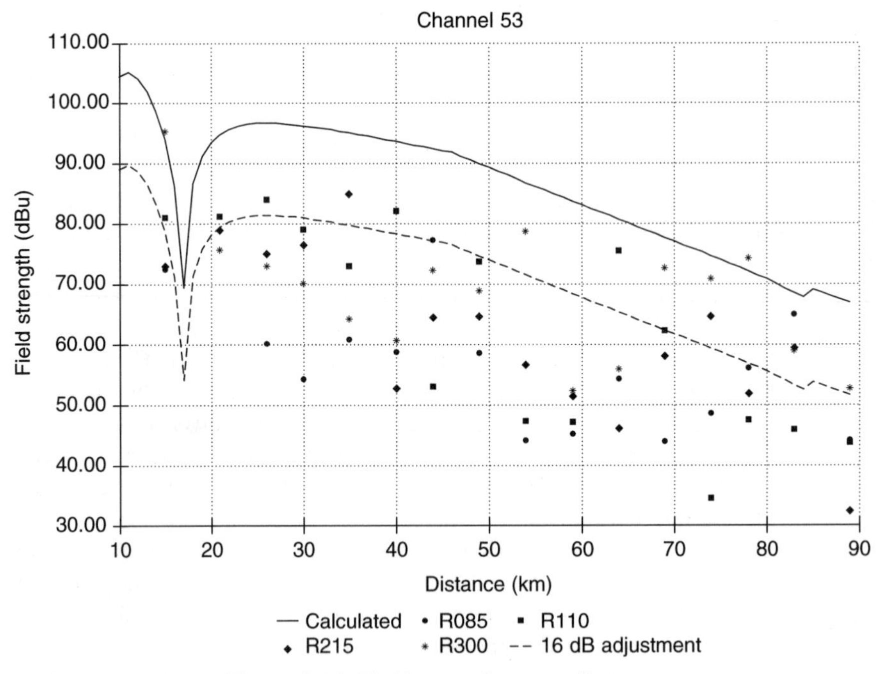

Figure 8-14. Field strength versus distance.

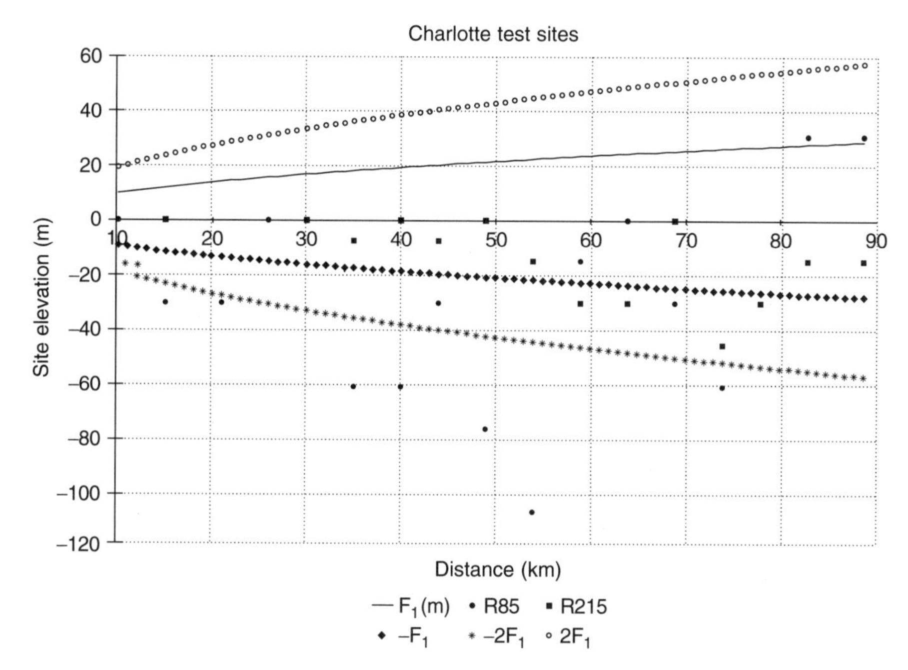

Figure 8-15. Site elevation versus distance.

conservative estimate. Radial R215 has the smoothest terrain.[19] Figure 8-15 shows the elevation of each site relative to the transmitter site. For R215, the average deviation of the measured field is only 3 dB relative to the calculated field. All other radials are classified to some degree as irregular terrain. The poorest match between measurements and calculations is on R300, for which the average deviation is 8 dB. This radial is relatively smooth out to 45 km but becomes very irregular at greater distances. Approximately ±0.7 dB variation is due to the circularity of the omnidirectional antenna.

For channel 53, the measured field strength is well below the calculated curve, except at very short and long ranges. Radial R085 deviates the most from the calculated field with an average deviation of 27 dB. This radial is rougher than either R110 or R215. Overall, the measurements and calculations match best for R300, for which the average deviation is 17 dB. The best evidence of specular reflection from a smooth earth is the measured field strength at a distance of 15 km on R300. Inspection of the terrain on this radial reveals a high flat plateau in the vicinity of this receiving test site. Overall, however, the measured data indicate significant losses, evidently due to diffused reflection from a rough surface. The earth's surface, which appears quite smooth to the channel 6 signal, appears to be very rough at the higher frequency. Approximately ±0.5 dB

[19] G. Sgrignoli, *Summary of the Grand Alliance VSB Transmission System Field Test in Charlotte, N.C.*, June 3, 1996, App. C.

variation in the channel 53 data is due to the circularity of the omnidirectional antenna.

When a surface roughness adjustment is introduced, the calculated field can be made to match the measured data much better. With a 16-dB adjustment, the measured field on R215 deviates above the calculated by about 6 dB. As shown in Figure 8-15, F_1 for the geometry of the Charlotte station ranges from 12 to 28 m. This is approximately the same as the surface roughness over much of R215. From Figure 8-9, knife-edge diffraction over a single obstacle with a height equal to F_1 produces a loss of about 16 dB. The overall roughness of R085 is approximately twice as great over much of the distance. The roughness of R110 is intermediate to R085 and R215. The measured field for R110 and R085 deviate above the calculated by about 10 and 12 dB, respectively.

The peak-to-peak variation of signal strength is much more severe and occurs at a higher spatial frequency for channel 53 than for channel 6. This indicates a greater multipath effect, which may be correlated with the effective roughness of the surface.

The subjective nature of the foregoing adjustments for surface roughness is obvious. With measured data in hand, it is relatively easy to analyze the terrain profile and conclude that "the surface roughness is approximated by F_1, etc." Making such a judgment without the benefit of measured data is much more difficult. It is interesting, however, that the average loss due to surface roughness along a radial may be approximated by the diffraction loss of a single knife edge.

The severity of the multipath is further indicated by the equalizer tap energy ratio.[20] When there is no channel distortion, only the main equalizer tap is on and the weighted tap energy ratio, E_t/E_m, is zero ($-\infty$ in dB). As the multipath becomes more severe, the tap energy increases. Analysis of the tap energy has shown that nearly 40% of channel 6 sites had a tap energy of -16 dB or greater while almost 50% of channel 53 sites had tap energies at or above this level. The tap energy may also tend to increase for the roughest radials. For example, radials R215 and R300 had tap energy for channel 6 of -16 dB or greater on only 6% of locations; the roughest radial, R305, had tap energy at or above this level at 18% of sites. However, the data are not as convincing for channel 53.

Multipath seems to become more severe with increasing path length. A plot of equalizer tap energy versus distance on several radials is shown for channel 53 in Figure 8-16. Although there is considerable variation at all locations, all energies at or above -11 dB are located beyond 60 km. Radials R050, R185, and R305 included knife-edge obstructions that could affect the field strength at distant sites. The most prominent of these is a sharp peak on R305 at a distance of approximately 50 miles (80 km), with an altitude of about 1650 ft (500 m) above mean sea level (AMSL). The free-space field strength at this site would be about 64 dBu at channel 6. The test sites at 83 and 89 km are approximately 550 and 450 ft (170 m and 135 m) below this peak, respectively. The calculated and

[20] Sgrignoli, op. cit., p. 17.

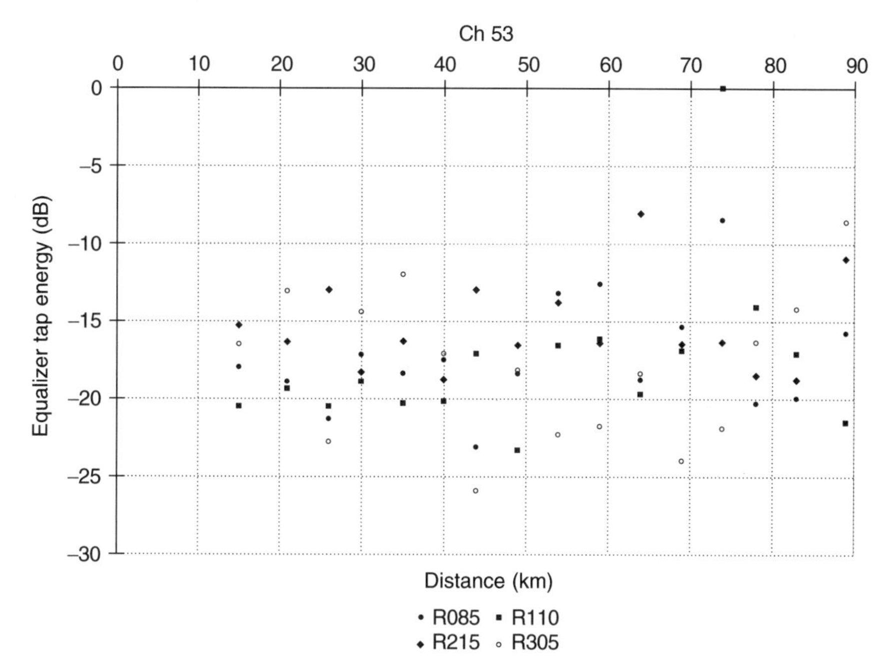

Figure 8-16. Equalizer tap energy.

measured field strength as a function of distance for channel 6 using the flat-earth model is plotted in Figure 8-17. As on the other radials, the calculated curve represents a good fit to the measured data; at some sites it is clearly conservative. The terrain for this radial, shown in Figure 8-18, is rougher than those previously considered and tends to slope upward with increasing distance. Most of the elevation variation is confined to a value between F_1 and $2F_1$. This might lead to the conclusion that the diffraction loss should be that due to shadowing by one Fresnel zone radius. However, a 16-dB adjustment would result in all measured points falling above the calculated curve. This situation again highlights the difficulty of accurately estimating the impact of surface roughness.

For comparison, the computed field strength is plotted along with predictions from FCC curves in Figures 8-19 and 8-20. At channel 6, the computed values match the FCC(50,90) within approximately 2 dB. Recall that the FCC curves are empirical in nature and published for the median frequency of 69 MHz. An adjustment of 1.9 dB is included for loss due to surface roughness. Measured data for R110 are repeated for comparison.

For channel 53, the computed values match the FCC(50,50) curve best at long range. For UHF, the FCC curves are published for the median frequency of 615 MHz. An adjustment of 4.8 dB is built into the FCC curves for loss due to surface roughness. Approximately 3 dB should be subtracted from the FCC curves to treat the Charlotte terrain properly. Measured data for R110 are repeated for comparison.

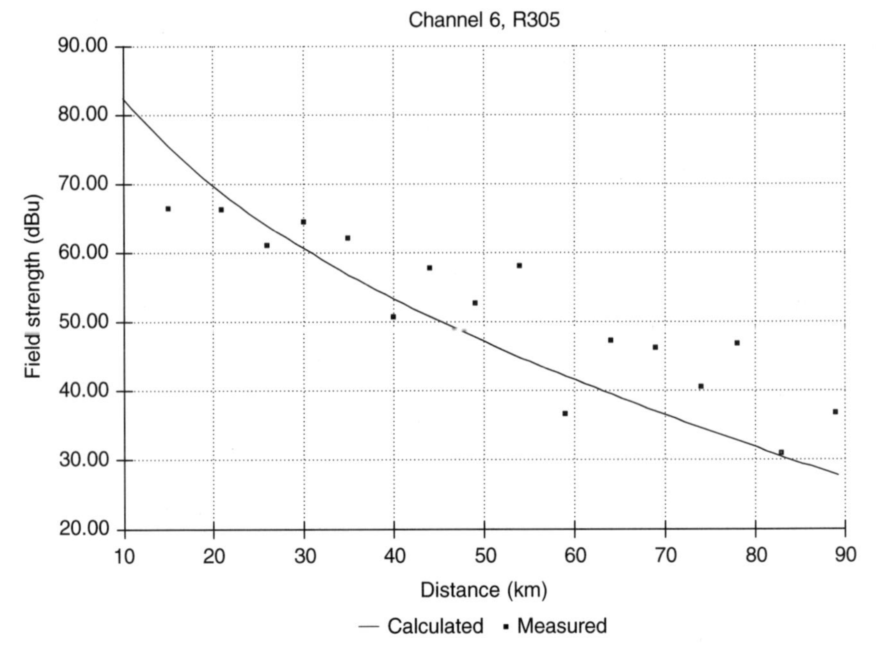

Figure 8-17. Field strength versus distance.

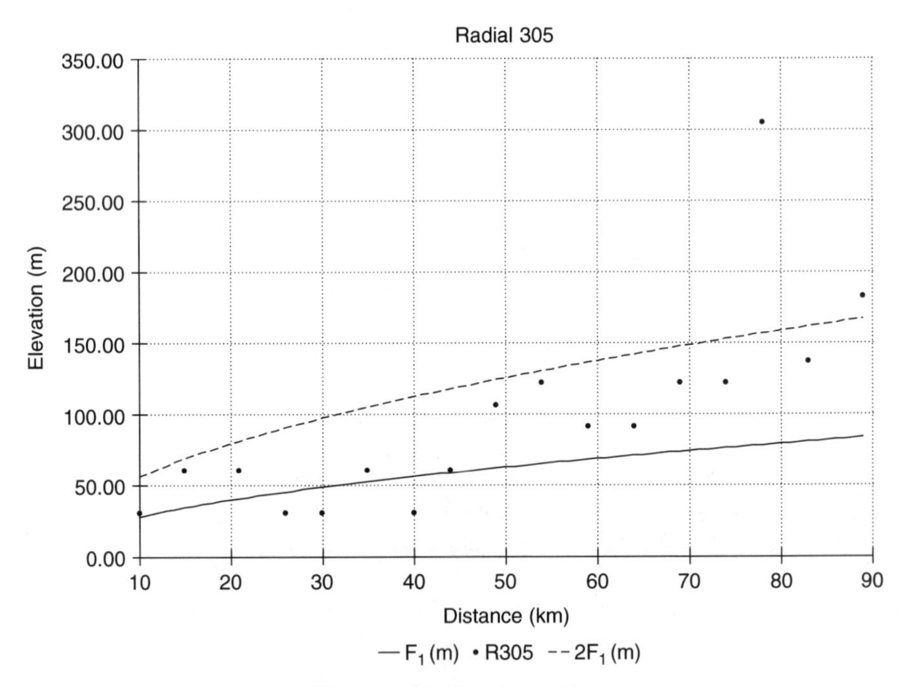

Figure 8-18. Terrain profile.

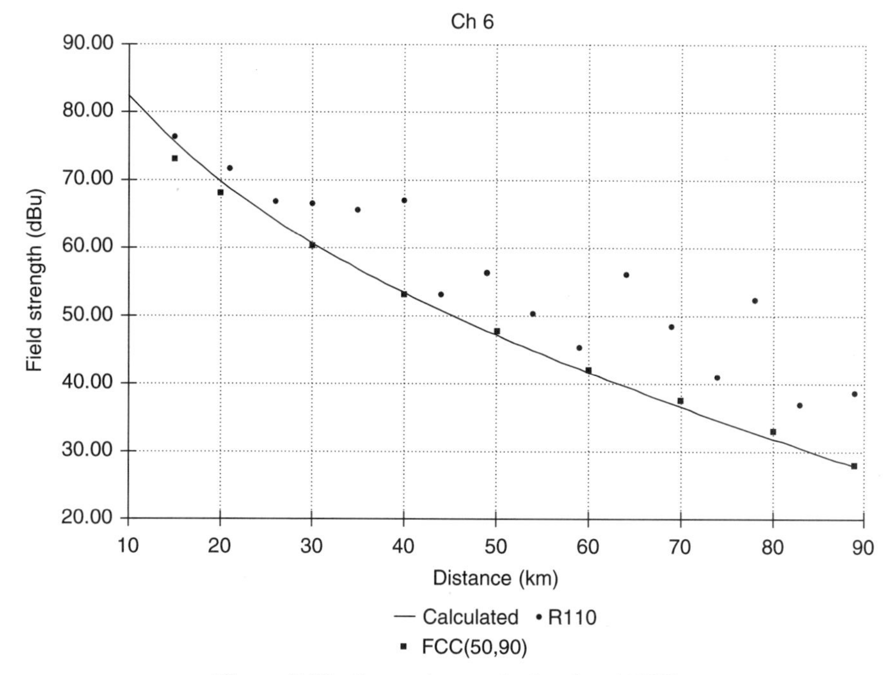

Figure 8-19. Comparison: calculated and FCC.

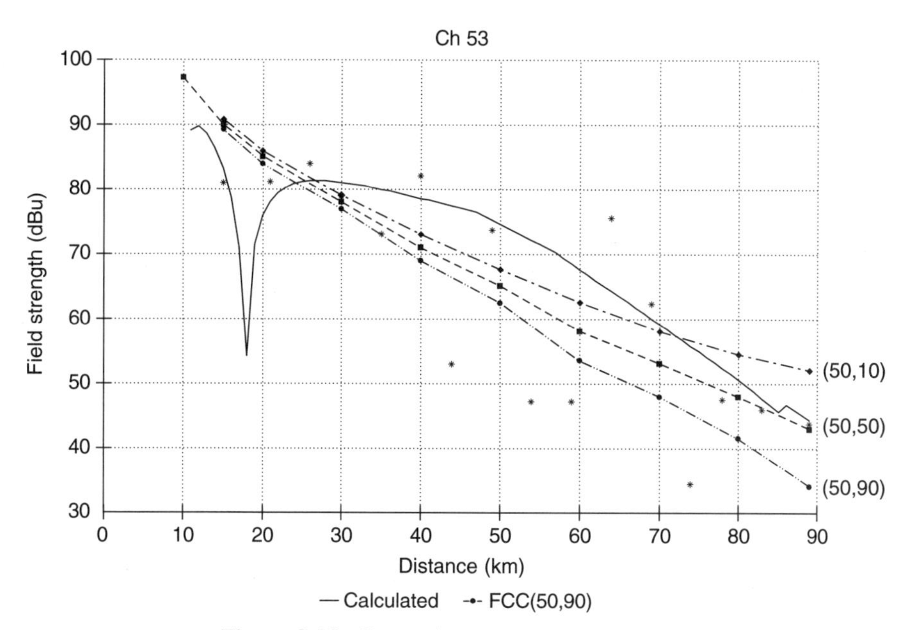

Figure 8-20. Comparison: calculated and FCC.

A limited number of indoor antenna tests at private homes were performed; tests were made at only eight sites on channel 6 and 10 sites on channel 53. The outdoor measurements were made as close as possible to the homes. Defying reasonable expectations, the relative signal strength at indoor antennas varied from 23 dB higher to 6 dB lower compared to outdoor antennas on channel 6; on channel 53 the indoor signal strength was from 6 dB higher to 6 dB lower. The tap energy was significantly higher for the indoor antennas. At channel 6, the tap energy difference varied from 1 to 12 dB worse for the indoor locations; at channel 53 the variation was from 3 to 13 dB. These data would indicate that the effect of multipath is worse for the less directive indoor receiving antennas.

CHICAGO, ILLINOIS

In Chicago, tests were performed on U.S. channel 20. The transmitting antenna was located on the east tower of the John Hancock Building at an approximate HAAT of 366 meters. The peak AERP was 284 kW (24.5 dBK). The AERP on radial R338 was about 12 dB lower due to the directional azimuth antenna pattern.[21] The tolerance on the AERP at all radials ranges from ± 2 to ± 4 dB due to interference from the west tower. Tests were made with a receive antenna height of 9 m.

For this analysis, a $\frac{4}{3}$ earth's radius is assumed and the distance to the radio horizon is 79 km. The free-space attenuation and attenuation factor due to ground reflections, assuming a ground-reflection coefficient of -1, were calculated. The divergence factor was also calculated. In addition, the diffraction loss due to a spherical earth was computed. The diffraction parameter, M, is 91. The resulting field strength is plotted in Figure 8-21. To obtain the curve, the available power at the receive site was computed using the AERP and relevant attenuation factors; the received power was then converted to field strength. Also shown is the measured field strength data for radials R251, R270, and R305. (Measurements from one of the special sites are included on radial R305 and from a home site on R270.)

All but one point of the measured data are well below the calculated curve. Radial R270 deviates the most from the calculated field, with an average deviation of 19 dB. This radial is one of two selected by the test engineers for its short delay multipath characteristics. Overall, the measurements and calculations match best for R305, on which the average deviation is 12 dB. This radial was selected for its long delayed multipath from reflections off the Sears Tower and the Amoco Building. Due to the large downtown buildings, 10- to 15-story apartment buildings and suburban housing and industry, none of the radials can be considered good examples of specular reflection from a smooth earth. At best, the measured data indicate significant losses due to diffused reflection

[21] M. McKinnon, M. Drazin, and G. Sgrignoli, "Tribune/WGN Field Test," *IEEE Trans. Broadcast.*, Vol. 44, No. 3, September 1998, pp. 261–273.

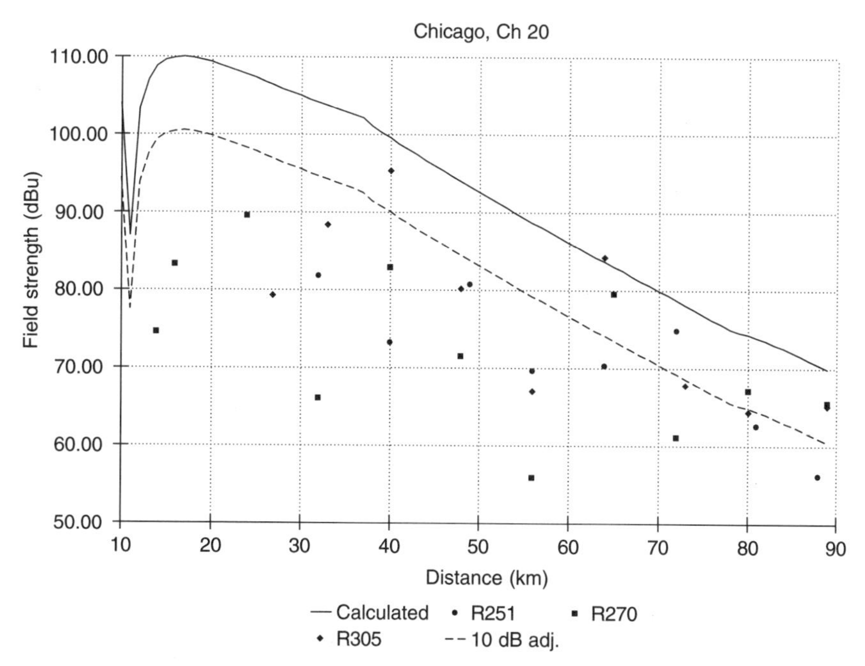

Figure 8-21. Field strength versus distance.

from a very rough surface plus the azimuthal reflections from the skyscrapers. Approximately ±3 dB variation in the data is due to the rapid azimuthal changes in the directional antenna pattern.

When an adjustment of approximately 6 to 10 dB is introduced, the calculated field is better matched to the measured data. With a 6-dB adjustment, the measured field for R305 deviates above the calculated by about 6 dB; with a 10-dB adjustment, the measured field for R270 deviates above the calculated by about 6 dB. Overall, a 10-dB adjustment results in a reasonable average of all measurements. Even so, the data nearest the tallest buildings are well below calculated values.

Analysis of the equalizer tap energy shows that nearly 57% of sites had a tap energy of −16 dB or greater. A plot of equalizer tap energy versus distance on several radials is shown in Figure 8-22. Multipath is most severe close to downtown, lowest at midranges, increasing to intermediate levels at the longest path lengths. Although there is considerable variation at all locations, all tap energies at or above −13 dB are at less than 30 km distance. The reflection from the west tower produced an echo with an approximate magnitude of −13 dB below the direct signal and delay of 0.1 to 0.2 μs. An echo from Sears Tower was clearly seen on R305 and R338 with a magnitude of about −14 dB and delays of 9.5 and 13.7 μs, respectively. An echo from the Amoco Building was clearly seen on R251 and R338 with similar magnitude and delay of 4.6 and 9.7 μs, respectively.

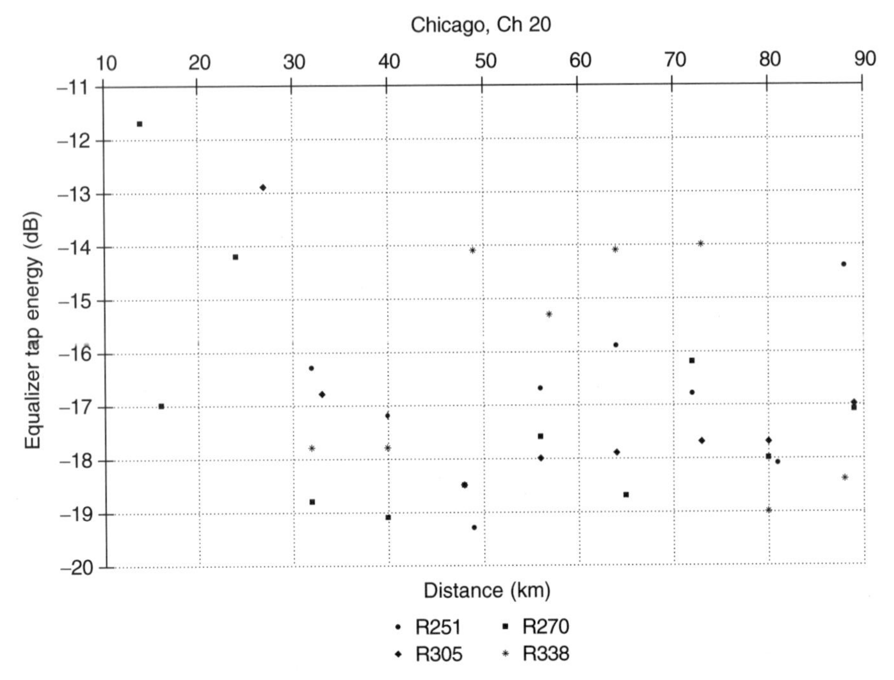

Figure 8-22. Tap energy versus distance.

The field strength calculated using the curved-earth model plus the effect of the three prominent reflections is plotted in Figure 8-23. The net effect of these reflections is less than a 4-dB variation about the curved-earth model. Most of the variation is due to the reflection from Sears Tower. Although the multipath from these structures is clearly important, evidently, local reflections and blockages in the urban and suburban areas contribute even more to the measured field strength at any given site. The calculated curve matches the measured data fairly well at distance beyond 50 km.

For comparison, the FCC curves are also plotted in Figure 8-23. The computed values match the FCC(50,10) curve within 1 dB or so at close and long range; at mid range the calculated values exceed the FCC(50,10) by about 10 dB and the FCC(50,50) by 11 dB. The FCC(50,90) is a reasonably good match to the low field strength measured at close range.

The reflections have a profound effect on the linear distortions within the channel. The magnitude and phase of the frequency response at a distance of 32 km from the transmitter on R338 due to the two identifiable reflections is plotted in Figures 8-24 and 8-25. Amplitude swings of several decibels and phase variations on the order of 0.6 rad are present. The latter correspond to group delay variations on the order of 1 μs. These distortions must be compensated by the adaptive equalizer. These results are similar in magnitude to those observed on

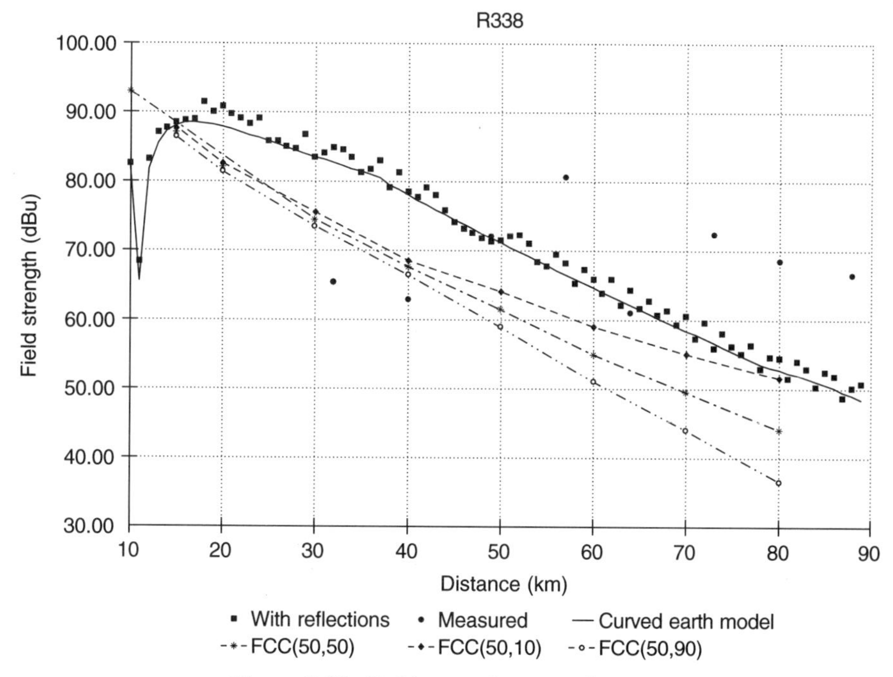

Figure 8-23. Field strength versus distance.

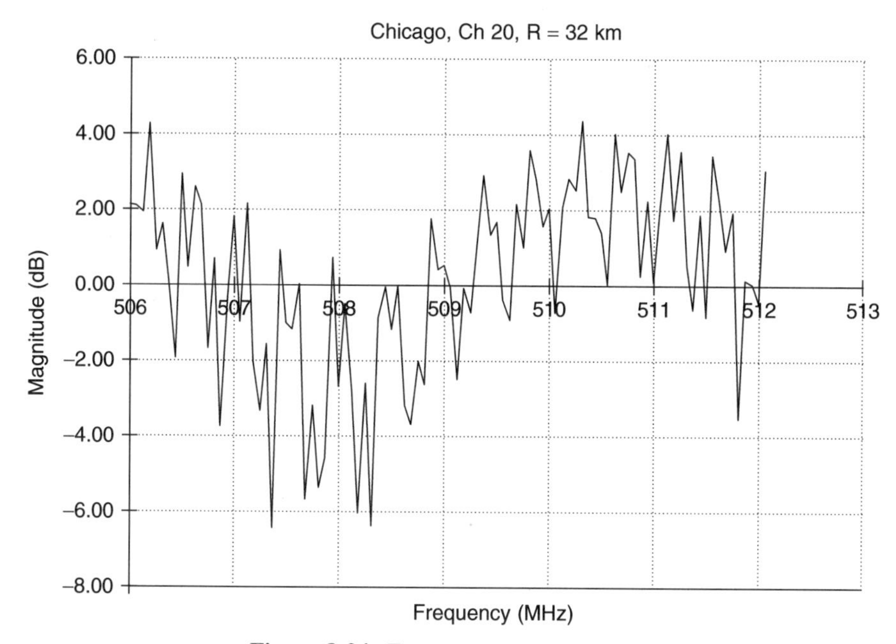

Figure 8-24. Frequency response, R338.

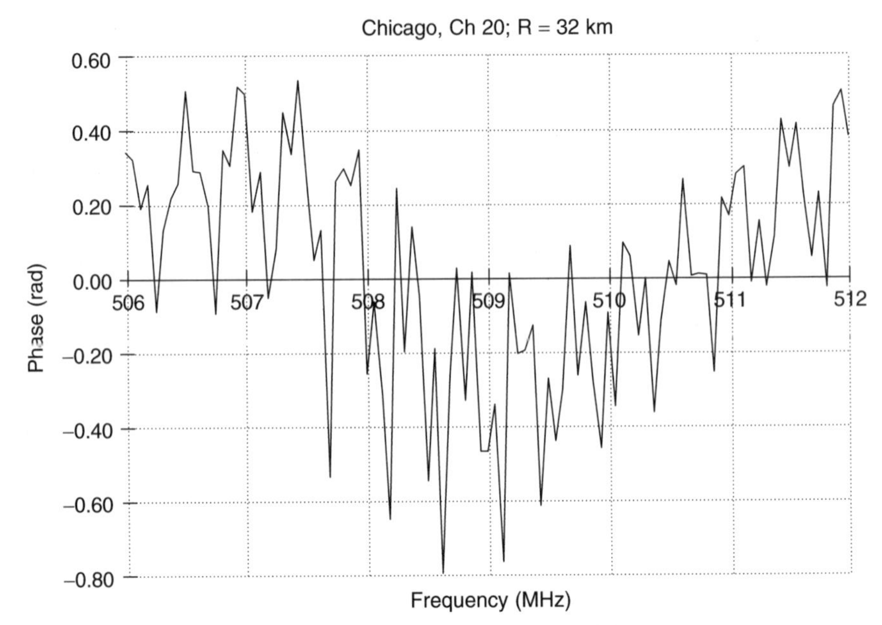

Figure 8-25. Phase response, R338.

channel characterization tests in downtown Ottawa, Canada.[22] On R305, only one identifiable echo was observed. The effect on linear distortion is less pronounced as seen in Figures 8-26 and 8-27.

A limited number of indoor antenna tests were performed. Although the tests were made at only 10 sites, it was confirmed that the signal strength is substantially lower and the tap energy significantly higher than for the outdoor sites. The loss in signal strength ranged from 3 to 18 dB and included the effect of height loss, building penetration loss, and a less directive receiving antenna. Considerable variation was found due to various types of construction and the location of the receiver within the building. Homes with metallic walls, such as aluminum siding, mesh-reinforced plaster, or foil-backed insulation, were especially lossy. These construction techniques tended to increase multipath and standing waves.

RALEIGH, NORTH CAROLINA

At Raleigh, tests were performed on U.S. channel 32. The antenna height was approximately 529 m above ground level (AGL). The AERP was 106 kW (20.3 dBK). Tests were made with a receiving antenna height of 9 m.

[22] B. Ledoux, "Channel Characterization and Television Field Strength Measurements," *IEEE Trans. Broadcast.*, Vol. 42, No. 1, March 1996, pp. 63–73.

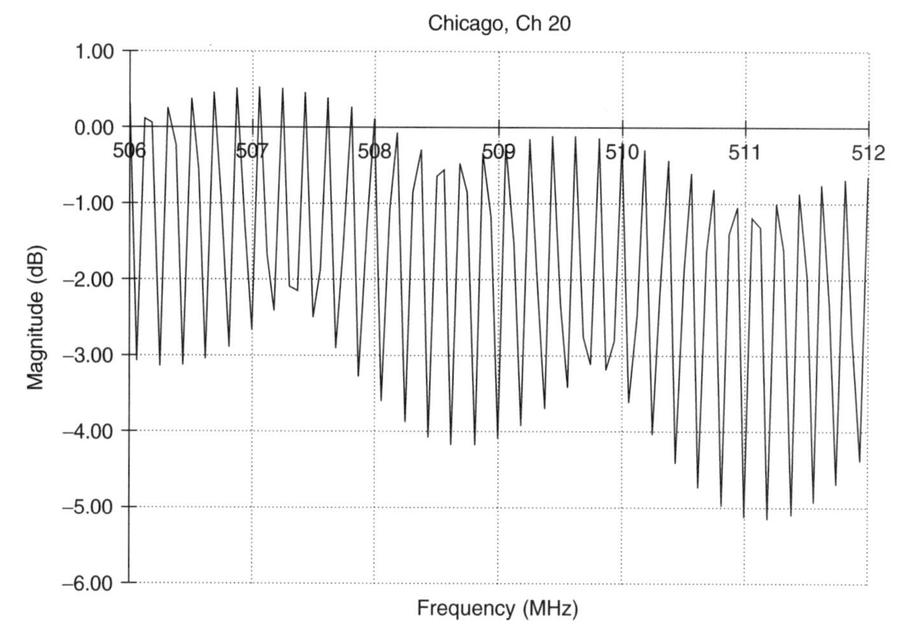

Figure 8-26. Frequency response, R305.

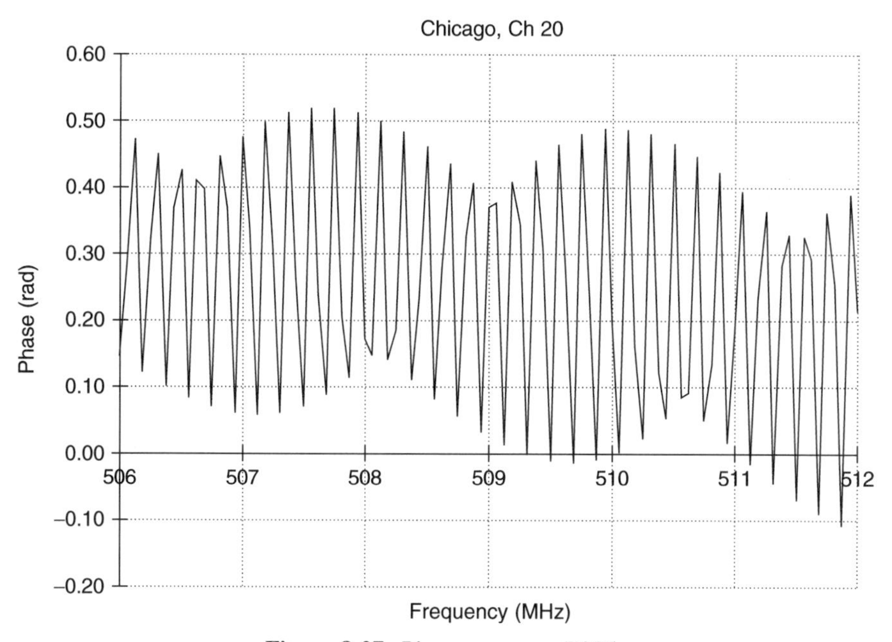

Figure 8-27. Phase response, R305.

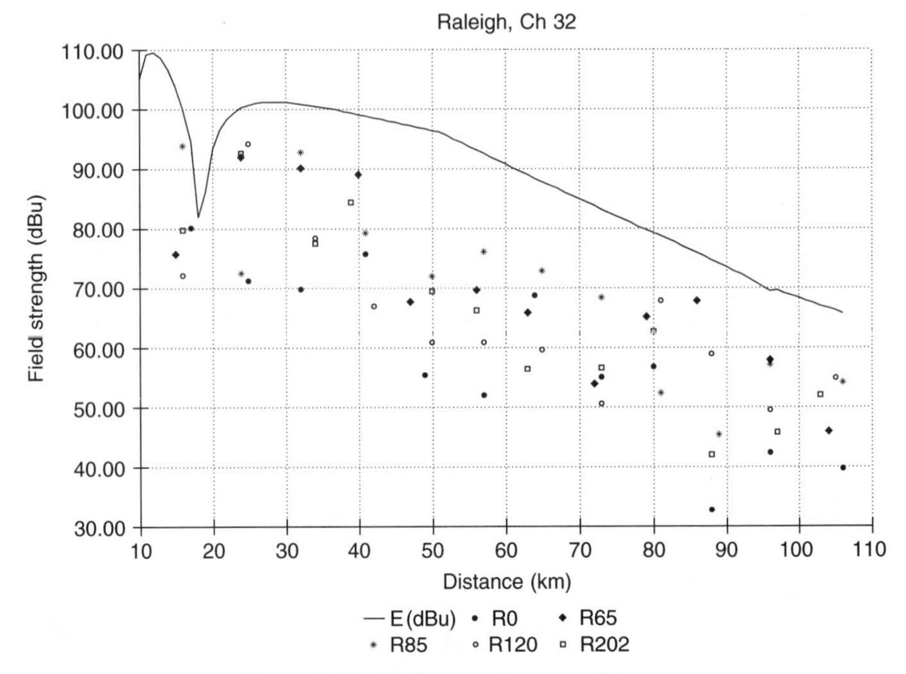

Figure 8-28. Field strength versus distance.

The climate is classified Continental Temperate and the K factor used in this analysis is 1.33. The distance to the radio horizon of a spherical earth is 94 km. The free-space attenuation and attenuation due to ground-reflections assuming a ground-reflection coefficient of -1 was calculated. The divergence factor was also calculated. In addition, the diffraction loss due to a spherical earth was computed. The diffraction parameter, M, is 139. The resulting field strength is plotted versus distance in Figure 8-28. To obtain these curves, the available power at the receive site was computed using the AERP and relevant loss factors; the received power was then converted to field strength. Also shown is the measured field strength data for selected radials.

As with channel 53 at Charlotte, the measured channel 32 field strength is well below the calculated curve. Radial R0 deviates the most from the calculated field with an average deviation of 25 dB. This radial is the roughest of all for which the data are plotted. The measurements and calculations match best for R065 and R085 for which the average deviation is 15 dB. Approximately ± 7 dB variation in the data is due to the circularity of the side-mounted antenna. Overall, however, the measured data indicate significant losses due to diffused reflection from a rough surface.

When a surface roughness adjustment of approximately 16 to 26 dB is introduced, the calculated field is a much better match to the measured data. With a 16-dB adjustment, the measured field for R85 deviates above the calculated

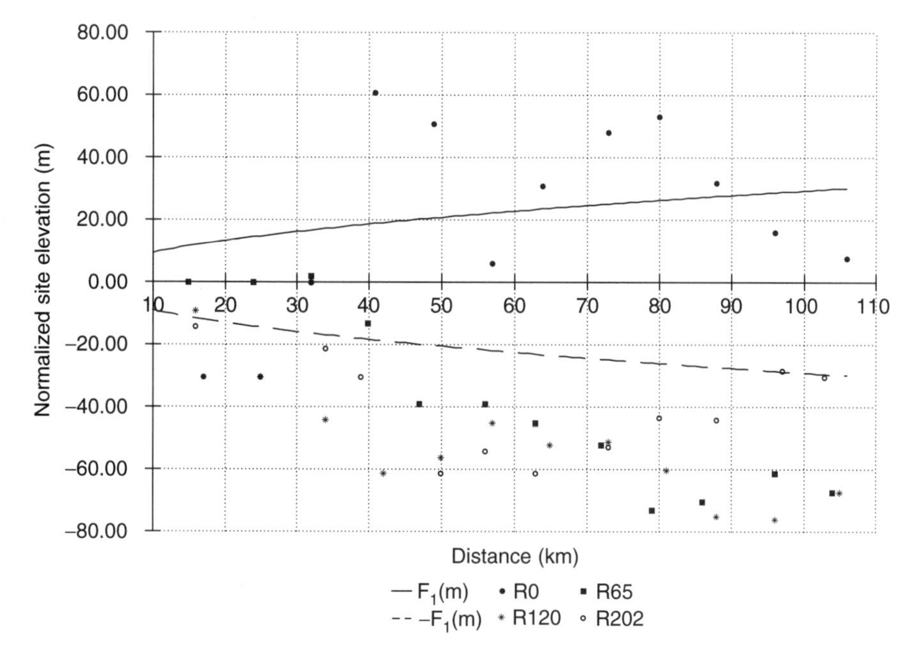

Figure 8-29. Normalized site elevation.

value by about 10 dB. This radial is the smoothest of all those measured. As shown in Figure 8-29, F_1 for the geometry of the Raleigh station ranges from 9 to 30 m. Over most of the distance, the elevation on R085 varies within a range bounded by $-3F_1$ and $-2F_1$.[23] This might justify a loss due to surface roughness equivalent to a clearance of F_1, or 16 dB. The overall roughness of R0 is approximately three times as great over much of the distance. Correspondingly, the apparent loss due to roughness is approximately 10 dB greater. The surface roughness and associated loss of R065, R120, and R202 are intermediate to R085 and *R*0.

Analysis of the equalizer tap energy shows that approximately 50% of sites had tap energies at or above −16 dB. This is almost identical to the results at Charlotte. The tap energy may also tend to increase for the roughest radials. For example, radials R085 had tap energy of −16 dB or greater at only 41% of locations; the roughest radial, R0, had tap energy at or above this level at 67% of sites. The highest tap energy, +2.3 dB, was observed on this radial.

A plot of equalizer tap energy versus distance on several radials is shown for channel 53 in Figure 8-30. Although there is considerable variation at all locations, all tap energies at or above −3 dB are located beyond 50 km. The correlation with distance is not strong, but the effect is to produce an increasing

[23] The author wishes to thank Luther Ritchie of WRAL-TV for kindly supplying a copy of the WRAL field test report, which is the original source of the data discussed herein.

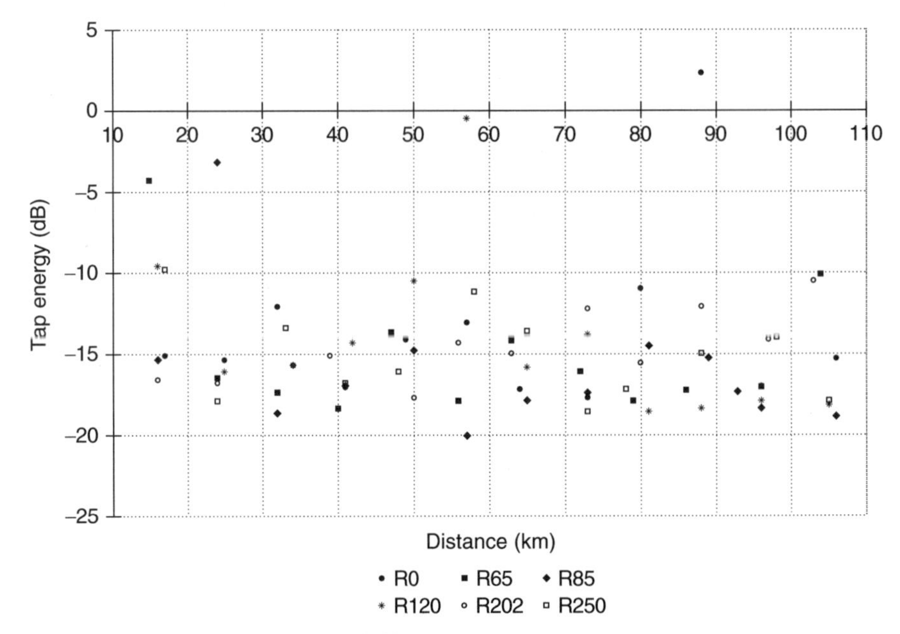

Figure 8-30. Equalizer tap energy.

number of failed reception sites at long distances, even though the field strength is above the minimum required by the planning factors.

The multipath produced a large amount of linear distortions within the channel. The measured frequency response at a couple of sites on R085 are shown in Figures 8-31 and 8-32. At site 6, the tap energy is only −20.1 dB and the frequency-response variation is 2.5 dB. At site 2, the tap energy is −3.2 dB and the frequency-response variation is 15 dB. In general, the greater the tap energy, the more the linear distortion, as shown in Figure 8-33. There is some scatter in the data because the shape of the frequency-response curve varies from site to site. Also, the phase response contributes to tap energy. These results are similar to those obtained in Chicago, where the multipath was due to urban clutter.

Radials R0, R275, and R315 included knife-edge obstructions that could affect the field strength at more distant sites. One of the more prominent of these is a sharp peak on R275 at a distance of approximately 53 miles (85 km) and an elevation of about 700 feet (210 m) AMSL. The free-space field strength at this site would be about 87 dBu. The test sites at 88, 96, and 105 km are approximately 50 and 100 ft (15 and 30 m) below this peak, respectively, and are blocked by this single obstruction. The theoretical diffraction loss is 20 dB or more, depending on the sharpness of this peak. Thus, the field strength at these sites should not exceed 67 dBu. The measured field strength as a function of distance is plotted in Figure 8-34. The measured values for these three sites are 47, 53, and 44 dBu, respectively.

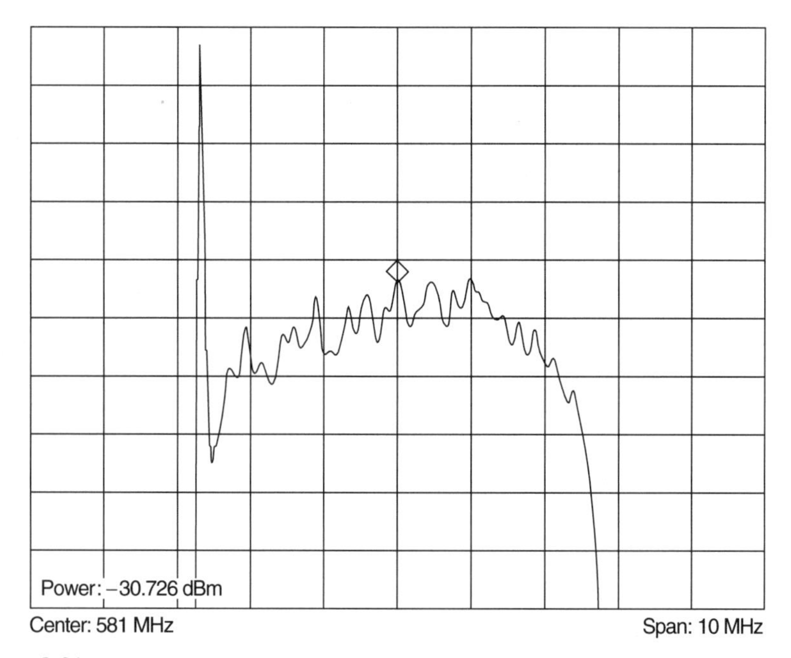

Figure 8-31. Frequency response at R85, site 6 (1 dB/division). (Data courtesy of WRAL-HD; used with permission.)

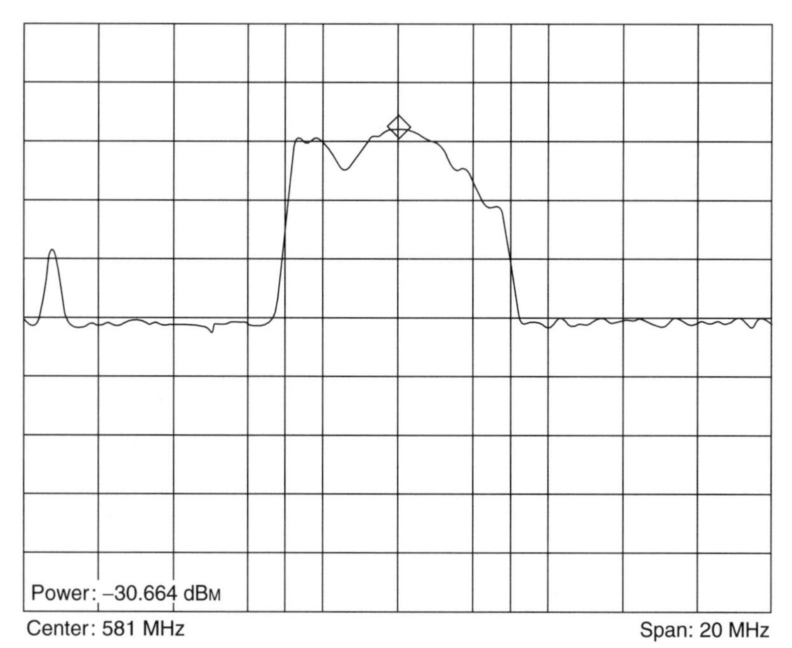

Figure 8-32. Frequency response at R85, site 2 (10 dB/division). (Data courtesy of WRAL-HD; used with permission.)

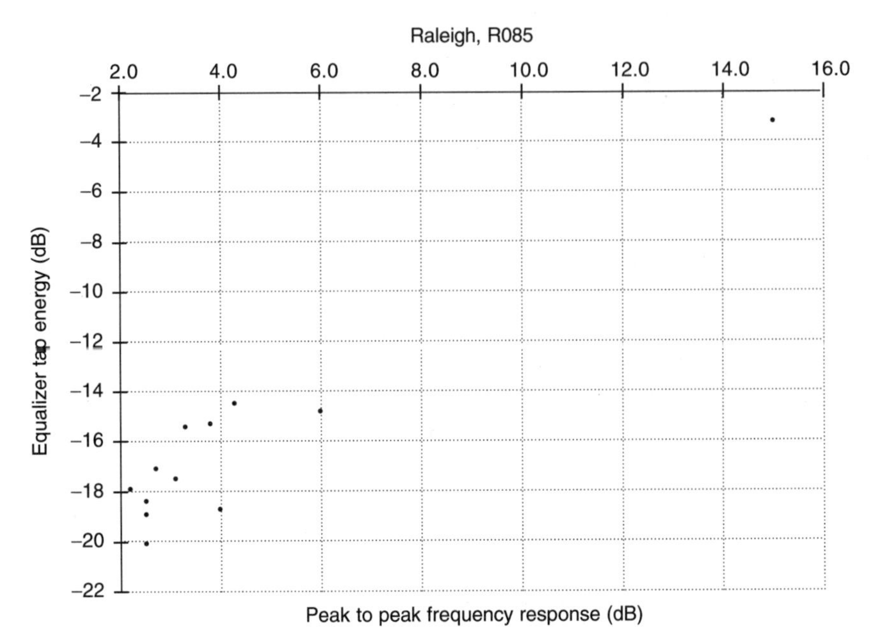

Figure 8-33. Tap energy versus response.

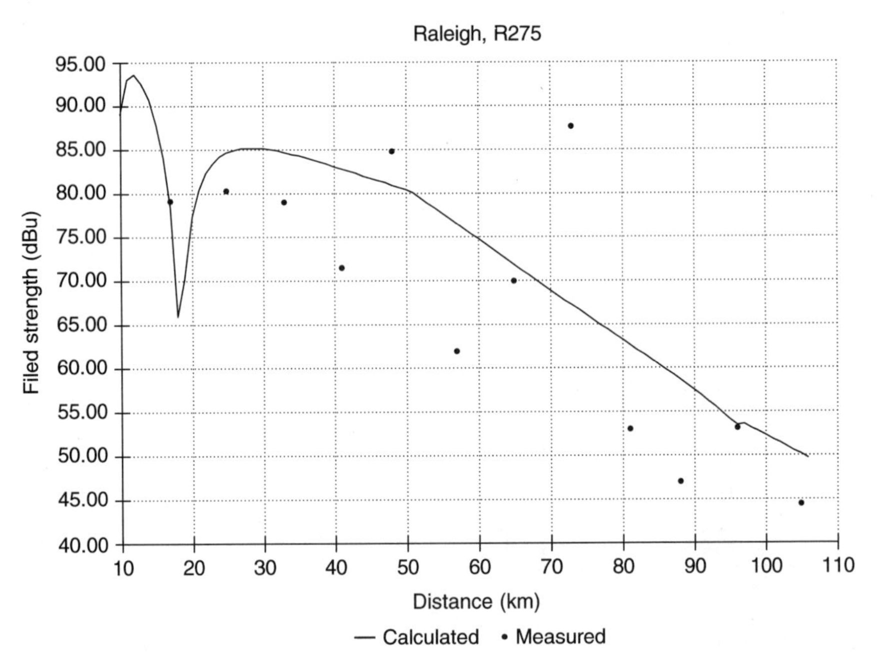

Figure 8-34. Field strength versus distance.

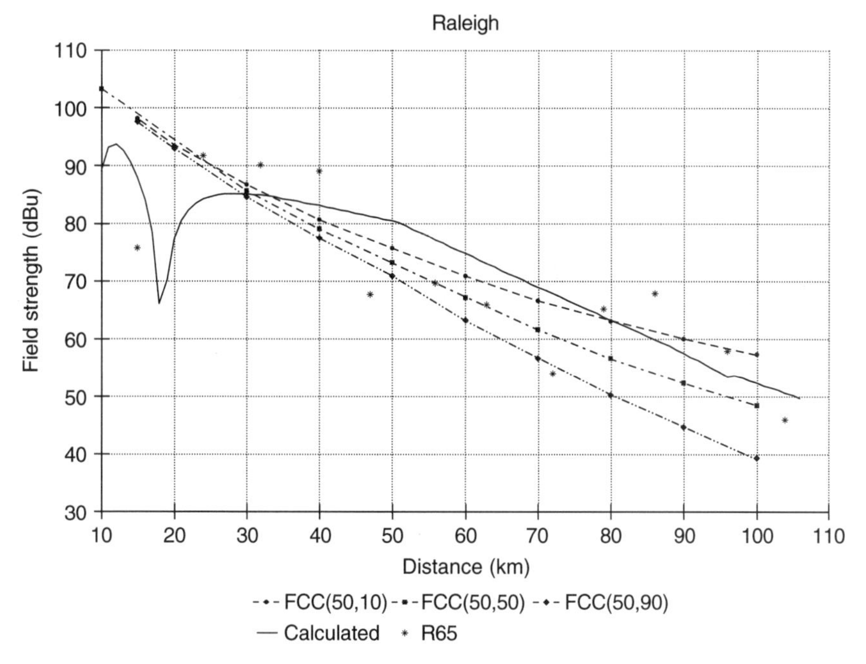

Figure 8-35. Comparison with FCC.

The computed field strength is plotted along with predictions from FCC curves in Figure 8-35. The computed curve matches the FCC(50,50) curve best at 30 km and at long range. Up to 0.5 dB should be subtracted from the FCC curve to treat the Raleigh terrain properly. Measured data for R065 are repeated for comparison.

Indoor antenna tests were performed at 36 sites. Three types of indoor antenna was tested: a loop, a single bowtie, and a dual bowtie over a ground plane. A usable signal with the indoor antennas was observed at all but three sites. At these sites, the median signal strength on the indoor antennas was lower than the outdoor measurements by 9.1, 6.8, and 11.1 dB, respectively. The loss in signal strength included the effect of height loss, building penetration loss, and a less directive receiving antenna. The equalizer tap energy was significantly higher than for the outdoor measurements. The average tap energy on the indoor antennas was about −6 dB compared to −15 dB on the outside antennas. This would indicated significantly higher multipath indoors.

SUMMARY

The factors that affect the propagation of digital television signals at VHF and UHF have been considered along with various means of estimating signal strength and frequency response. It is evident that the means do not exist to predict with precision the field strength or frequency response at any location and time. This is due to the nature of the propagation environment. Free-space attenuation, ground reflections from a plane or spherical earth, refraction by an ideal atmosphere, and diffraction over spherical earth and well-defined obstacles lend themselves to precise calculations. However, the real world is much different. The effect of the earth's rough surface, the temperature, humidity, and pressure variations of the atmosphere, and the locations, shapes, and reflection coefficients of natural and man-made obstacles are difficult to estimate. Nevertheless, it is important to understand the contribution of each of these factors.

Understanding these factors is useful in assessing the difference between propagation at VHF and UHF. Both free-space attenuation and losses due to surface roughness are much higher for UHF. These losses are partially offset by the effect of ground reflections from smooth earth. In addition, diffraction losses are generally lower at UHF since fixed clearances are greater when measured in terms of Fresnel zone radii. Nevertheless, overall propagation losses are almost always greater for UHF.

9

TEST AND MEASUREMENT FOR DIGITAL TELEVISION

Although there are many tests and measurements for the transmission of digital television that are similar to those made for analog television, some are distinctly different. These will be the focus of this chapter. These tests include the measurement of power as well as linear and nonlinear distortions. Frequency measurements are also discussed. This discussion is not meant to be exhaustive. There are many tests that may be made in connection with the subsystems discussed in previous chapters. There are other tests that may be made at the systems level. The purpose of this chapter is to highlight a few of the key tests that may be used to characterize the RF performance of a digital television system.

POWER MEASUREMENTS

The measurement of power is fundamental to all digital TV transmission tests. Power output establishes the transmitter operating point and thus determines the level of nonlinear distortions at the source. The stress on high-power RF filters, transmission lines, and antennas is determined by incident and reflected peak and average power. At the receiver, the available signal power relative to noise and interference determines the availability of a viewable picture.

Although the concept of power was discussed earlier, it is important that it be defined clearly as it relates to measurement. As noted earlier, both average and peak power are important to the transmission of digital TV. The average power must be known in relation to the dissipation and temperature rise in transmission equipment as well as the signal power available at the receiver. Average power refers to the product of the RMS signal voltage and current, integrated over the modulated signal bandwidth. Since the transmitted data stream is random in nature, the average power is constant if the average is taken over a sufficiently long time. This is in contrast to the analog television signal, for which the average power varies with video content.

Even though the average power is used to establish TPO, system ERP, and C/N, it is often desirable to measure peak power. Nonlinear distortions may lead to degraded system performance. This most often is due to overdrive somewhere within the system; the ability to measure peak power is a valuable tool for troubleshooting. The peak power must also be known in relation to the rating of transmission components.

The peaks of the RF envelope are determined statistically by the random pattern of the data and the bandlimiting of the system. Thus the peak power levels must be described by both their magnitude and the percent of time they occur.[1] For these statistics, the peak envelope power (PEP) is defined as the average power contained in a continuous sine wave with peak amplitude equal to the signal peak. Thus the PEP for a digital TV signal is defined in the same manner as for analog TV. The contrast is in the regular recurring peaks of the analog sync pulses at a constant amplitude versus the random occurrence of the digital peaks at random amplitudes. It is customary to state the peak power relative to the average power. Usually, this is a logarithmic ratio and is given in decibels.

Since the peak power is statistical in nature, the peak-to-average power ratio is often presented in the form of a cumulative distribution function (CDF). This is a concept borrowed from the mathematics of probability that permits the description of the relative frequency of occurrence (probability) of a particular peak power level (the random variable). The RF power is sampled at regular intervals, and the power level measured at each interval is collected in one of many incremental ranges or "bins." The number of times the measured level falls into a particular bin relative to the total number of measurements is computed for each bin and may be plotted as a histogram. Thus the histogram is a record of the frequency at which a particular incremental power range is measured. When properly constructed with sufficiently small power increments and a large number of measurements, the histogram approximates a probability distribution function (PDF).

The probability of the peak-to-average power ratio exceeding a particular level is the usual parameter of interest to the engineer. This may be determined from the CDF, which is obtained by integrating the PDF from the maximum peak to average ratio down to unity. The peak and average powers are equal approximately 50% of the time; as the peak-to-average power ratio increases, the frequency of occurrence approaches but never becomes zero. A typical CDF for the 8 VSB signal is as shown in Figure 2-7.

A variety of instruments are used to measure power. Some of these measure only average power. Others are capable of measuring peak power, from which

[1] G. Sgrignoli, "Measuring Peak/Average Power Ratio of the Zenith/AT&T DSC-HDTV Signal with a Vector Signal Analyzer," *IEEE Trans. Broadcast.*, Vol. 39, No. 2, June 1993, pp. 255–264.

average power and the relevant statistics are computed. In either case, it is important that the measuring device provide sufficient bandwidth and accuracy over the range of power levels to be measured.

AVERAGE POWER MEASUREMENT

Compared to peak power, average power is much easier to measure. Just as with an analog television signal, the high average power at the transmitter may be measured using one of two methods: water-flow calorimetry or a precision probe in the transmission line connected to a power meter. The power meter may also be used at the receiving site, provided that there is adequate properly calibrated low-noise amplification.

CALORIMETRY

Measurement of power by means of calorimetry is a direct measurement of the amount of heat energy dissipated in a liquid per unit time. For the purpose of discussion, it is assumed that the liquid is water, although it is common to use water containing glycol in many systems. In either case, the principle is the same; only the specific heat of the liquid is affected.

Water is an excellent medium for the conversion of RF energy to heat. It is well known that for every kilocalorie of added heat, the temperature of 1 kg of water rises by 1°C. Since power is simply energy per unit time (1 watt is 1 joule per second), the power dissipated in a water load may be computed if the temperature rise, ΔT, and rate of flow, R_f, of the water are known. Thus

$$\text{TPO} \propto \Delta T R_f$$

The flow rate is often measured in gallons per minute, so that the constant of proportionality (specific heat of water) is 0.264.[2] Disadvantages of calorimetry are that this measurement must be made while the transmitter is off-air, and it is not accurate for very low power measurements.

POWER METERS

Average power may be measured at the output of the transmitter or RF filter with a power meter if a suitable calibrated probe or coupler is available. For example,

[2] "Transmitter for Analog Television," in J.G. Webster (ed.), *Encyclopedia of Electrical and Electronic Engineering*, Wiley, New York, 1999, Vol. 22, p. 489.

a 60-dB coupler provides approximately 15 mW (11.8 dBm) to a power meter if the expected power output is in the range of 15 kW. Power is sensed at the output of the coupler by a thermocouple or diode detector. Thermocouples measure true average power by detecting the voltage generated in the metallic sensor due to a temperature gradient. Diode sensors use resistive–capacitive loads with long time constants to produce a voltage proportional to the average power. When using a diode sensor, care must be taken to avoid driving it above its square-law characteristic. Otherwise, calibration errors are introduced by the transient peaks. Measuring average power by this method has the advantages of providing on-air data and being suitable for high- and low-power systems.

PEAK POWER MEASUREMENT

A variety of instruments, including peak power meters, spectrum analyzers, and the vector signal analyzer, are available to measure peak power. Calorimeters and conventional power meters are not suitable since their output is the average of the signal power. Peak power meters detect the time-varying signal envelope by means of a fast diode sensor which provides a voltage output that is proportional to the RF envelope. The output of the sensor is amplified and digitized so that the appropriate digital signal processing (DSP) computations can be made. The peak power distribution is integrated over a specified time limit so that peak power, average power, and their ratio can be displayed. Similar features are provided in the vector signal analyzer and some spectrum analyzers with DSP capability.

The CDF of the peak-to-average power ratio may be measured using a simple setup that includes equipment available at most analog TV stations and manufacturers' laboratories. The major pieces of equipment include a frequency counter, average reading power meter, and calibrated attenuator.[3] Although the method is described for the VSB signal, it is applicable for any digitally modulated system. The frequency counter responds to the signal peaks that exceed the calibrated power levels set by attenuator. The resulting data may be combined with the measured average power to determine peak power. Techniques for assuring accurate measurement of average power are also described.

MEASUREMENT UNCERTAINTY

It is important to recognize that RF measurements, especially absolute power measurements, always include a certain amount of uncertainty. These uncertainties may arise from many factors, including instrument and coupler calibration, the efficiency of the power sensor, and mismatches within the system.[4] Thermocouple sensors must be operated in a suitable range above the noise level.

[3] C.W. Rhodes, "Measuring Peak and Average Power of Digitally Modulated Advanced Television Systems," *IEEE Trans. Broadcast Technol.*, December 1992.

[4] *HP Application Note AN 64-1A*, "Fundamentals of RF and Microwave Power Measurements," pp. 37–61.

The effect of any non-square-law characteristic of diode sensors must be known. For calorimetric measurements, errors are present in the measurement of both temperature and flow rate. Unfortunately, the effects of these sources of uncertainty are often overlooked or completely ignored. However, small errors may represent large amounts of power. For example, an error of just 0.1 dB in the measurement of the output of a 25-kW transmitter represents 525 W. In many cases it is likely that the measurement error is even greater.

It is also important to distinguish between the accuracy and precision of the measurement. Although these words are often consider synonyms, in a technical sense measurement accuracy refers to the difference between the measured power level and the true power expressed in either decibels or percent. Precision or resolution refers to the numerical ambiguity or number of significant digits that may be assigned to a measurement. With the availability of digital instruments, calculators, and computers capable of displaying numbers with many significant digits, it is tempting to assume that such numbers are useful in their entirety. Unless adequate attention is given to sources of error, the result may be an inaccurate number known to great precision.

TESTING DIGITAL TELEVISION TRANSMITTERS

The key measurements required for a digital television transmitter proof of performance include average output power, frequency response, pilot frequency, error vector magnitude, intermodulation products, and harmonic levels. The first four of these primarily evaluate the in-band performance of the transmitter; the last two are out-of-band parameters. Some of the in-band and out-of-band parameters are related, however.

The most critical of these measurements is average output power, pilot frequency, in-band frequency response, and adjacent channel spectrum. These parameters should be checked periodically to assure proper transmitter operation. In every case, they can be measured while the transmitter is in service with normal programming using a power meter and/or spectrum analyzer. Experience has shown that when these parameters are satisfactory, peak power and system EVM are usually satisfactory. Thus it may be necessary to measure peak power and EVM only at the time of initial setup and whenever nonlinear performance is suspected.

The pilot frequency (or frequencies) may be measured with a frequency counter or spectrum analyzer. For the ATSC system, the results should be the frequency of the lower channel edge plus $309{,}440.6 \pm 200$ Hz, unless precise frequency control is required and/or a frequency offset is employed. The frequency response of the transmitter and output filter can be measured directly with a spectrum analyzer. This measurement is fundamental because poor in-band response will result in intersymbol interference, degraded C/N, bit errors, symbol errors, and degraded EVM. Frequency-response measurements also are required to demonstrate compliance with the emissions mask.

In practice, it is difficult to measure full compliance with the DTV or DVB-T emissions masks directly. For near-in, out-of-band spectral components, the best procedure may be to (1) measure the output spectrum of the transmitter without the high-power filter using a spectrum analyzer, (2) measure the filter rejection versus frequency using a network analyzer, and (3) add the filter rejection to the measured transmitter spectrum. The sum should equal the transmitter spectrum with the filter. It is recommended that the transmitter IP level be measured with the resolution bandwidth set for about 30 kHz throughout the frequency range of interest. This setting results in an adjustment to the FCC mask by 10.3 dB. Under this test condition, the measured shoulder breakpoint levels should be at least −36.7 dB from the midband level.

Output harmonics may be determined in the same manner as the rest of the out-of-band spectrum. For the ATSC system, they should be at least −99.7 dB below the midband power level. Once the output filter response is measured by the manufacturer, it should not be necessary to remeasure unless detuning has occurred.

EVM is the key numerical parameter indicating the status of the transmitted signal constellation. For this reason, once a transmitter is set up at the correct frequency and power with good spectral characteristics, it is often desirable to measure EVM as a final check. A vector signal analyzer is necessary for this measurement. If the EVM is satisfactory, both bit error and symbol error performance will be satisfactory.

In addition to EVM, the vector signal analyzer provides several qualitative and quantitative measures of system performance. The symbol errors may be displayed as a function of time along with the symbol table. The signal constellation in the I–Q plane and/or eye diagram may be displayed to indicate distortion due to compression (AM/AM and AM/PM), noise, and timing errors. A satisfactory I–Q diagram for 8 VSB will exhibit eight narrow vertical columns of dots. Spreading of the columns indicates the presence of excessive white noise. If the columns are slanted with respect to the vertical, phase distortion is indicated. Similar diagnostics may be performed on the I–Q diagrams of the DVB-T and ISDB-T constellations.

The eye diagram should display the distinct signal levels at the correct sampling time. The in-band and out-of-band spectrum may also be displayed by the vector signal analyzer along with a computation of adjacent channel power. C/N may also be displayed and correlated with EVM. All measurements made with the vector signal analyzer may be done while the transmitter is in or out of service. For out-of-service measurements, it should be possible to generate pseudorandom data simply by creating an open or short circuit at the exciter input.

SYMBOLS AND ABBREVIATIONS

CHAPTER 1

α_N	Nyquist filter shape factor
AERP	average effective radiated power
ATSC	Advanced Television Systems Committee
BST-OFDM	band-segmented transmission–OFDM
COFDM	coded orthogonal frequency-division multiplex
D/A	digital to analog
DiBEG	Digital Broadcasting Experts Group (Japan)
DQPSK	differential quadrature-phase shift keying
DVB-T	digital video broadcast–terrestrial
ETSI	European Telecommunications Standards Institute
FCC	Federal Communications Commission
FEC	forward error correction
f_{frame}	data frame rate
f_{seg}	segment rate
HAAT	height above average terrain
η_s	spectral efficiency
HDTV	high-definition television
I	in-phase component
IDFT	inverse discrete Fourier transform
IF	intermediate frequency
ISDB	Integrated Services Digital Broadcasting
ISDB-T	Integrated Services Digital Broadcasting–Terrestrial
ISO	International Standards Organization
ITU-R	International Telecommunications Union, Radio Sector

LO local oscillator
MPEG Motion Pictures Expert Group
PN pseudorandom number
Q quadrature
QAM quadrature amplitude modulation
QPSK quadrature phase-shift keying
R/S Reed–Solomon
SDTV standard definition television
SFN single-frequency networks
S/N signal-to-noise ratio
STL studio-to-transmitter link
T symbol time
T_F frame duration
TMCC transmission and multiplex control
TPS transmission parameter signaling
8 VSB eight-level vestigial sideband

CHAPTER 2

α_r attenuation of receive antenna transmission line
ATTC Advanced Television Test Center
AWGN additive white Gaussian noise
B channel bandwidth
BER bit error rate
BPS bits per second
C average carrier power
CDF cumulative distribution function
C/N carrier-to-noise ratio
$C/(N+I)$ carrier-to-noise plus interference ratio
D_a actual constellation vector
D_i ideal constellation vector
D/U desired-to-undesired ratio
E_b/N_0 ratio of average energy per bit to noise density
e_i error signal
E_s energy per symbol
ERP effective radiated power
EVM error vector magnitude
F receiver noise factor
g gain of amplifier in linear region of transfer function
g_3 coefficient of third-order nonlinearity
g_{3I} in-phase component of third-order nonlinearity
g_{3Q} quadrature component of third-order nonlinearity
G_r receive antenna gain in decibels
IP intermodulation products
ISI intersymbol interference

k	Boltzmann's constant $= 1.38 \times 10^{-23}$ joules/Kelvin
L	transmission line loss in decibels
M	number of levels
N	noise power
NF	receiver noise figure
N_s	number of samples
N_t	thermal noise limit for perfect receiver at room temperature
PAR	peak-to-average ratio
P_{ma}	threshold average power at antenna
P_{mr}	threshold average power at receiver
P_r	average power of received signal
R_b	transmission rate in bits per second
SER	symbol or segment error rate
S_i	input signal
S_o	output signal
T_0	ambient temperature
T_a	antenna noise temperature in Kelvin
T_s	receive system noise temperature in Kelvin
TOV	threshold of visibility
TPO	total average transmitter output power
V	center-to-center distance between symbol levels

CHAPTER 3

a0, a1	output vectors of OFDM bit interleaver
b	dc level
b0, b1	pair of substreams at output of OFDM demultiplexer
C_c	channel capacity
d_i	series of pulses representing symbols
δ	Dirac delta or impulse function
Δ	guard interval
$\Delta C/N$	change in C/N
d_m	minimum distance between sequences of encoded signal
f_b	block code data rate
f_c	channel center frequency
FDM	frequency-division multiplex
f_p	payload data rate
f_t	trellis code data rate
IFFT	inverse fast Fourier transform
k	carrier number
k_b	length of R/S block before coding
k_t	length of trellis code word before coding
n_b	length of R/S block after coding
NRZ	non return to zero

n_t	length of trellis code word after coding
P_a	average power
$P_k(f)$	power spectral density of *k*th OFDM carrier
P_t	transmitted power
$S_f(t)$	mathematical representation of frequency-division multiplex signal in time domain
SMPTE	Society of Motion Picture and Television Engineers
$S_n(f)$	power spectral density of noise or interference
$S_v(t)$	mathematical representation of VSB signal in time domain
SSB-SC	single-sideband suppressed carrier modulation
$S_x(f)$	power spectral density of transmitted signal
t	time
t_b	maximum number of byte errors a R/S code is capable of correcting
TPO	transmitter power output
T_u	active symbol interval
VSB	vestigial sideband modulation
$x(t)$	baseband signal in time domain
$x_i(t)$	in-phase signal in time domain
$x_q(t)$	quadrature signal in time domain
Y	output vector of OFDM symbol interleaver

CHAPTER 4

AGC	automatic gain control
ALC	automatic level control
AVR	automatic voltage regulator
DSP	digital signal processing
FET	field-effect transistor
$f_0(v_c)$	polynomial representing power amplifier nonlinearities
$H_0(\omega)$	complex frequency response of power amplifier and filters
$H_{eq}(\omega)$	complex frequency response of equalizer
HPA	high-power amplifier
$H_s(\omega)$	system transfer function
IOT	inductive output tube
IPA	intermediate power amplifier
LDMOS	lateral diffused MOSFET
MTBF	mean time between failures
PA	power amplifier
PFC	precise frequency control
PLL	phase-locked loop
SiC	silicon carbide
v_c	complex output voltage of precorrector

v_i input voltage to precorrector
v_0 complex output voltage of power amplifier

CHAPTER 5

α_c cavity attenuation in nepers per unit length
A_{pb} attenuation at passband edge frequency
A_{sb} attenuation at stopband edge frequency
Δ_ω radian frequency difference between half-power points
ε passband ripple
ε_r relative dielectric constant
f_0 center frequency; frequency at which transmission line is 1/4 wavelength long
f_1 lower band edge frequency
f_2 upper band edge frequency
f_{pb} passband edge frequency
f_{sb} stopband edge frequency
h_c half length of cavity
h_c/a cavity length-to-radius ratio
Γ reflection coefficient function
λ_c cutoff wavelength of waveguide
λ_g waveguide wavelength
M_{mn} coupling factors
n number of poles or filter order
P_a partial pressure of dry air in millimeters of mercury
P_w partial pressure of water vapor in millimeters of mercury
P_l power delivered to load
Q quality factor
Q_u unloaded Q
Q_l loaded Q
ω_0 angular resonant frequency
R_n ratio of polynomials defining filter poles and zeros
S complex frequency variable
S_{dB} cutoff slope
T_a absolute temperature in Kelvin
t_f filter transmission function
Z_0 characteristic impedance
Z_{sc} input impedance of short-circuited lossless transmission line

CHAPTER 6

α attenuation constant
A conductor loss factor
a_i inside width of rectangular waveguide

β	transmission line phase constant
B	dielectric loss factor
b_i	inside height of rectangular waveguide
BW	bandwidth ratio
C	capacitance per unit length
D_{io}	inside diameter of outer conductor of coaxial line or circular waveguide
d_o	outside diameter of inner conductor
D_s	shroud diameter
f	frequency in megahertz
f_{co}	cutoff frequency in megahertz
FOM	figure of merit
g_a	antenna gain
γ	complex propagation constant, $\gamma = \alpha + j\beta$
η_l	transmission line efficiency
Γ_i	current reflection coefficient
I_l	total current on transmission line
I'	direct-wave current
I''	reflected-wave current
λ_c	waveguide cutoff wavelength
λ_g	guide wavelength
L	inductance per unit length
M_α	increase in line loss due to temperature
N_l	length of transmission line in standard units
P_d	power dissipated
P_i	input power
P_o	output power
T_1	ambient temperature
T_2	maximum allowable inner conductor temperature
V'	direct-wave voltage
V''	reflected-wave voltage
V_l	total voltage on transmission line
v_p	velocity of propagation
VSWR	voltage standing wave ratio

CHAPTER 7

α_e	phase shift from element to element in radians
α_n	current phase of nth array element relative to center of array
a	radius of circular array
AF	array factor
CP	circular polarization
d	distance between array elements
d_h	horizontal pattern directivity
d_v	vertical pattern directivity

η	antenna efficiency
E_θ	theta component of electric field
EP	elliptical polarization
G	antenna gain in decibels
h	distance between dipole and ground plane
h_t	transmitting antenna height
H_ϕ	phi component of magnetic intensity
$H_n^{(2)'}(ka)$	first derivative of Hankel function
ϕ	azimuth coordinate in spherical coordinate system
ϕ_n	angular position of nth array element
I_{eff}	effective current
I_m	maximum dipole current
I_n	current amplitude of nth array element
I_2/I_1	ratio of antenna driving currents
l	length of a dipole antenna
L_a	antenna length
λ	wavelength
N_r	number of radiating elements
ω_h	upper edge angular frequency
ω_l	lower edge angular frequency
P_{rad}	power radiated
r	radial distance in a spherical coordinate system
R	radius of earth
R_r	antenna input resistance
R_{rad}	radiation resistance
θ	elevation coordinate in spherical coordinate system
θ'	angle referenced to z-axis in spherical coordinate system
Θ_3	half-power beamwidth
Θ_t	beam tilt angle
ζ_0	characteristic impedance of free space
Z	antenna input impedance
Z_{11}	antenna self-impedance
Z_{12}	mutual impedance between pair of antennas

CHAPTER 8

α_{gr}	ground reflection attenuation factor
a	major axis of first Fresnel zone
A_a	effective area of antenna
AGL	above ground level
A_i	effective area of isotropic antenna
AMSL	above mean sea level
A_n	amplitude of nth wave,
b	minor axis of first Fresnel zone

B_n	net amplitude of nth wave due to troposcatter and transmission through partially opaque objects
c	wave velocity in vacuum
δR	incremental distance traveled by reflected wave
δR_n	incremental distance traveled by nth wave
D	divergence factor
ΔF	loss in signal strength relative to perfectly smooth earth due to surface roughness
Δh	height difference between peaks and valleys
$\Delta\phi$	relative bearing of echo and receiver
ΔT	temperature rise
E	field intensity
E_d	direct-wave field intensity
E_r	reflected-wave field intensity
E_t/E_m	weighted tap energy ratio
F_1	first Fresnel zone radius
F_2	second Fresnel zone radius
F_n	radius of nth Fresnel zone
GD	group delay
Γ	complex reflection coefficient
h	altitude
H	clearance height
H_h	total height of hill
h_r	receive antenna height
k	propagation constant
K	equivalent earth radius factor
L_d	diffraction loss due to spherical earth
L_{ke}	knife-edge diffraction loss
LOS	line of sight
L_s	free-space path loss
n	index of refraction
N	total number of waves arriving by other than direct path
N_r	modified index of refraction or refractivity
ν	height parameter; height measured relative to first Fresnel zone radius in absence of hill
$\mathcal{P}$	power density
p_h	contour parameter; sharpness of peak of hill
ψ	grazing angle
R	distance from transmitter to receiver
R_1, R_2	radii of concentric spheres
R_{eff}	effective earth radius
R_h	radius of a cylinder over pedestal representing hill

R_r distance from transmitter to echo
ρ radius of curvature of propagation path
θ_i angle of incidence
θ_r angle of reflection
v wave velocity in medium other than vacuum

CHAPTER 9

PDF probability distribution function
PEP peak envelope power
R_f flow rate

AUTHOR INDEX

SUBJECT INDEX